W9-DFN-256

Peacock / Calhoun
**Polymer Chemistry**

Andrew J. Peacock
Allison Calhoun

# Polymer Chemistry

**Properties and Applications**

## HANSER

Hanser Publishers, Munich • Hanser Gardner Publications, Cincinnati

QD
381
.P43
2006

*The Authors:*
Dr. Andrew J. Peacock, Tredegar Film Products, 1100 Bouldres Parkway, Richmond VA 23225, USA
Dr. Allison Calhoun, Department of Chemistry, Whitman College, Walla Walla, WA 99362, USA

Distributed in the USA and in Canada by
Hanser Gardner Publications, Inc.
6915 Valley Avenue, Cincinnati, Ohio 45244-3029, USA
Fax: (513) 527-8801
Phone: (513) 527-8977 or 1-800-950-8977
www.hansergardner.com

Distributed in all other countries by
Carl Hanser Verlag
Postfach 86 04 20, 81631 München, Germany
Fax: +49 (89) 98 48 09
www.hanser.de

Fairleigh Dickinson
University Library

Teaneck, New Jersey

The use of general descriptive names, trademarks, etc., in this publication, even if the former are not especially identified, is not to be taken as a sign that such names, as understood by the Trade Marks and Merchandise Marks Act, may accordingly be used freely by anyone.
While the advice and information in this book are believed to be true and accurate at the date of going to press, neither the authors nor the editors nor the publisher can accept any legal responsibility for any errors or omissions that may be made. The publisher makes no warranty, express or implied, with respect to the material contained herein.

Library of Congress Cataloging-in-Publication Data
Peacock, Andrew J., 1959-
  Polymer chemistry : properties and applications / Andrew J. Peacock,
Allison Calhoun.-- 1st ed.
      p. cm.
  Includes bibliographical references and index.
  ISBN-13: 978-1-56990-397-1
  ISBN-10: 1-56990-397-2
  1. Polymers--Textbooks. 2. Polymerization--Textbooks. I. Calhoun, Allison
R. II. Title.
  QD381.P43 2006
  668.9--dc22
                          2006008756
Bibliografische Information Der Deutschen Bibliothek
Die Deutsche Bibliothek verzeichnet diese Publikation in der Deutschen Nationalbibliografie;
detaillierte bibliografische Daten sind im Internet über <http://dnb.ddb.de> abrufbar.

ISBN-10: 3-446-22283-9
ISBN-13: 978-3-446-22283-0

All rights reserved. No part of this book may be reproduced or transmitted in any form or by any means, electronic or mechanical, including photocopying or by any information storage and retrieval system, without permission in writing from the publisher.

© Carl Hanser Verlag, Munich 2006
Production Management: Oswald Immel
Typeset by Manuela Treindl, Laaber, Germany
Coverconcept: Marc Müller-Bremer, Rebranding, München, Germany
Coverdesign: MCP · Susanne Kraus GbR, Holzkirchen, Germany
Printed by Druckhaus »Thomas Müntzer «, Bad Langensalza, Germany

*This book is dedicated to the memory of*
*William Lindsay Peacock*

# Preface

This book provides a comprehensive introduction to the field of polymer science, covering all relevant topics in a single volume. It systematically develops the various branches of polymer science, starting from basic chemical structures, working through the scientific principles that describe polymer behavior and culminating with a description of their conversion to usable products. The final section of the book is devoted to case studies that discuss the most important classes of synthetic polymers. It is with regret that the authors decided to omit a discussion of elastomeric materials. It was felt that a single volume could not do justice to both plastics in the general sense and the wide ranging field of elastomers and rubbers.

The opening chapter of this book introduces the basic chemical features of synthetic polymers that set them apart from the majority of compounds encountered in classic organic and inorganic chemistry. The importance of the distribution of chemical structure is discussed in detail, with particular relevance to how this influences properties and end uses. Subsequent chapters enlarge upon this theme describing the relationships between molecular structure and molten and solid state properties. The principal polymerization processes are surveyed with respect to the range of molecular structures available therefrom. A chapter is devoted to the rheological behavior of polymer melts, particular as it pertains to non-Newtonian flow and how this affects processing. This theme continues with a discussion of the solidification of polymers from the molten state and the formation of the anisotropic structures that characterize most polymer articles. The physical attributes of polymers in their solid state are then described in terms of mechanical, thermal, optical, and electrical characteristics. Analytical methods for characterizing polymers are briefly outlined as appropriate. Chemical degradation and the importance of stabilization are briefly explored. Commercial processes for converting raw polymers to useful products are described and their application to various classes of polymer is discussed. The principle recycling processes as applied to polymers are outlined. The final section of this book is dedicated to a series of case studies that describe the most important classes of polymer. Each case study outlines the chemistry, properties and applications of a particular polymer type in its own right and compared to competing materials, both polymeric and non-polymeric. The book is illustrated throughout with figures and tables that serve to further explain the principles under discussion.

This work is intended to introduce the field of polymer science to undergraduates in the areas of physical sciences and engineering. It is also aimed at professionals who find themselves working with polymers for the first time and need to rapidly familiarize themselves with guiding principles. In keeping with its role as a comprehensive text, this book assumes no prior knowledge of the field of polymers. A bibliography is provided at the end of the book for those wishing to delve further into a particular topic.

*Andrew Peacock* and *Allison Calhoun*

# Acknowledgements

No work of this type could be written without the help, encouragement and support of many practitioners in the field. The authors are indebted to their various teachers (both formal and informal) over the years who instilled in them a love of knowledge, science in general, and polymers in particular.

Dr. Peacock is grateful to his father Lindsay Peacock (who was also a chemist and did not live to see the completion of this work) for starting him off on the right foot and encouraging his scientific studies. Other memorable teachers include Dr. Pat Hendra, who inspired all his graduate students to exercise their imagination to solve problems in creative ways, and Dr. Leo Mandelkern who demanded rigorous scientific analyses and thorough experimental programs. Last, but not least Dr. Peacock would like to thank his wife, Shavon, for allowing him to devote a substantial portion of his "free" time to creating a work that he hopes will be valuable to future generations of scientists and engineers.

Dr. Calhoun is grateful to Dr. Allen King, Jr. and Dr. Darwin Smith – mentors who act as a source of inspiration by their own joy for learning, their expertise in physical chemistry and their exceptional skills at communicating their extensive knowledge with others. Their patience, high expectations, academic rigor and faith were, undoubtedly, the greatest influences in her career choices. Dr. Calhoun offers her sincerest gratitude to the three most important people in her life – her husband, William, and two daughters, Kaitlyn and Jordan, for their love and support. Finally, she thanks her colleagues at Whitman College, her family and friends for their encouragement.

Both authors thank Dr. Christine Strohm, the publisher and editor of this text, for her hard work and unfailing kindness.

# Contents

# 1 Introduction to Synthetic Polymers

## 1.1 Definition of Polymers

In our everyday lives we are surrounded by a seemingly endless range of synthetic polymers. Synthetic polymers include materials that we customarily call plastics and rubbers. These include such commonplace and inexpensive items as polyethylene grocery sacks, polyethylene terephthalate soda bottles, and nylon backpacks. At the other end of the scale are less common and much more expensive polymers that exhibit specialized properties, including Kevlar™ bullet proof vests, polyacetal gears in office equipment, and high-temperature resistant fluorinated polymer seals in jet engines. Manufacturers are constantly introducing new grades of polymers and expanding the range of finished products in which they are used. It is our intention in this book to provide an introduction to the wide range of materials and properties available. We shall systematically describe the principles that control a polymer's behavior in terms of its chemical and physical nature.

At the most basic level, a polymer molecule consists of hundreds, thousands, or even millions of atoms joined together to form a chain with an extended length at least an order of magnitude greater than its thickness. In practice, polymers are much more complex than this; any given polymeric material consists of molecules with a distribution of many chemical variations. These include variations of molecular weight (length), branching, steric configuration, interconnections, and chemical defects of many kinds. In addition to their chemical variety, the molecules that make up a polymeric material adopt a range of physical conformations. Physical variants include the organization and alignment of neighboring chains to various degrees at various scales of size. The properties of any polymeric item represent the sum total of the statistical distribution of the chemical and physical configurations of its component molecules. Throughout this book we shall be examining these chemical and physical variations in light of their specific effects on tangible properties.

The overwhelming majority of synthetic polymers is organic in nature, and it is on these that we will concentrate. The simplest and most common synthetic polymer is polyethylene, which will be our first example. Figure 1.1 shows the basic chemical structure of polyethylene. Pairs of hydrogen atoms are attached to the carbon atoms that make up the backbone. The repeat unit in this structure contains two carbon atoms and is derived from the ethylene monomer. In the case of polyethylene, the number of monomer residues, which is known as the polymerization

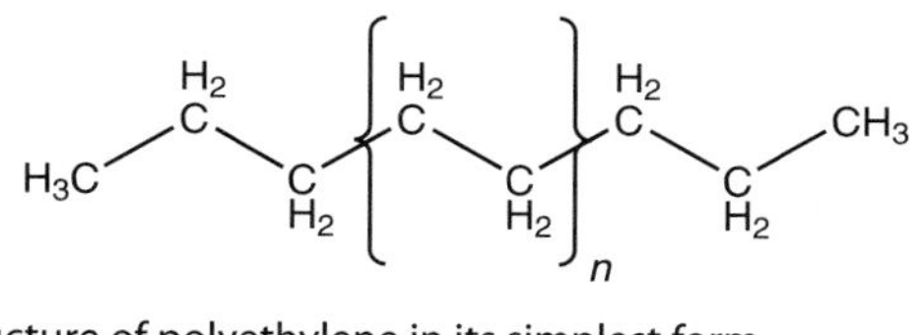

**Figure 1.1** Chemical structure of polyethylene in its simplest form

number, can range from approximately 100 to well over 100,000. This equates to molecular weights ranging from approximately 1,400 to over 1,400,000 Daltons. If these molecules were stretched out fully, they would have lengths ranging from 12.5 nanometers ($12.5 \times 10^{-9}$ m) to 12.5 μm ($12.5 \times 10^{-6}$ m). The terminal groups of polymer molecules are generally some variant of the repeat units that comprise the majority of the chain.

The number of monomer units that are connected to form a molecular chain strongly influences it properties. When the polymerization number is small, only a few monomers are linked, forming molecules known as oligomers. As the polymerization number increases, oligomers increase in molecular weight, changing from gases to liquids and then to solids that may be waxy or brittle. As the molecular weight rises further, the chains become long enough to entangle with each other. When this length is reached, groups of molecules begin to exhibit the physical properties that we typically associate with polymers. There is no explicit demarcation between oligomers and polymers. In practice, all plastics and rubbers consist of a statistical distribution of molecular lengths that may incorporate oligomers or even monomers.

There are numerous variations on the basic linear structure of polymers. Returning to our example of polyethylene, we find short chain branches and long chain branches, as shown in Figs. 1.2 and 1.3, respectively. The number and type of these branches strongly influences the way that the molecules pack in the solid state, and hence affect the physical properties. Long

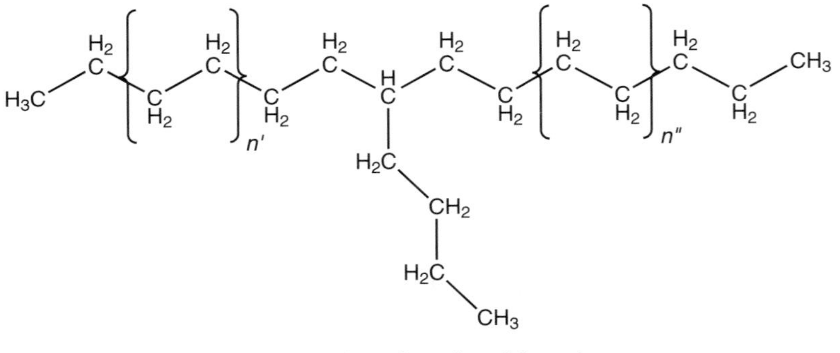

**Figure 1.2**  Polyethylene molecule with a short chain (butyl) branch

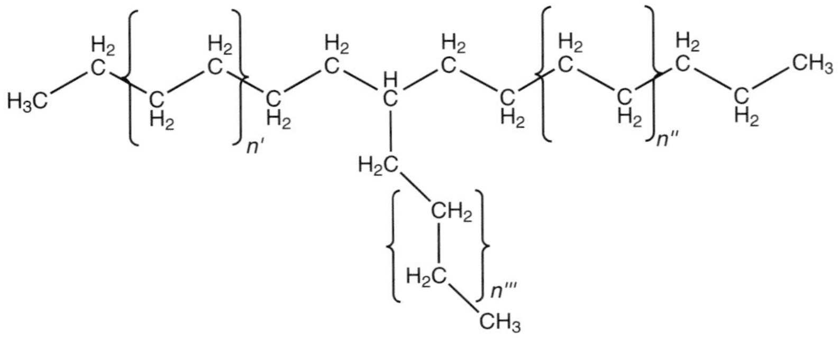

**Figure 1.3**  Polyethylene molecule with a long chain branch

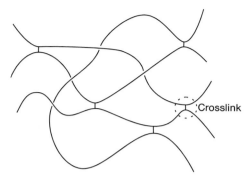

**Figure 1.4**   Schematic representation of a crosslinked structure

chain branches may themselves be branched, and these branches may in turn be branched and so forth. The properties of an assembly of linear molecules are very different from those of an assembly of highly branched chains. Taking the concept of branching to its extreme, each chain can be connected directly or indirectly to every other chain via a series of crosslinks, to form a three-dimensional network, as shown in Fig. 1.4.

Most of the polymers that we will discuss in this book are commonly known as plastics or rubbers. Plastic materials exhibit a wide range of physical states, ranging from hard and brittle, such as polycarbonates and polymethyl methacrylate (Plexiglas™), to soft and ductile, such as low density polyethylene and plasticized polyvinyl chloride. Rubbers are polymers that exhibit a high degree of elastic recovery following significant deformation. Automobile tires obviously fall into this class, but so do the elastic films used in diaper fastening systems and the soft overmolding on toothbrush handles. The wide range of chemical and physical structures adopted by polymers gives rise to a spectrum of properties that defies categorization within precise demarcation lines. It is the nature of the relationships between polymer properties and polymer structure that we will explore throughout this book.

## 1.2     Chemical Categories of Polymers

The vast majority of polymers is made by either addition or condensation polymerization. During addition polymerization, monomers that contain double bonds are added sequentially to the growing end of a chain, in a process known as "chain growth". Examples of monomers that participate in addition polymerization include ethylene, styrene, vinyl acetate, and vinyl chloride, which are shown in Fig. 1.5. During condensation polymerization, monomers, oligomers, and polymer chains combine by the elimination of small molecules to form longer chains in a process known as "step growth". Each monomer that contributes to the growing chain during step growth contains a minimum of two functional groups. Examples of such monomers include diacids, diols, diamines, and triamines, such as those shown in Fig. 1.6.

We can further characterize polymers into "thermoplastics" and "thermosets". Thermoplastics consist of linear or lightly branched chains that can slide past one another under the influence of temperature and pressure. These polymers flow at high temperatures which facilitates their molding into useful products. Thermosets consist of a network of interconnected chains whose positions are fixed relative to their neighbors. Such polymers do not flow when heated.

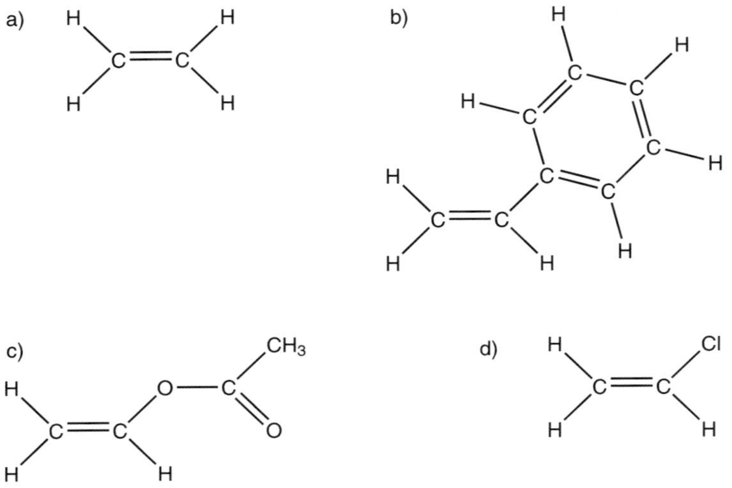

**Figure 1.5**  Examples of monomers that participate in addition polymerization: a) ethylene, b) styrene, c) vinyl acetate, and d) vinyl chloride

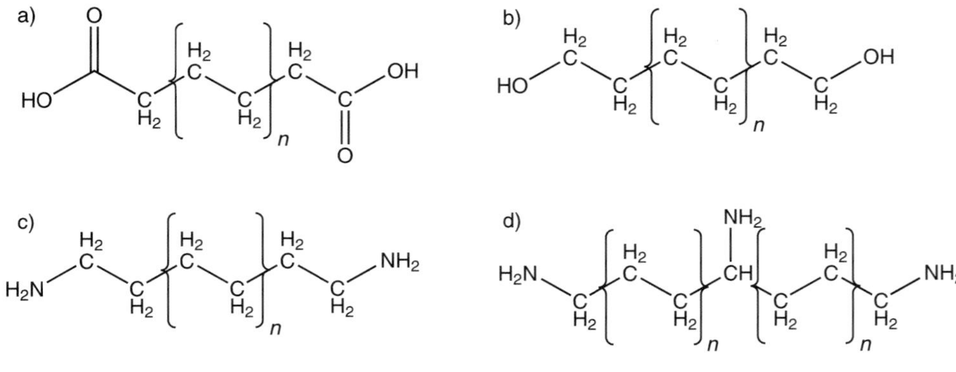

**Figure 1.6**  Examples of monomers that participate in condensation polymerization: a) diacid, b) diol, c) diamine, and d) triamine

## 1.2.1      Addition Polymers

Addition polymers, which are also known as chain growth polymers, make up the bulk of polymers that we encounter in everyday life. This class includes polyethylene, polypropylene, polystyrene, and polyvinyl chloride. Addition polymers are created by the sequential addition of monomers to an active site, as shown schematically in Fig. 1.7 for polyethylene. In this example, an unpaired electron, which forms the active site at the growing end of the chain, attacks the double bond of an adjacent ethylene monomer. The ethylene unit is added to the end of the chain and a free radical is regenerated. Under the right conditions, chain extension will proceed via hundreds of such steps until the supply of monomers is exhausted, the free radical is transferred to another chain, or the active site is quenched. The products of addition polymerization can have a wide range of molecular weights, the distribution of which depends on the relative rates of chain growth, chain transfer, and chain termination.

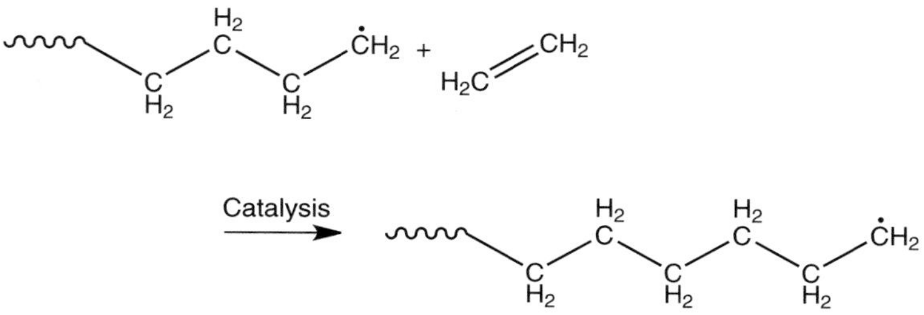

**Figure 1.7**   Addition polymerization as typified by the polymerization of ethylene

### 1.2.1.1      Homopolymers versus Copolymers

The monomers used to make an addition polymer need not be identical. When two or more different monomers are polymerized into the same chain, the product is a copolymer. For instance, we routinely copolymerize ethylene with small percentages of other monomers such as $\alpha$-olefins (e.g., 1-butene and 1-hexene) and vinyl acetate. We call the products of these reactions linear low density polyethylenes and ethylene-vinyl acetate copolymer, respectively. We encounter these copolymers in such diverse applications as cling film, food storage containers, natural gas distribution pipes, and shoe insoles.

Relatively small changes in comonomer content can result in significant changes in physical or chemical properties. Polymer resin manufacturers exploit such relationships to control the properties of their products. The composition of a copolymer controls properties such as stiffness, heat distortion temperature, printability, and solvent resistance. For example, polypropylene homopolymer is brittle at temperatures below approximately 0 °C; however, when a few percent ethylene is incorporated into the polymer backbone, the embrittlement temperature of the resulting copolymer is reduced by 20 °C or more.

Comonomers can be incorporated into the growing chain at random or as groups of identical monomers, known as blocks. The polymerization catalyst and reaction conditions control whether random or block copolymerization occurs. The relative lengths of blocks play a significant role in controlling the physical properties of copolymers.

### 1.2.1.2   Tacticity

The term "tacticity" refers to the configuration of polymer chains when their constituent monomer residues contain a steric center. Figure 1.8 illustrates the three principal classes of tacticity as exemplified by polypropylene. In isotactic polypropylene, the methyl groups are all positioned on the same side of the chain, as shown in Fig. 1.8 a). In syndiotactic polypropylene, the methyl groups alternate from one side to the other, as shown in Fig. 1.8 b). Random placement of the methyl groups results in atactic polypropylene, which is shown in Fig. 1.8 c). We can readily observe the effects of tacticity on the properties of polypropylene: isotactic polypropylene is hard and stiff at room temperature, syndiotactic polypropylene is soft and flexible, and atactic polypropylene is soft and rubbery.

In practice, there is no such thing as a pure isotactic or syndiotactic polymer. Once again, we find that polymers comprise a statistical distribution of chemical structures. Polymers that contain steric centers inevitably incorporate a certain number of steric defects that prevent us from obtaining 100% isotacticity or syndiotacticity. Polymer manufacturers vary the catalyst type and reaction conditions to control the tacticity level and the resulting properties.

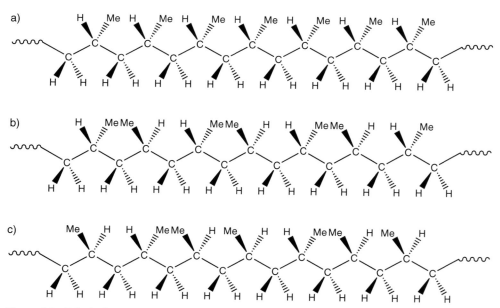

**Figure 1.8**   Principal types of tacticity found in polypropylene:
a) isotactic, b) syndiotactic, c) atactic

### 1.2.1.3    Branching

There are two ways that we can introduce branching into step growth polymers; either by copolymerization or by "backbiting". Copolymerization creates well defined branches, such as the butyl branch illustrated in Fig. 1.2, in which 1-hexene is copolymerized with ethylene. Long chain branches, as shown in Fig. 1.3, are created by backbiting. Backbiting occurs when a growing polymer chain transfers its terminal free radical to a site further back along the chain, from which a new chain proceeds to grow. Long chain branches do not have a fixed length; again we encounter a distribution of molecular compositions.

## 1.2.2    Condensation Polymers

Condensation polymers, which are also known as step growth polymers, are historically the oldest class of common synthetic polymers. Although superseded in terms of gross output by addition polymers, condensation polymers are still commonly used in a wide variety of applications; examples include polyamides (nylons), polycarbonates, polyurethanes, and epoxy adhesives. Figure 1.9 outlines the basic reaction scheme for condensation polymerization. One or more different monomers can be incorporated into a condensation polymer.

**Figure 1.9**   Generalized reaction scheme for condensation polymerization

### 1.2.2.1    Linear Condensation Polymers

Linear condensation polymers are produced when the constituent monomers contain two functional groups each. When a single monomer is polymerized, the product is made of chains whose repeat unit corresponds to the monomer. An example of this type is nylon 6, the structure of which is shown in Fig. 1.10. If two different monomers are polymerized, the result most often is a chain whose repeat unit corresponds to the two different monomers arranged alternately. An example of this type is nylon 66, the structure of which is shown in

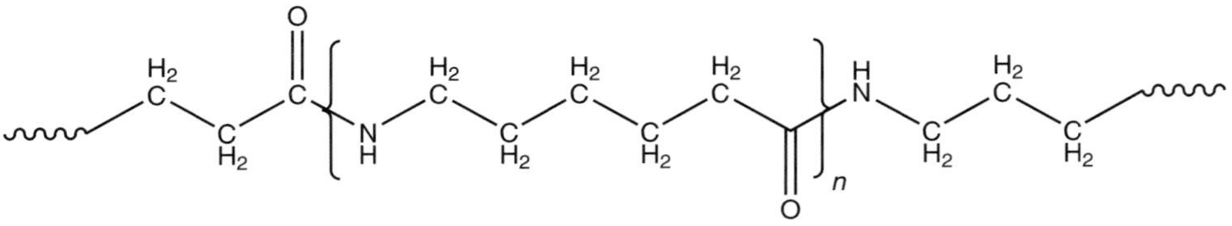

**Figure 1.10**   Structure of nylon 6 showing single monomer unit

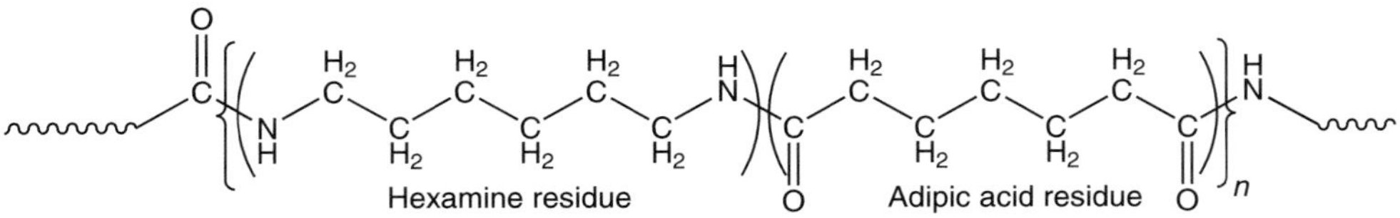

**Figure 1.11**   Structure of nylon 66 showing alternating monomer units

Fig. 1.11. The molecular weight of linear condensation polymers is generally controlled by the addition of a small percentage of monofunctional molecules.

### 1.2.2.2   Branched and Crosslinked Condensation Polymers

Branched and crosslinked condensation polymers are produced when the reaction mixture includes tri-functional monomers as well as bi-functional ones. The incorporation of a single tri-functional monomer into a chain generates a branch point. As we increase the fraction of tri-functional monomers, branching becomes more prevalent and the resulting molecules more complex. When sufficient tri-functional monomers are present we create a three-dimensional crosslinked network. Figure 1.12 shows the general outline of the effects of tri-functional monomers on condensation polymers.

### 1.2.3   Thermoplastic Polymers

A thermoplastic polymer can be repeatedly softened by heating, molded to a new shape, and then cooled to harden it. Thermoplastic polymers consist of chains that have no permanent chemical bonds to their neighbors. When we heat them, their molecules take on the properties of a viscous liquid that flows when we apply pressure. When we cool them, they solidify to take on a shape that remains constant until they are once again subjected to heat and pressure. We can dissolve thermoplastic polymers in solvents without destroying any chemical bonds.

We can create thermoplastic polymers by chain growth or step growth reactions. In either case the polymer chains consist of a string of monomer residues, each of which is attached to two other monomer residues. The polyethylene molecule shown in Fig. 1.1 is an example of a thermoplastic polymer made via chain growth polymerization, as shown in Fig. 1.7.

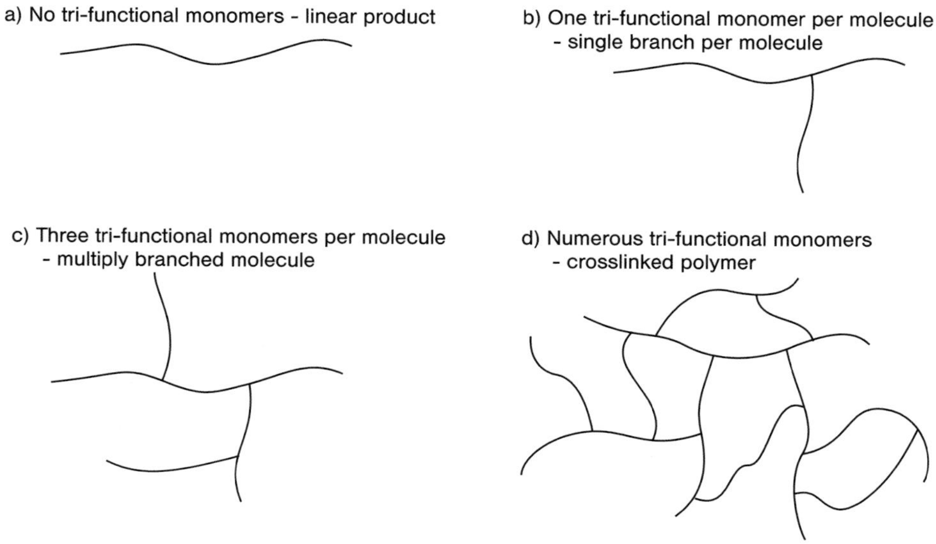

a) No tri-functional monomers - linear product

b) One tri-functional monomer per molecule
   - single branch per molecule

c) Three tri-functional monomers per molecule
   - multiply branched molecule

d) Numerous tri-functional monomers
   - crosslinked polymer

**Figure 1.12**   General outline of the effects of tri-functional monomer content on condensation polymers

The nylon 66 molecule shown in Fig. 1.11 is a thermoplastic polymer, created by the step growth polymerization of hexamine and adipic acid. The majority of commercial polymers are thermoplastics, which permits us to readily mold them to many useful shapes.

## 1.2.4    Thermoset Polymers

A thermoset polymer does not flow when it is heated and subjected to pressure. Thermoset polymers consist of an interconnected network of chains that are permanently chemically connected to their neighbors, either directly or via short bridging chains, as shown in Fig. 1.4. We refer to such networks as being crosslinked. Thermoset polymers do not dissolve in solvents, but they can soften and swell.

Commercially, we create thermoset polymers directly by step growth reactions or by crosslinking a thermoplastic polymer to form a thermoset material in a separate reaction. To create a thermoset by step growth, we need to incorporate a significant proportion of tri- (or higher) functional monomers, which creates the type of network illustrated in Fig. 1.12 d). We can crosslink linear polymers to form thermosets via various reactions, depending on the chemical structure of the base polymer. For instance, we can crosslink polyethylene by blending it with an organic peroxide at a temperature below that at which the peroxide decomposes. Subsequently, we heat the polymer sufficiently for the peroxide to decompose into peroxy radicals, which abstract hydrogen atoms from the chains to leave radicals which combine with each other to form covalent carbon-carbon bonds between adjacent chains.

## 1.3    Physical Categories of Polymers

The chains that make up a polymer can adopt several distinct physical phases; the principal ones are rubbery amorphous, glassy amorphous, and crystalline. Polymers do not crystallize in the classic sense; portions of adjacent chains organize to form small crystalline phases surrounded by an amorphous matrix. Thus, in many polymers the crystalline and amorphous phases co-exist in a "semicrystalline" state.

### 1.3.1    Amorphous State

The least organized of the polymer phases is known as the amorphous state. Polymer chains in this state exhibit no crystalline ordering. That is, they are not packed into structures that have a well defined repeat unit. We can envisage such chains as a pot full of cooked spaghetti that we have vigorously stirred. On a larger scale, the molecules may exhibit some degree of ordering. This typically takes the form of preferential alignment of chain segments. Such alignment is inevitable following most commercial forming processes. Polymer molecules in the molten state are stretched out and oriented during viscous flow. As these oriented chains cool, they solidify in their aligned state.

Amorphous polymers convert reversibly between the rubbery and glassy states as their temperature rises or falls. Below their glass transition temperature, amorphous polymers exist in a glassy state. Above their glass transition temperature they are rubbery. We can demonstrate this easily with a racquet ball, which is made of an amorphous polymer. At room temperature, as we all know, the ball bounces; at this temperature it is in the rubbery state. If we immerse the ball in liquid nitrogen it becomes brittle and will shatter when we drop it, i.e., it has become a glass. If we were to allow the frozen ball to warm up to room temperature, it would become rubbery once more. We can freeze and thaw the same ball repeatedly with no loss of its properties at room temperature.

#### 1.3.1.1    Rubbery Amorphous

In the rubbery amorphous state, the groups of atoms that make up the polymer chains possess substantial local freedom of movement. That is, small groups of atoms can vibrate, twist, and rotate, within the general confines of the chains around them. This freedom of movement is possible because the polymer chains are relatively loosely packed together, and thus are surrounded by a comparatively large "free volume". The configuration of a chain segment, comprising a small number of backbone atoms and their associated atoms, can vary widely over a short period of time due to the cooperative motion of the surrounding molecular segments, which are also in a high state of agitation. The high degree of local molecular motion results in materials that are typically soft and flexible.

Rubbery amorphous polymers behave this way, and not like liquids, because their chains are not entirely free to slide past one another. The principal factors that limit long range slippage are chemical crosslinks, chain entanglements, long chain branching, short chain branching, polar interactions, and hydrogen bonds. Permanent crosslinks, such as those illustrated in Fig. 1.4, prevent wholesale rearrangement of the chains comprising a rubber. Chain entanglements exist in virtually all polymers simply by the nature of the polymerization process. Such entanglements act as temporary crosslinks, which impede molecular motion. We can think of entanglements as loose knots involving two or more molecular chains. If we were to pull rapidly on the end of the chain, the knot would tighten and we could not extract the chain. Short and long chain branches can act as temporary anchors that impede the slippage of a molecule between its neighbors. Naturally, the larger the branch, the more likely it is to catch upon its neighbors. Thus, long chain branches limit molecular translation more effectively than short chain branches. Hydrogen bonds and other interactions between polar substituents act as transient crosslinks, which impose a significant energy barrier to chain translation.

When a rubbery amorphous polymer is deformed by an external force, its molecules slide past one another. In the absence of permanent crosslinks and given sufficient time, the molecules in the rubbery state start to disentangle. The rate of disentanglement increases with temperature, as the polymer chains gain sufficient energy to overcome activation barriers to motion. At sufficiently high temperatures, the rubbery nature of amorphous polymers gradually gives way to viscous flow, characteristic of the molten state. Even in the molten state, polymers exhibit rubbery characteristics when the instantaneous rate of deformation exceeds the rate of chain disentanglement. For example, if we take a lump of molten polymer we can roll it into a ball that bounces quite nicely when it is thrown against a wall. In this case, the rate of deformation as the ball strikes the solid surface exceeds the rate of chain disentanglement; thus the polymer behaves as a rubber. On the other hand, if we were to tread on the same ball of molten polymer it would flatten permanently, because the rate of deformation is slow enough for the molecules to slide past one another.

Rubbery amorphous polymers do not hold their shape well unless they are permanently crosslinked. If automobile tires were not crosslinked, they would be a soft sticky mess that would flow under the weight of the car. For this reason, we rarely encounter rubbery amorphous polymers that are not crosslinked.

### 1.3.1.2   Glassy Amorphous

In the glassy amorphous state polymers possess insufficient free volume to permit the cooperative motion of chain segments. Thermal motion is limited to classical modes of vibration involving an atom and its nearest neighbors. In this state, the polymer behaves in a glass-like fashion. When we flex or stretch glassy amorphous polymers beyond a few percent strain they crack or break in a brittle fashion.

Glassy amorphous polymers exhibit excellent dimensional stability and are frequently transparent. Everyday examples include atactic polystyrene, polycarbonate, and polymethylmethacrylate (Plexiglas™), which we encounter in such applications as bus shelters, motorcycle windshields, and compact disc cases.

### 1.3.2    Semicrystalline State

Polymers do not crystallize in the same way that small molecules do. When small molecules crystallize, they form macroscopic structures that are large enough to be seen with the naked eye, such as sugar granules or grains of table salt. Such crystals contain thousands of discrete molecules or a regular array of cations and ions. In contrast, polymers crystallize on a microscopic basis to form tiny crystals that we call crystallites, which are surrounded by an amorphous matrix. Crystallites rarely grow large enough for us to see them with our naked eyes.

Polymer crystallites are made up of linear segments from numerous adjacent molecules. Figure 1.13 illustrates schematically how a single polymer chain can contribute to several crystallites. Polymer chains can double back on themselves, contributing several adjacent linear chain segments, linked by folds at the crystallite surface. On a larger scale, a single polymer molecule can contribute chain segments to several crystallites. The molecular segments between crystallites are in the amorphous phase, thus each molecule contributes to both the ordered and disordered phases. In this way, the crystalline and amorphous phases are intimately linked. Neighboring crystallites are linked via "tie chains". The linkages between crystallites and the crystalline and amorphous phases are very important. These links serve to tie the phases together, giving semicrystalline polymers high strength and toughness, which we take advantage of in such applications as drainage pipes, chemical storage tanks, and grocery sacks.

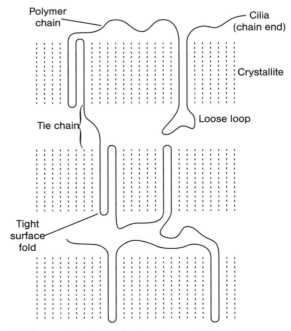

**Figure 1.13**   Schematic diagram showing a single polymer chain contributing to several crystallites

In addition to the chain segments within a crystallite being parallel with each other, the chains in neighboring crystallites may share a common orientation. When crystallite orientation occurs on a macroscopic scale, a polymer is said to be anisotropic. That is, its properties depend on the testing direction. We can see a good example of anisotropy in the humble grocery sack, which behaves quite differently depending on whether we stretch it lengthwise versus stretching it crossways.

The basic difference between the crystallization of polymers and small molecules lies in the fact that polymer molecules are entangled. When a molten polymer is cooled, its constituent molecules tend to pack better and will crystallize if it is thermodynamically favorable to do so. A polymer's molecular structure and the kinetics of its chain motion control the extent to which it crystallizes. Entanglements between neighboring molecules impede the slippage of chains that would be required for large scale molecular re-organization to form macroscopic crystals. Some stereo irregular chains, such as atactic polypropylene, are incapable of forming the regular repeat units required for crystallization. In other cases, branches hamper the formation of regular structures. When a polymer is cooled slowly, its chains have plenty of time to reorganize, which results in a higher degree of crystallinity than can form during quench cooling. The relative proportions of the crystalline and amorphous phases and the links between them determine the properties of a semicrystalline polymer. We can see this readily in polyethylene. High crystallinity polyethylene samples are rigid and tough, which makes them an excellent choice for milk crates and industrial pallets. On the other hand, low crystallinity polyethylene is soft and flexible, which suits it for medical tubing and many children's toys.

# 1.4    Statistical Distributions

There is no such thing as a "pure" polymer. All polymers comprise molecules that exhibit chemical and physical distributions of many variables; these include molecular weight, branching, steric defects, molecular configuration, preferential chain orientation, and crystallite size and shape. The properties and characteristics that we exploit in polymers are controlled by the overall balance of these distributions.

In addition to the statistical distributions inherent in an individual polymer, distributions are further broadened by the commercial practice of blending. We commonly blend two, three, four, or even more polymers of similar or dissimilar types in order to achieve the specific properties required.

## 1.4.1    Molecular Weight Distribution

Polymer scientists are well aware of the importance of molecular weight distribution and have learned to precisely measure and control it. This enables us to control many important

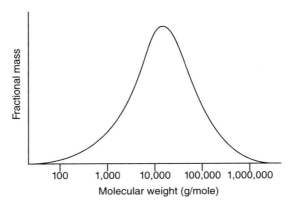

**Figure 1.14**    Generalized plot of molecular weight distribution

properties, such as the flow of molten polymers, and their solid state toughness and extensibility. The relative rates of chain growth, chain transfer, and chain termination during polymerization primarily determine a polymer's molecular weight distribution. Subsequent events, such as chemical attack, heating, and shearing during processing, or exposure to ultraviolet radiation can radically modify a polymer's molecular weight distribution.

We commonly portray molecular weight distributions as a plot of fractional mass versus the logarithm of the molecular weight of the constituent molecules, as shown in Fig. 1.14. The range of molecular lengths in a polymer may cover many orders of magnitude, ranging from traces of monomer and oligomers up to molecules containing hundreds of thousands of atoms.

We typically use a set of shorthand values to describe a polymer's molecular weight distribution. These values are a series of averages that are increasingly weighted to the longer chains of the distribution. The most common molecular weight averages that we encounter are the number average and weight average molecular weights.

The number average molecular weight $(M_n)$ is calculated from Eq. 1.1.

$$M_n = \frac{\sum M_i N_i}{\sum N_i} = \frac{\sum W_i}{\sum N_i} \tag{1.1}$$

Where:

$M_i$ = Molecular weight of chains in fraction i

$N_i$ = Number of chains in fraction i

$W_i$ = Weight of chains in fraction i

Similarly, we can calculate the weight average molecular weight $(M_w)$ using Eq. 1.2.

$$M_w = \frac{\sum M_i^2 N_i}{\sum M_i N_i} = \frac{\sum M_i W_i}{\sum W_i} \tag{1.2}$$

We often use the ratio between the weight and number average molecular weights as a guide to summarize a polymer's overall molecular weight distribution.

In some cases, two or more catalysts are present during polymerization. Inevitably, the catalysts exhibit different polymerization kinetics, which results in different populations of molecules. In such cases, we produce polymers with a bimodal molecular weight distribution.

Different lengths of chains play different roles in controlling polymer properties. For instance, shorter chains flow more readily in the molten state and are more readily incorporated into crystallites because they have fewer entanglements to impede their motion. Conversely, longer chains tend to resist flow and impede crystallization.

## 1.4.2    Short Chain Branching Distribution

Short chain branches are frequently introduced into polymers by copolymerization. The chemical structure of the comonomer controls the type and length of the short chain branch. The polymerization catalyst, reaction conditions, and comonomer content in the reaction medium determine the probability of finding a branch at any particular location along a chain. Comonomers, and hence the short chain branches derived from them, can be introduced at random or as blocks.

By using two or more polymerization catalysts simultaneously, polymer chemists can produce copolymers with a bimodal composition distribution. This is made possible by the fact that no two catalysts incorporate monomers at exactly the same rate. The net result is that short chain branches may be preferentially incorporated into either the higher or lower molecular weight fractions. Polymer manufacturers can obtain a similar result by operating two polymerization reactors in series. Each reactor produces a resin with a different copolymer distribution, which are combined to form a bimodal product. Copolymers with a bimodal composition distribution provide enhanced toughness when extruded into films.

## 1.4.3    Long Chain Branching Distribution

Polymer chemists can introduce long chain branches into polymers by various means. During addition polymerization they exploit the phenomenon known as "backbiting" to create the long chain branches. Alternatively, they can copolymerize a mixture of bifunctional and trifunctional monomers during condensation polymerization. Long chain branches provide various beneficial effects, including improved melt flow characteristics and increased toughness. In general, long chain branches are introduced at random locations along the length of a chain.

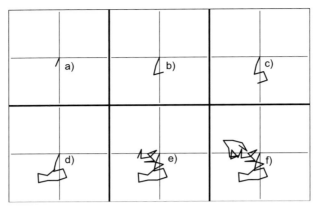

**Figure 1.15**   Two-dimensional freely jointed random walk with:
a) one bond, b) three bonds, c) five bonds, d) ten bonds, e) twenty bonds,
and f) thirty bonds

## 1.4.4    Shape Distributions

Due to their length and flexibility, polymer molecules adopt an infinite diversity of molecular configurations. One way to visualize the shapes that polymer chains can adopt is by using the concept of a "three-dimensional random walk", which is illustrated schematically in Fig. 1.15. We start by placing one end of a polymer chain at a fixed point in space. We place the next atom of the backbone one bond length away in a random direction relative to the point of origin. Likewise, we place the third atom one bond length away from the second in a random direction. We progress similarly for the remainder of the atoms in the backbone. For simplicity in the figure, we have shown a very short random walk in only two dimensions. In reality, random walks occur in three dimensions and the bond angles are limited by the rules governing molecular structure and the requirement that chains must avoid themselves and others with which they overlap. Each polymer molecule adopts a unique configuration.

## 1.4.5    Orientation Distribution

Polymer molecules in fabricated items rarely adopt a true random walk. Manufacturing processes stretch out molecules and then freeze them in an extended configuration before they have time to relax to the random state. Manufacturers exploit orientation in order to control physical properties.

Polymer items may exhibit uni-axial or multi-axial molecular orientation, depending on their design and manufacturing conditions. For instance, extruded polymer fibers exhibit preferential orientation along their length; this gives them high tensile strength. In contrast, a plastic margarine tub lid, which is produced by injection molding, exhibits radial orientation from the injection point at its center. Plastic bottles, which are produced by inflating a molten precursor, exhibit bi-axial orientation within their walls. Other, more complicated, molded items can have even more complex orientation patterns.

## 1.5 Properties and Applications

When we consider the wide range and vast numbers of polymer items that surround us, it is difficult to imagine how modern life could continue if they were not available. In this section, we are going to present a few examples that illustrate the range of properties available to us from polymers. We will describe items in terms of their molecular characteristics and how this relates to their end usage.

The properties of a plastic or rubber item are a combination of its intrinsic molecular properties as modified by the manufacturing process. For any given polymer there is a range of properties that can be accessed by changing the way in which it is processed.

### 1.5.1 Mechanical Properties

In order to understand the mechanical properties of polymers it is useful to think of them in terms of their viscoelastic nature. Conceptually we can consider a polymeric item as a collection of viscous and elastic sub-components. When a deforming force is applied, the elastic elements deform reversibly, while the viscous elements flow. The balance between the number and arrangement of the different components and their physical constants controls the overall properties. We can exploit these relationships to create materials with a broad array of mechanical properties, as illustrated briefly by the following examples.

Ultra-oriented polyethylene fibers consist of very high molecular weight molecules (weight average molecular weight greater than 1,000,000 Daltons) that are strongly aligned with the long axis of the fiber. In this arrangement, the molecules crystallize to a high degree. The net result is a material with an extraordinarily high elastic modulus that rivals steel on a weight-for-weight basis. Fabrics woven from these fibers are so tough that they are used in bullet proof vests.

At the opposite end of the polyethylene spectrum are the so-called ultra low density products with a high concentration of short chain branching that inhibits crystallization. These materials are soft, flexible, and transparent; we encounter these materials in applications such as medical tubing, meat packaging films, and ice bags.

Other polymers, such as polycarbonate and polymethylmethacrylate, are hard, tough, and transparent. These materials are ideal for applications that are likely to experience severe impact. We find these polymers in bus shelters, motorcycle helmet visors, and jet fighter canopies.

Polymers that are rigid at high temperatures are known as "engineering plastics". This class of polymers includes polyacetal and many nylons. These polymers are used in applications such as small gears in office equipment and under the hood of automobiles.

The properties of polyurethanes can be tailored by prudent selection of their constituent monomers. They can be converted into elastic foams, which are widely used in upholstery, and are used as covers for the handles of various tools and implements, such as the soft touch grips on ball-point pens and power tools.

Synthetic and natural rubbers are amorphous polymers, typically with glass transition temperatures well below room temperature. Physical or chemical crosslinks limit chain translation and thus prevent viscous flow. The resulting products exhibit elastic behavior, which we exploit in such diverse applications as hoses, automotive tires, and bicycle suspension units.

## 1.5.2    Barrier Properties

Many polymers are used in barrier applications, either to keep contents in or contaminants out. Food packaging is an excellent example of such usage. Plastic films and containers of many types are used to package food. Blow molded bottles often contain numerous layers, each of which provides specific benefits. Polyethylene layers are excellent water barriers, polyvinyl alcohol is a good oxygen barrier, and polyethylene terephthalate impedes the diffusion of carbon dioxide from carbonated drinks. Other barrier applications include toothpaste tubes, diaper backsheets, tarpaulins, and geomembranes, which are used to line containment ponds and landfill pits.

## 1.5.3    Surface Contact Properties

Many polymer items are designed specifically to make contact with other materials. Where surface contact is concerned, two key properties are coefficient of friction and abrasion resistance. Polymers used in such applications include ultra high molecular weight polyethylene, polyacetal, fluorinated polymers, and natural and synthetic rubbers. Examples that we routinely come across include furniture upholstery, bushings and gears in office equipment, and bicycle tires. Industrial uses include the outer cover of electrical cables, and pipes that convey abrasive liquids such as slurries and powders.

## 1.5.4    Optical Properties

Depending on the application, polymers are transparent, opaque, glossy, or matte. Transparent applications range from automotive tail lights and food packaging, to camera and contact lenses. Opaque applications include buckets, kitchenware, and diaper backsheets. We can tailor the surface characteristics of a polymer by the production process to be either glossy or matte, depending on aesthetic requirements. Computer housings and automotive dashboards are non-reflective, whereas household appliance housings are generally glossy.

### 1.5.5    Electrical Properties

Polymers are widely used as electrical insulators in applications such as wire and cable insulation, electrical appliance housings and capacitor films. Polymers used in these applications include polyvinyl chloride, polyethylene, and isotactic polypropylene.

## 1.6    Conclusions

In this chapter we have outlined the general concepts that we will be exploring in greater detail throughout this book. We have described the various classes of polymer, how we make them and their physical characteristics. In the following nine chapters we will discuss the chemistry and properties that govern polymer behavior. The next seven chapters are devoted to the processes by which we convert raw polymer into usable items. In the remaining eight chapters we will bring it all together to describe how and why specific polymers are used in particular applications.

## Review Questions for Chapter 1

1.    Explain the differences between monomers, oligomers and polymers. Draw examples of each.

2.    What is meant by the statement "plastics and rubbers comprise a statistical distribution of molecular lengths"?

3.    Explain the differences between the following sets of terms:
     a)  homopolymer and copolymer
     b)  atactic, isotactic and syndiotactic
     c)  rubbery amorphous state and glassy amorphous state

4.    What is meant by the term "preferred orientation" when talking about fabricated plastic parts? Why would this orientation result in anisotropic mechanical properties in the final product?

5.    What are the mechanisms that can produce short chain and long chain branching?

6.  What characteristic functionality is required in a monomer to produce a polymer via addition polymerization?

7.  What is the difference between a rubber and a plastic? Why are they both polymers?

8.  Name at least three characteristics of polymers that exhibit statistical distributions.

9.  Name five factors that would govern the choice of a polymer for a specific application.

10. What is meant by the term "molecular entanglement"?

# 2 Polymer Chemistry

## 2.1 Introduction

In this chapter, we will see how polymers are manufactured from monomers. We will explore the chemical mechanisms that create polymers as well as how polymerization methods affect the final molecular structure of the polymer. We will look at the effect of the chemical structure of monomers, catalysts, radicals, and solvents on polymeric materials. Finally, we will apply our molecular understanding to the real world problem of producing polymers on a commercial scale.

## 2.2 Thermoplastics and Thermosets

Thermoplastics consist of linear or lightly branched chains that can slide past one another under the influence of temperature and pressure. These polymers flow at high temperatures, allowing us to mold them into useful products. When we heat and/or shear thermoplastic polymers, we can change their shape. For example, polyethylene milk containers can be reprocessed back into the melt stage and then formed into a park bench. We find thermoplastics in a wide variety of commonly used items, such as pantyhose (polyamide), compact disks (polycarbonate), grocery bags (polyethylene), house siding (polyvinyl chloride), gas and water pipes (polyethylene, polyvinyl chloride, and polypropylene), and medical intravenous fluid bags (polyvinyl chloride).

Thermosets consist of a network of interconnected chains whose positions are fixed relative to their neighbors. Such polymers do not flow when heated. Instead, when exposed to high temperatures, thermosets degrade into char. Examples of thermosets include some polyurethanes and epoxy resins.

To understand the difference between thermoplastic and thermoset polymers, we must look towards the molecular structure of the polymers for insight. In thermoplastics, the individual polymer chains are chemically separate from one another while being physically entangled. The chains can slide over one another when heated and sheared, allowing the polymer to flow or become rubbery. This, in turn, allows the polymer to take on new shapes. The polymer chains in thermosets differ from thermoplastics because their chains are linked to one another through chemical crosslinks. The crosslinks create an extended network in which every chain is attached to every other chain. Therefore, the molecular weight of a fully crosslinked thermoset article is equal to its weight in grams. Thermoset polymers cannot flow because the crosslinks prevent large scale reorganization of their polymer chains.

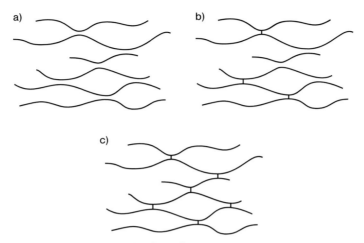

**Figure 2.1**  Effect of crosslinking on molecular architecture:
a) no crosslinking – linear molecules,
b) light crosslinking – long chain branching and
c) complete crosslinking – interconnected network of chains

As with most classification schemes, there exist grey areas between the thermoset and thermoplastic categories. The traditional usage of the term thermoset focused on chemicals that polymerized to create the interconnected network, such as urea-formaldehyde resins (the Formica® used in counter tops). This definition becomes blurred by our understanding that traditional thermoplastic materials are often the starting point for the crosslinking process. For example, gamma irradiation of standard polyethylene creates a crosslinked network which is, in essence, a thermoset. Crosslinked polyethylene is used in bulk liquid storage tanks, high voltage electrical insulation, and kayaks. Since the degree of crosslinking can vary, depending on process parameters, such as temperature and radiation dose, there is a continuum between true thermoplastics and true thermosets ranging from no crosslinks (thermoplastics), lightly crosslinked (long chain branched polymers), to mostly crosslinked (rubbers), to fully crosslinked (true thermosets) as shown in Fig. 2.1.

## 2.3    Chain Growth Polymerization of Thermoplastics

Chain growth polymers, which are often referred to as addition polymers, form via chain addition reactions. Figure 2.2 presents a generic chain addition mechanism. Chain addition occurs when the active site of a monomer or polymer chain reacts with an adjacent monomer molecule, which is added to the end of the chain and generates a new active site. The active site is the reactive end of a monomer or polymer that participates in the polymerization reaction.

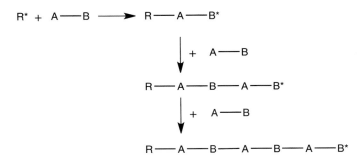

**Figure 2.2**   Schematic representation of generic chain addition mechanism

Chain growth polymers comprise most of the commodity polymers found in consumer products. Common examples include the polyethylene used in trash can liners, the polyvinyl chloride used as wire insulation, and the polypropylene used in food storage containers.

Chain growth polymerization begins when a reactive species and a monomer react to form an active site. There are four principal mechanisms of chain growth polymerization: free radical, anionic, cationic, and coordination polymerization. The names of the first three refer to the chemical nature of the active group at the growing end of the monomer. The last type, coordination polymerization, encompasses reactions in which polymers are manufactured in the presence of a catalyst. Coordination polymerization may occur via a free radical, anionic, or cationic reaction. The catalyst acts to increase the speed of the reaction and to provide improved control of the process.

Free radical, anionic, and cationic addition polymerization processes occur in four distinct steps: initiation, propagation, chain transfer, and termination. Figure 2.3 illustrates these steps as exemplified by the free radical polymerization of ethylene. In the first step, the initiator, R•, creates an active site on the monomer, as indicated by the unpaired electron. During propagation, the active site reacts with another monomer, thereby adding the monomer residue to the end of the chain and generating a new active site, causing the chain to grow. Chain growth is terminated when the active site becomes deactivated. Chain transfer is an alternate reaction of the active site. In this process, an active site transfers to another molecule creating one terminated species and a new activated species.

The choice of one polymerization method over another is defined by the type of monomer and the desired properties of the polymer. Table 2.1 lists advantages and disadvantages of the different chain growth mechanisms. Table 2.2 summarizes some well known addition polymers and the methods by which they can be polymerized.

**Table 2.1**   Advantages and Disadvantages of Chain Growth Polymerization Mechanisms

| Polymerization mechanism | Advantages | Disadvantages |
|---|---|---|
| Free radical polymerization | • Relatively insensitive to trace impurities<br>• Reactions can occur in aqueous media<br>• Can use chain transfer to solvent to modify polymerization process | • Structural irregularities are introduced during initiation and termination steps<br>• Chain transfer reactions lead to reduced molecular weight and branching<br>• Limited control of tacticity<br>• High pressures often required |
| Anionic polymerization | • Narrow molecular weight distribution<br>• Limited chain transfer reactions<br>• Predictable molecular weight average<br>• Possibility of forming living polymers<br>• End groups can be tailored for further reactivity | • Solvent-sensitive due to the possibility of chain transfer to the solvent<br>• Can be slow<br>• Sensitive to trace impurities<br>• Narrow molecular weight distribution |
| Cationic polymerization | • Large number of reactive monomers<br>• High reactivity of active site, therefore quick reaction times<br>• Reactions proceed rapidly even at low temperatures | • High reactivity leads to undesirable side reactions<br>• Requires very high purity reaction medium<br>• Chain transfer leading to low molecular weight and high branching<br>• Kinetic mechanisms are poorly understood<br>• Reaction often does not go to completion |
| Coordination polymerization | • Can engineer polymers with specific tacticities based on the catalyst system<br>• Can limit branching reactions<br>• Polymerization can occur at low pressures and modest temperatures<br>• Otherwise non-polymerizable monomers (e.g., propylene) can be polymerized | • Mainly applicable to olefinic monomers |

**Table 2.2**  Selected Homopolymers and the Potential Methods of Polymerization for Each

| Polymer | Monomer | Methods of polymerization available |
|---|---|---|
| Polyethylene | Ethylene | Free radical |
| | | Coordination* |
| Polypropylene | Propylene | Coordination* |
| Polyvinyl chloride | Vinyl chloride | Free radical* |
| Polystyrene | Styrene | Free radical* |
| | | Cationic |
| | | Anionic |
| | | Coordination |
| Polymethyl methacrylate | Methyl methacrylate | Free radical* |
| | | Anionic |
| Polyisobutylene | Isobutylene | Cationic* |

*    Indicates the most frequently used method of polymerization

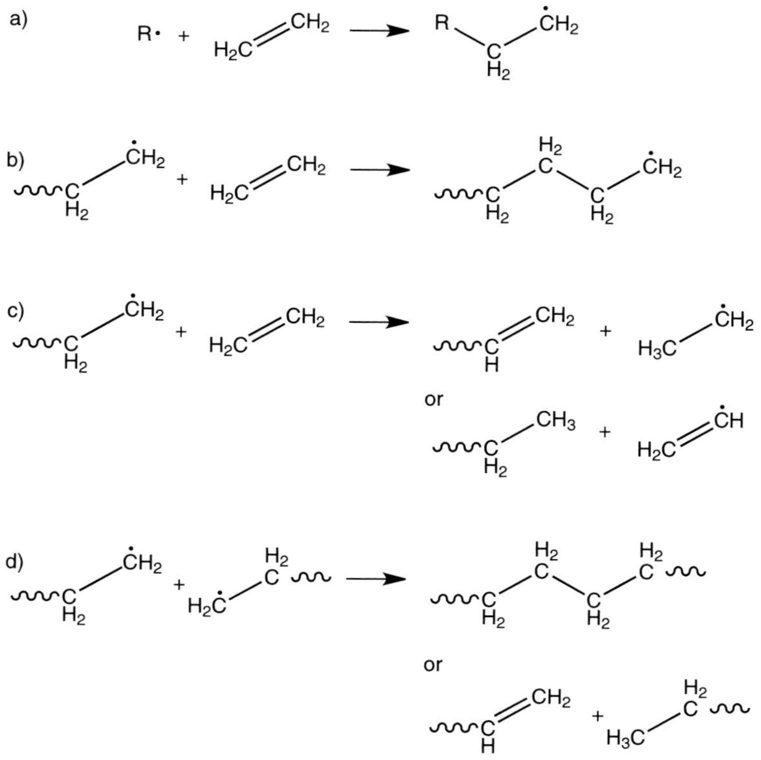

**Figure 2.3**   Chain growth polymerization exemplified by free radical polymerization of polyethylene: a) initiation, b) propagation, c) chain transfer, and d) termination

## 2.3.1     Free Radical Polymerization

A monomer that can undergo free radical polymerization must be able to accept a radical from an initiating species and then transfer that radical to another monomer to create a polymer chain. Figure 2.4 shows several monomers that meet this requirement. During initiation, as shown in Fig. 2.3 a), a free radical attacks the double bond of a monomer and transfers a high energy radical electron to the end of the monomer. The modified end of the monomer is highly reactive, which leads to the second step, propagation, which is shown in Fig. 2.3 b). The unpaired electron of the activated monomer reacts with another monomer creating an activated dimer, with $n$ (the polymerization number) equaling one. Repeated propagation builds the polymer, one monomer at a time, increasing the value of $n$.

During chain transfer, shown in Fig. 2.3 c), the active end of a growing polymer chain reacts with a terminated chain, a monomer, a deliberately added chain transfer agent (such as hydrogen), or a solvent molecule. Radical chain growth terminates while the inactive species gains the free radical and polymerization continues at this site. If the active site is transferred to an existing polymer chain, a branch is created. If it is transferred to a small molecule, a new polymer chain is initiated. During chain transfer, the number of active sites remains constant. By deliberately adding chain transfer agents we can limit the average molecular weight of the polymer.

Termination occurs when the active sites of two growing chains meet, as shown in Fig. 2.3 d). The unpaired electrons form a bond that couples the ends of the chains. Alternatively, disproportionation may occur. This happens when one chain transfers a hydrogen atom to the other and the electrons on both species rearrange themselves to satisfy the octet rule.

There are many varieties of free radical initiators. Chemical initiators decompose to create radicals; examples include organic peroxides, azo compounds, or even oxygen. More rarely we initiate polymerization via a physical condition, such as heat or high energy radiation, to create free radicals directly from the monomers.

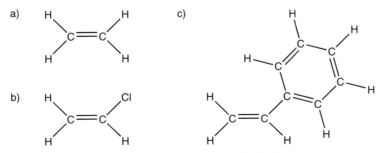

**Figure 2.4**   Examples of monomers that can undergo free radical polymerization: a) ethylene, b) vinyl chloride and c) styrene

## 2.3.2 Anionic Polymerization

Figure 2.5 illustrates anionic polymerization. Polymerization is initiated when a monomer reacts with a negative ion to create an anionic active site. The negative end of the monomer residue, which is now a carbanion, forms an ion pair with the cation from the initiating ionic species, as shown in Fig. 2.5 a). Only monomers that contain a highly electrophilic (electron withdrawing) group adjacent to the vinyl group can be activated to create a carbanion. Some examples of monomers that undergo anionic polymerization are shown in Fig. 2.6. The uneven sharing of charge creates an overall dipole, allowing the partially positive end of the monomer to react with the anion in the catalyst. In a similar manner, activated anionic monomer residues react with monomers, extending the polymer chain and creating a new carbanion in preparation for further monomer addition, as shown in Fig. 2.5 b). Termination, which is illustrated in Fig. 2.5 c), occurs when the active site encounters a species that contains an exchangeable proton (such as an alcohol or water) or with carbon dioxide followed by an acid to form a carboxylic acid terminal group. Chain transfer rarely occurs in anionic polymerization, but is favored by increased reaction temperatures. Due to the limited number

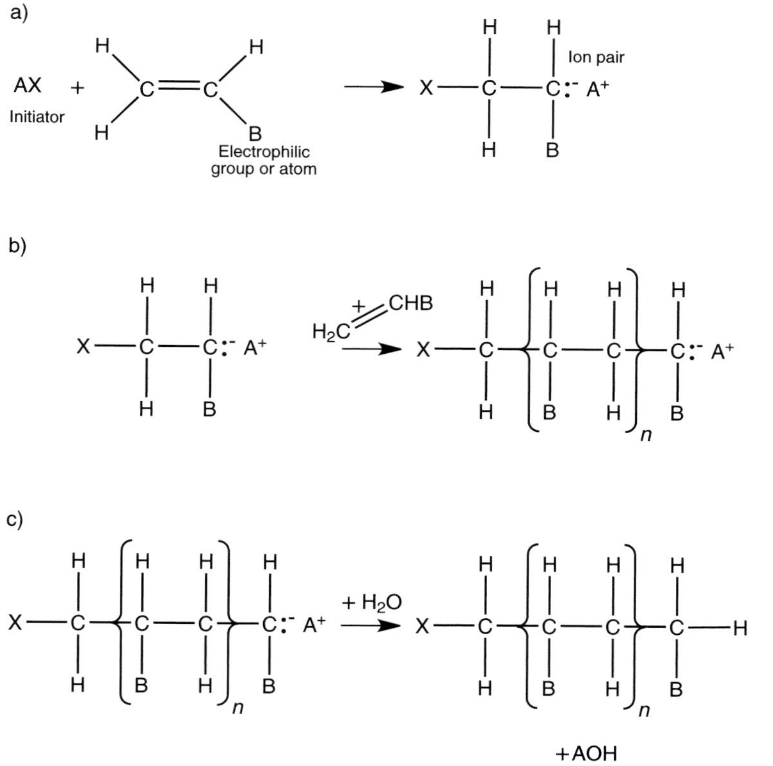

**Figure 2.5** Anionic polymerization:
a) initiation, b) propagation, and c) termination

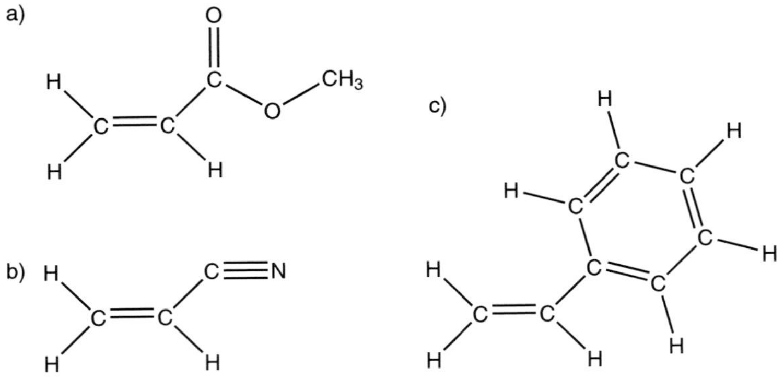

a)

b)

c)

**Figure 2.6**   Examples of monomers that can undergo anionic polymerization:
a) methyl methacrylate, b) acrylonitrile and c) styrene

of side reactions, anionic polymers typically have narrower molecular weight and branching distributions than polymers produced via free radical mechanisms.

In both anionic and cationic polymerization it is possible to create "living polymers". In this process, we starve the reacting species of monomer. Once the monomer is exhausted, the terminal groups of the chains are still activated. If we add more monomer to the reaction vessel, chain growth will restart. This technique provides us with a uniquely controllable system in which we can add different monomers to living chains to create block copolymers.

### 2.3.3   Cationic Polymerization

Figure 2.7 illustrates the principal steps of cationic polymerization. Monomers that undergo cationic polymerization have nucleophilic substituents adjacent to the vinyl group. Examples of monomers that can undergo cationic polymerization include isobutylene, vinyl methyl ether, and styrene, which are shown in Fig. 2.8. These monomers all contain a functional group that can react with a free proton to create a positively charged active site. The free proton can be introduced through either a strong acid or a Lewis acid co-catalyzed with a species containing an exchangeable proton. The strongly electrophilic proton reacts with the weakly nucleophilic double bond electrons of the vinyl group to form a carbon-hydrogen bond, as shown in Fig. 2.7 a). The adjacent carbon atom then becomes positively charged, forming a carbocation, and the negatively charged counter ion remains in proximity to maintain charge neutrality. During propagation, as illustrated in Fig. 2.7 b), monomers are added sequentially to the growing chain. Carbocations are highly reactive which results in a fast reaction rate as well as a high occurrence of chain transfer reactions. These events limit the length of the polymer chains resulting from this reaction mechanism. Termination, as shown in Fig. 2.7 c) occurs when the cationic proton transfers back to the initiating catalyst, thereby deactivating the polymer chain. Commercial polymers manufactured by cationic polymerization include butyl rubber, polymethyl styrene, and polystyrene.

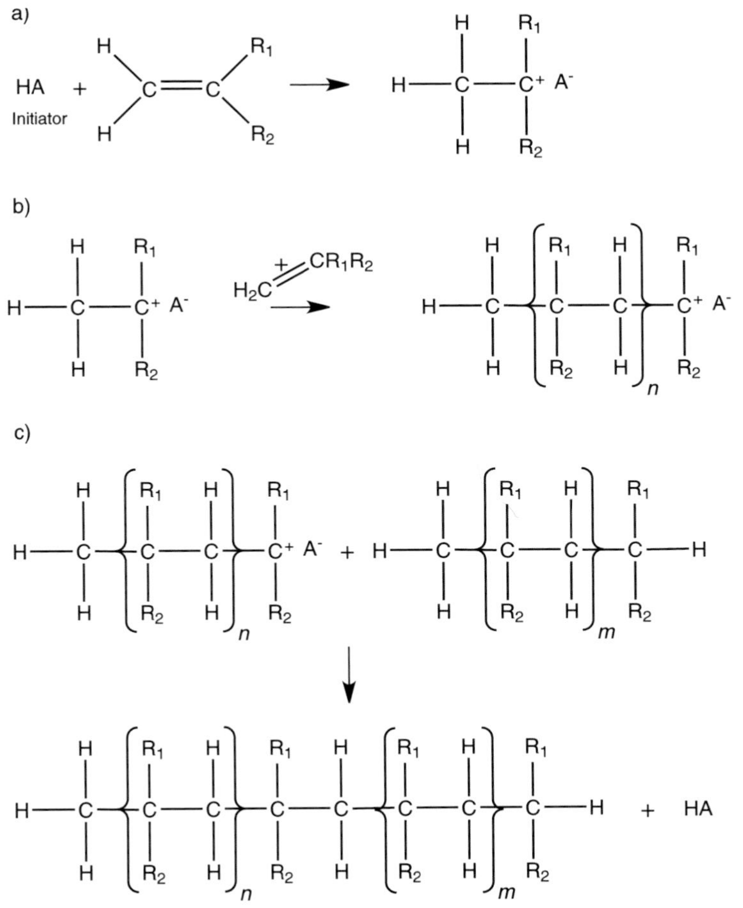

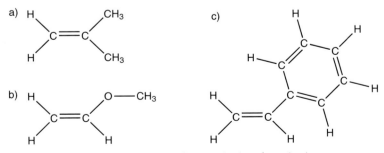

**Figure 2.7**   Cationic polymerization initiated by Bronsted-Lowry acid:
a) initiation, b) propagation, c) termination

**Figure 2.8**   Examples of monomers that can undergo cationic polymerization:
a) isobutylene, b) vinyl methyl ether, and c) styrene

## 2.3.4    Coordination Polymerization of Thermoplastics

Coordination polymerization is another common method of polymer synthesis. During this process, a catalyst forms a coordinated complex with the monomer, as shown in Fig. 2.9, allowing a rearrangement of the electrons that make up the bonds in the monomer. The catalyst holds the intermediate species in place, making it available to react with an incoming monomer. When an incoming monomer approaches the complex with the correct orientation, rearrangement takes place. A bond forms between the approaching monomer and the monomer residue held by the catalyst; the newly attached monomer shifts into the complexation site. The catalyst allows a polymerization reaction to proceed at lower temperatures and pressures than seen in most standard free radical polymerization processes. Coordination catalysts also allow for the stereo-specific addition of monomers by hindering any monomer from approaching the catalytic site that is not correctly oriented. This becomes important for monomers such as propylene and styrene that can form isotactic, syndiotactic, or atactic chains (as shown in Fig. 1.8). The actual mechanism of reaction can be anionic, cationic, or free radical, depending on the monomer and reaction conditions.

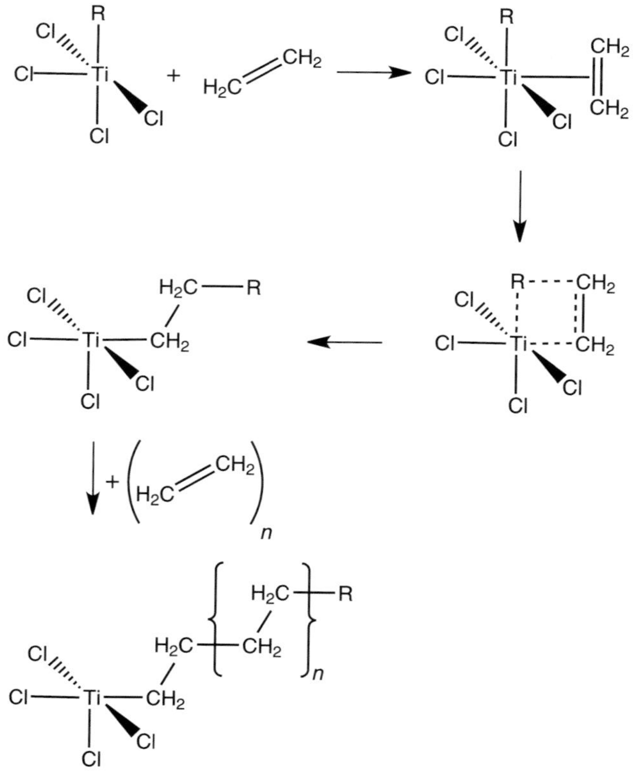

**Figure 2.9**   Polymerization of ethylene via coordination catalysis

In order to enhance the activity of coordination catalysts we typically add a cocatalyst. The cocatalyst works synergistically with the catalyst to allow us to tailor the tacticity and molecular weight of the product while also enhancing the rate of the reaction. An example of a commercially used cocatalyst is methylaluminoxane used in conjunction with metallocene catalysts.

Two well known and widely used families of coordination catalysts are Ziegler-Natta and metallocene catalysts. Ziegler-Natta catalysts consist of a metal chloride (typically titanium) complexed with an organo-metallic salt. Metallocene catalysts, as shown in Fig. 2.10 a) and b), contain a group IV transition metal ion sandwiched between two aromatic five member rings. The metal center acts as the sole polymerization site, creating what is often referred to as a single-site catalyst. The rings may be modified with different functional groups, depending on the desired polymer properties. The nature of the functional groups modifying the cyclopentadienyl rings defines the tacticity of the final polymer. By creating steric interferences between the growing chain and the incoming monomer, the isotactic or syndiotactic structure can be favored. Other commercially important single site catalysts incorporate a single cyclopentadienyl ring bridged to the transition metal, as shown in Fig. 2.10 c), to create a constrained geometry.

Polymers manufactured via single site catalyst technologies, because of the unique chemical catalytic environment, exhibit a more controlled molecular weight distribution and tacticity than seen with Ziegler-Natta catalyst systems.

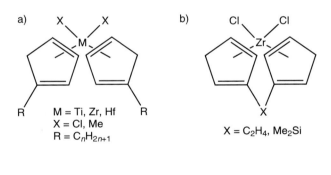

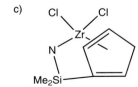

**Figure 2.10**   Examples of single-site catalysts:
a) generic metallocene catalyst,
b) bridged metallocene catalyst and
c) constrained geometry catalyst

## 2.4    Step Growth Thermoplastic Polymers

We can contrast the addition polymerization mechanism with that of step growth polymerization. In step growth polymerization, of which condensation polymerization is the most common mechanism, monomers combine with one another through chemical reactions between the end groups. With the exception of monomers that are incorporated at the ends of polymeric chains, the building blocks of these polymers must be at least difunctional. The growing chain can react with the functional groups of monomers, dimers, trimers, oligomers, or polymers as illustrated in Fig. 1.9. The name "condensation polymer" arose from the low molecular weight byproduct released by each polymerization step, which was seen to condense on the inside of the reaction vessel. Some examples of these byproducts are water, ethanol, and hydrogen chloride. In step growth polymerization, each reactive group is available to react with any other compatible reactive group, regardless of the current length of the chain.

The functional groups that typically participate in this type of polymerization are carboxyl, amine, and alcohol groups. Examples of step growth polymers include polyesters and nylons, which are often spun into fibers used to manufacture carpeting and fabrics, and polycarbonates, which are converted into compact discs, jewel cases, and the large bottles used in water coolers.

### 2.4.1    Mechanisms of Step Growth Polymerization

Figure 2.11 illustrates several examples of step growth polymerization. For most polymers created in this manner, a low molecular weight species such as water, ethanol, or hydrogen chloride is evolved. A minor portion of step growth polymers arise through the rearrangement of the bonds in monomers creating a linkage without losing a low molecular weight species. Table 2.3 lists different step growth polymers and the chemical reaction associated with their manufacture.

One interesting observation regarding the polymerization of many step growth polymers is that the length of the chain does not affect the probability of that chain reacting. Instead, the probability of reaction of any single terminal function group is proportional to its concentration. The theory of equal reactivity of functional groups states that the rate constant of the reaction is independent of the chain length of the reactive species. This relationship does not hold true for the first few polymerization steps when n (the polymerization number) is less than or equal to approximately four. This deviation occurs because the functional groups are so close to one another that they interfere with each other's reactivities. Also, monomers whose intramolecular functional groups are near one another have the same problem.

Monomers that participate in step growth polymerization may contain more or fewer than two functional groups. Difunctional monomers create linear polymers. Trifunctional or polyfunctional monomers introduce branches which may lead to crosslinking when they are present in sufficiently high concentrations. Monofunctional monomers terminate polymerization by capping off the reactive end of the chain. Figure 2.12 illustrates the effect of functionality on molecular structure.

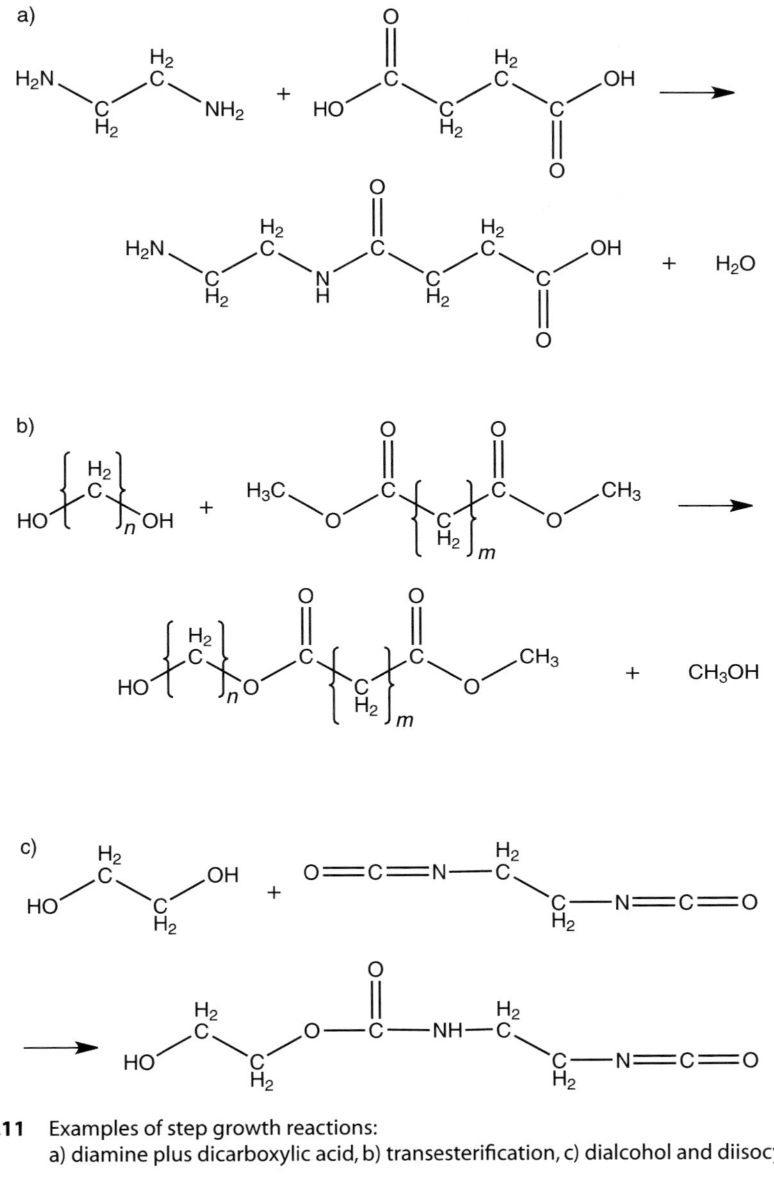

a)

b)

c)

**Figure 2.11**  Examples of step growth reactions:
a) diamine plus dicarboxylic acid, b) transesterification, c) dialcohol and diisocyanate

**Table 2.3**   Polymers Formed via Step Growth Polymerization and Their Reaction Mechanisms

| Type of reaction | Polymer |
|---|---|
| Diamine + dicarboxylic acid | Polyamides (e.g., nylon 66; nylon 610) |
| Diester + diamine | Polyamides (nylon 7, nylon 66) |
| Transesterification<br>-dialcohol + diester<br>-dialcohol + dicarboxylic acid | Polyesters |
| Dialcohol + diisocyanate | Polyurethanes |

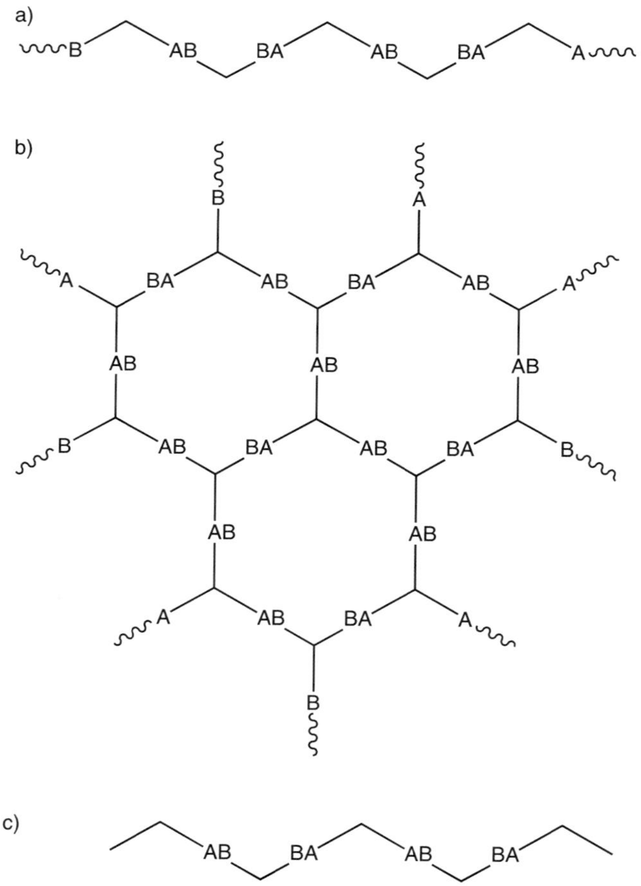

**Figure 2.12**   Effect of monomer functionality on molecular structure:
a) difunctional monomers forming a linear molecule,
b) trifunctional monomers forming a network and
c) monofunctional monomers terminating a linear chain

## 2.5    Comparison of Chain Growth and Step Growth Polymers

We can explain the differences between step growth and chain growth polymers in terms of the mechanism of their manufacture. During addition polymerization, the monomer is essentially consumed at a constant rate as polymer is formed. This fact arises because there is a very low concentration of growing chains in the reaction vessel, relative to the overall concentration of monomer. Since the relative rates of propagation, termination, and chain transfer remain constant during the process, the average molecular weight remains constant throughout polymerization. Additionally, chain growth polymerization is rarely taken to completion. During step growth polymerization, all of the monomers present react quickly at the beginning of the polymerization and the molecular weight gradually increases as a function of time. Figure 2.13 shows how the concentration of monomer and the average molecular weight vary as a function of time for step growth and chain growth mechanisms.

Chain growth and step growth polymers exhibit different molecular weight distributions. Chain growth polymers typically have a broad distribution of molecular weights, ranging from monomers to very high molecular weight species. The products of step growth

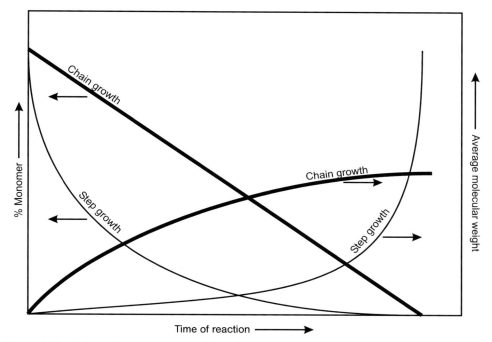

**Figure 2.13**   Schematic representation of the effect of polymerization mechanism on the rate of monomer usage and the average molecular weight as a function of the time of the reaction

**Table 2.4**    Differences in Chain Growth and Step Growth Polymerization Mechanisms

|  | Chain Growth Polymers | Step Growth Polymers |
|---|---|---|
| Molecular weight during polymerization | Average molecular weight changes very little once polymerization has begun as long polymer chains form early in the process | Average molecular weight increases with reaction time |
| Species present during polymerization | Monomer, polymer and propagating chains (at very low concentration) | Monomer, dimer, trimer, oligomer, polymer |
| Rate of monomer utilization | Constant consumption of monomer during polymerization process | Monomer consumed early in the reaction |
| Effect of long reaction times | High conversion of monomer to polymer | High molecular weight species |

polymerization exhibit narrow molecular weight distributions. This is because the average molecular weight increases during the course of the reaction until all the functional groups have been consumed.

A summary of properties of chain growth and step growth polymers can be found in Table 2.4.

# 2.6    Commercial Methods of Thermoplastic Polymerization

When we design commercial polymerization plants we must consider the characteristics of both the monomer and the final product. This allows us to define the optimum configuration to produce a specific polymer. Polymerization reactions can take place in homogeneous solutions or heterogeneous suspensions. For homogeneous processes, the diluted or pure monomer(s) are added directly to one another and the reaction occurs in the media created when mixing the reactants. When the reactants are added directly to one another, the process is referred to as a bulk process. With heterogeneous processes, a phase boundary exists which acts as an interface where the reaction occurs.

## 2.6.1    Bulk Polymerization

Step growth polymers, such as polyesters, are often manufactured via bulk polymerization. The reactive species are mixed together in a stirred reactor designed to promote intimate contact between the reactants. Variables such as temperature and pressure are used to control the molecular properties of the final polymer.

We rarely use bulk polymerization methods to produce chain growth polymers. Bulk polymerization finds little use for these materials, because the process generates relatively few growing chains. These chains, though, grow at a fast rate, often evolving a great deal of heat. Therefore, there is a propensity for "hot spots" in the reaction vessel where the temperature, viscosity, and molecular weight are locally very high. The high viscosity and temperature alter the localized chemical kinetics and heat transfer at that site relative to the rest of the media, making the system difficult to control. The worst case scenario is the runaway reaction where the polymerization is no longer controllable. One catastrophic result is a reactor plugged with very high molecular weight polymer. Chain saws and chisels are required to remove it from the reactor. Even worse, though, is when the heat generated volatilizes components in the reactor, thereby rupturing the pressure relief diaphragm. A mixture of hot, inflammable solvent and monomer is released into the atmosphere, where it is likely to explode, making the neighbors unhappy! To address these issues, engineers work towards creating reaction vessels which promote excellent mixing during polymerization.

## 2.6.2    Solution Polymerization

In solution polymerization, monomers mix and react while dissolved in a suitable solvent or a liquid monomer under high pressure (as in the case of the manufacture of polypropylene). The solvent dilutes the monomers which helps control the polymerization rate through concentration effects. The solvent also acts as a heat sink and heat transfer agent which helps cool the locale in which polymerization occurs. A drawback to solution processes is that the solvent can sometimes be incorporated into the growing chain if it participates in a chain transfer reaction. Polymer engineers optimize the solvent to avoid this effect. An example of a polymer made via solution polymerization is poly(tetrafluoroethylene), which is better known by its trade name Teflon®. This commonly used commercial polymer utilizes water as the solvent during the polymerization process.

## 2.6.3    Types of Interfacial Polymerization

Many polymers can be produced via interfacial polymerization. These multiphase reactions occur in emulsions, suspensions, slurries, or at the interface between a gas and a solid.

### 2.6.3.1    Emulsion Polymerization

In emulsion polymerization, a solution of monomer in one solvent forms droplets, suspended in a second, immiscible solvent. We often employ surfactants to stabilize the droplets through the formation of micelles containing pure monomer or a monomer in solution. Micelles assemble when amphiphilic surfactant molecules (containing both a hydrophobic and hydrophilic end) organize at a phase boundary so that their hydrophilic portion interacts with the hydrophilic component of the emulsion, while their hydrophobic part interacts with the hydrophobic portion of the emulsion. Figure 2.14 illustrates a micellized emulsion structure. To start the polymerization reaction, a phase-specific initiator or catalyst diffuses into the core of the droplets, starting the polymerization.

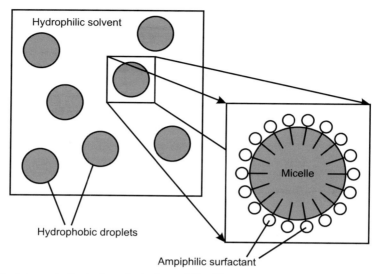

**Figure 2.14**   Schematic illustration of micelles stabilized by amphiphilic surfactant

In suspension polymerization, the monomer is agitated in a solvent to form droplets, and then stabilized through the use of surfactants to form micelles. The added initiator is soluble in the solvent such that the reaction is initiated at the skin of the micelle. Polymerization starts at the interface and proceeds towards the center of the droplet. Polystyrene and polyvinyl chloride are often produced via suspension polymerization processes.

Suspension and emulsion polymerization processes are very similar in that they both require an interface. The main difference is *where* the reaction takes place, on the skin of the suspended droplet or in the center of the micelle.

### 2.6.3.2   Interfacial Polymerization at a Phase Boundary

When the reactants involved in a step growth polymerization process are mutually immiscible, we can employ an interfacial polymerization method. Two solutions, each containing one of the monomers, are layered one on top of the other. This creates a phase boundary that forms with the least dense liquid on top. The different monomers can then meet and polymerize at the interface. A commonly demonstrated example of this is the manufacture of nylon 610 by the interfacial reaction between an aqueous solution of hexamethylenediamine with sebacoyl chloride dissolved in carbon tetrachloride. Because the reaction only occurs at the interface, it is possible to pull the products from this interface to isolate the final product.

Gaseous monomers can polymerize in the gas phase in the presence of a fluidized catalyst bed. As polymer forms, hot gas forces the newly made material out of the reactor to a collector. Figure 2.15 shows a simplified schematic diagram of a generic polymerization reactor.

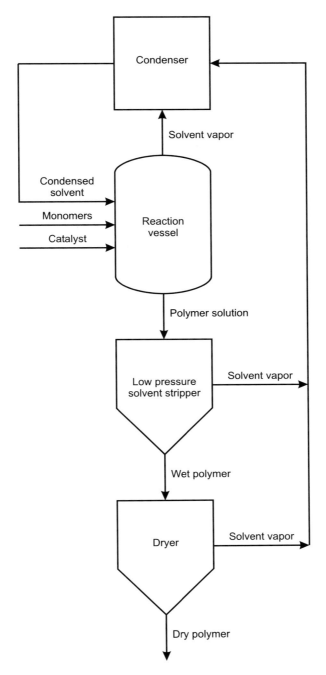

**Figure 2.15**   Simplified diagram showing the principal components of a polymerization system

## 2.6.4     Synthesis of Copolymers

We make copolymers by incorporating two or more different monomers into a single polymer. We can make copolymers via either chain growth or step growth polymerization methods. Copolymers are characterized based on the ordering of their monomers in the final chain. Figure 2.16 illustrates several of the more common classes of copolymer: random, alternating, block, and branched block.

During chain growth polymerization, comonomers incorporate into the growing chain at random or as series of identical monomers known as blocks. The polymerization catalyst and reaction conditions control whether random or block copolymerization occurs. Random copolymers are formed when the probability of incorporating a particular monomer is proportional to its concentration in the reaction medium. Alternatively, the probability that a particular type of comonomer will incorporate into the growing chain may depend on the nature of the active site at the end of the chain. Thus, a given comonomer residue comprising the growing end of a chain may enhance the probability that a similar monomer will be incorporated from the available comonomers. When this happens, the product is a block copolymer.

The type of copolymer formed during step growth polymerization depends on the reactivity of the functional groups and the time of introduction of the comonomer. A random copolymer forms when equal concentrations of equally reactive monomers polymerize. The composition of the copolymer, then, will be the same as the composition of the reactants prior to polymerization. When the reactivities of the monomers differ, the more highly reactive monomer reacts first, creating a block consisting predominantly of one monomer in the chain; the lower reactivity monomer is added later. This assumes that there is no chain transfer and no monofunctional monomer present. If either of these conditions were to exist,

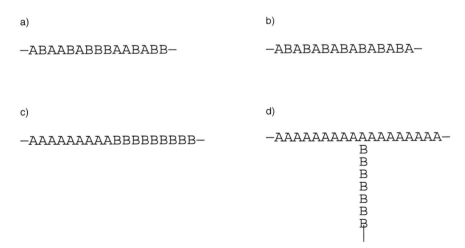

a)

−ABAABABBBAABABB−

b)

−ABABABABABABABA−

c)

−AAAAAAAAABBBBBBBBB−

d)

−AAAAAAAAAAAAAAAAA−
B
B
B
B
B
B
B

**Figure 2.16**    Examples of copolymers:
a) random, b) alternating, c) linear diblock and d) branched block

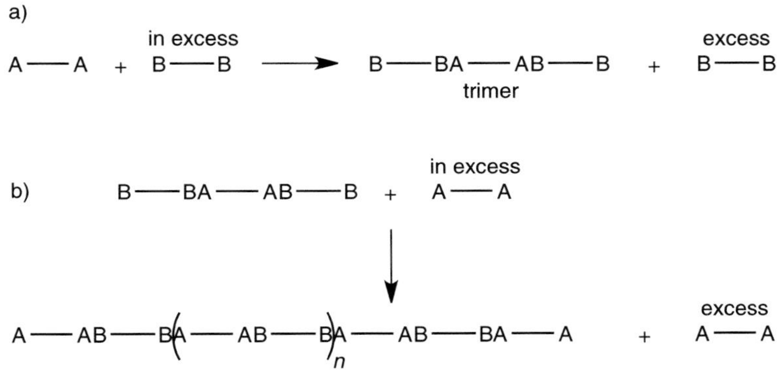

**Figure 2.17**   Formation of an alternating copolymer via a two-step condensation process: a) formation of precursor trimer and b) formation of alternating copolymer

we would produce a bimodal mixture of two copolymers, one of which being predominantly one monomer and the other predominantly the other.

We cannot make block copolymers in single-step batch processes; instead we must use a two-stage process. In the first stage, we create short chains of each of the different monomers, called prepolymers. Polymerization of the different prepolymers together creates the blocks in the second stage, forming the block copolymer. Alternating copolymers can be manufactured in an analogous way. For example, we can create a trimer by reacting a pair of monomers, one of which is present in great excess as shown in Fig. 2.17. The monomer in excess will form the ends of the trimer, while the limiting reagent will be found in the center. This batch of trimers can then react with another species added later to create an alternating structure.

## 2.7   Polymerization of Thermosets

Thermosets differ molecularly from thermoplastics in that their individual chains are anchored to one another through crosslinks. The resulting network creates cohesive materials that demonstrate better thermal stability, rigidity, and dimensional stability than thermoplastics. Some examples of traditional thermosets are melamine-formaldehyde resins, which are used to treat fabrics to make them wrinkle-free, and Bakelite (a phenol-formaldehyde resin), a historically important polymer used in many applications, such as costume jewelry, electrical switches, and radio casings.

We can create crosslinks during chain growth polymerization by copolymerizing dienes with vinyl monomers. When the two vinyl functions of the diene are incorporated into separate chains, a crosslink is formed. This process is shown in Fig. 2.18. When we use a low concentration of dienes, we produce a long chain branched polymer, while high concentrations of dienes create a highly crosslinked polymer network.

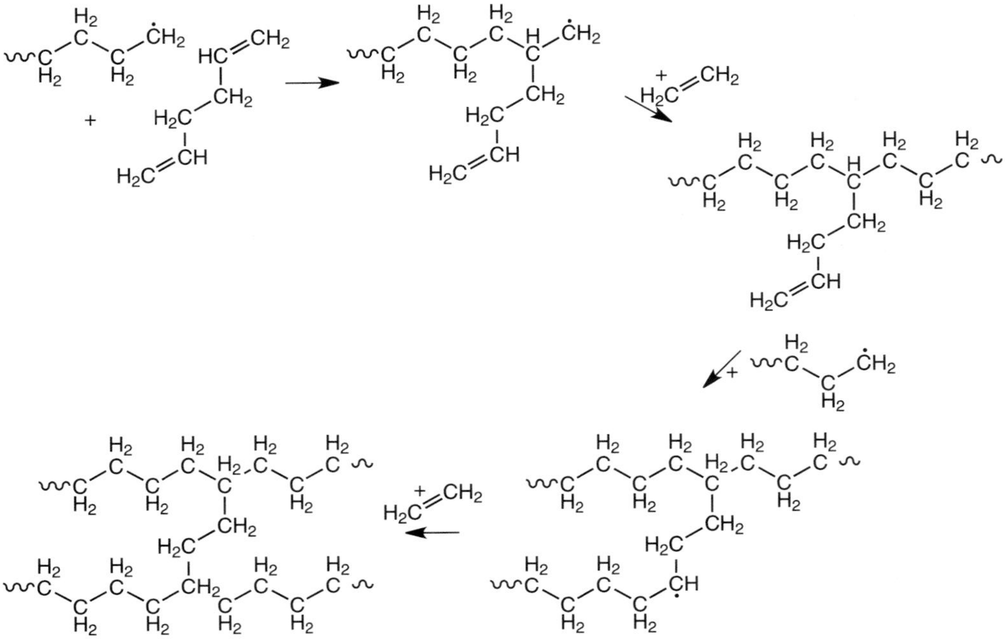

**Figure 2.18**   Crosslink formation in chain growth polymers by the incorporation of dienes with monomers

Secondary processes are normally employed to crosslink chain growth polymers. In one example a linear thermoplastic, such as polyethylene, is compounded with an organic peroxide that is thermally stable at standard processing temperatures but decomposes to chemically react with the polymer chain at higher temperatures creating crosslinks.

Traditionally, we create thermoset polymers during step growth polymerization by adding sufficient levels of a polyfunctional monomer to the reaction mixture so that an interconnected network can form. An example of a network formed from trifunctional monomers is shown in Fig. 2.12 b). Each of the functional groups can react with compatible functional groups on monomers, dimers, trimers, oligomers, and polymers to create a three-dimensional network of polymer chains.

Step growth polymerization can also yield highly crosslinked polymer systems via a prepolymer process. In this process, we create a prepolymer through a step growth reaction mechanism on two of the sites of a trifunctional monomer. The third site, which is chemically different, can then react with another monomer that is added to the liquid prepolymer to create the crosslinked species. We often use heat to initiate the second reaction. We can use this method to directly create finished items by injecting a mixture of the liquid prepolymer and additional monomer into a mold where they polymerize to create the desired, final shape. Cultured marble countertops and some automotive body panels are created in this manner.

## 2.8     Conclusions

In this chapter we have discussed methods of polymerization, the resulting molecular weight distribution, and the interplay between the chemistry of the monomer and the type of polymer that will be produced. We also briefly introduced some of the commercial methods of producing polymers and the role that the type of polymerization has on the choices made in commercial applications. In the following chapters we will build on this framework to explore the role of physical chemical processes, such as the thermodynamic and kinetic processes involved in polymer manufacture. We will also gain an understanding of structural properties of polymers and the means to explore these properties.

## Review Questions for Chapter 2

1.  Give the two types of polymerization and their alternate names. State the characteristics of the monomers used for each type of polymerization.

2.  Describe the fundamental differences between the chemical mechanisms of these two types of polymerization.

3.  Name three polymers made by each of the two types of polymerization. Draw the structure of their monomers.

4.  Name the four sub-types of chain growth polymerization. Give two examples of polymers made by each of the types of addition polymerization.

5.  Name the four mechanistic stages of chain growth polymerization. Draw a general example of each.

6.  Name the two principal types of coordination catalyst. Describe their general characteristics.

7.  Draw the general progression of step growth polymerization. Draw a specific example of the reactions to make one step growth polymer.

8.  Name three industrial types of polymerization reactor.

9.  Name three classes of copolymer. Name three physical properties that are affected by copolymer composition.

10.  How do thermosets differ from thermoplastics?

11.  What is a "living polymer"? What sort of copolymer can be made from it?

# 3    Thermodynamics, Statistical Thermodynamics, and Heat Transfer

## 3.1    Introduction

Thermodynamic studies explore the role of energy changes in a process. Polymerization reaction energies, phase transitions, energies of dissolution, intermolecular forces, molecular conformations and polymer degradation can all be studied using the tools of thermodynamics. The power of thermodynamics lies in its ability to predict what will happen under a given set of conditions. For example, from thermodynamic data we know that when hexamethylene diamine and adipic acid are placed together in a reaction chamber, a reaction will proceed to create a polymer, nylon 66. When we look at the enthalpies of formation for the bonds that are formed and apply some simple thermodynamic laws, we can predict that the reaction vessel will get hot. We can even refine the calculations to determine how much the temperature of the reaction vessel will rise as a function of the reaction and predict how much unreacted monomer will remain once the reaction is complete. Thermodynamics provides us with a powerful set of tools to assist us in the development of polymerization and polymer conversion processes. There are limits though: thermodynamic studies can predict *what* will happen, but they cannot predict *the timescale* on which the events will happen. To understand how quickly reactions occur, we rely on kinetics, which we shall explore in Chapter 4.

Thermodynamic studies are also limited in that they provide information only about the bulk process; they cannot provide information about the behavior of individual molecules. For that level of detail, we rely on quantum mechanics and statistical thermodynamics.

Heat transfer, which is a separate field, explains how quickly energy, in the form of heat, will move through a medium. We know from experience that heat passes more quickly through a solid metal handle on a pot than through a plastic-coated handle. With thermodynamics we can predict which way the heat will flow (from the burner to the handle) but not how long it will take for the energy to make its trip. We use heat transfer equations, details of the type of media, geometrical considerations, and temperature differences to predict the rate at which heat will flow, and the expected temperature gradients. An understanding of heat transfer helps us design polymerization reactors, processing equipment and final products, based on the needs of the application. For example, when we injection mold a polymer, we need to know how long the part has to cool before it can be safely removed from the mold. If we eject the part too early, it may be distorted or damaged by the ejection pins. The time it takes for the part to cool depends on how well the polymer transfers heat. Also, in the previous chapter we mentioned that chain addition polymers are rarely manufactured in a bulk process. This is because heat transfer from the reaction site through the viscous media created by the reaction can be slow, leading to "hot spots". These hot spots create non-uniform reaction conditions resulting in an inhomogeneous polymer.

This chapter will describe how we can apply an understanding of thermodynamic behavior to the processes associated with polymers. We will begin with a general description of the field, the laws of thermodynamics, the role of intermolecular forces, and the thermodynamics of polymerization reactions. We will then explore how statistical thermodynamics can be used to describe the molecules that make up polymers. Finally, we will learn the basics of heat transfer phenomena, which will allow us to understand the rate of heat movement during processing.

### 3.1.1    Energy

Any discussion of thermodynamics must begin with the definition of the system that we are studying. Anything outside of the system is called the surroundings. Although this seems rather simplistic, the concept of a system becomes complicated by the many different types of systems that we can explore. Figure 3.1 depicts various types of systems. In an isolated system neither energy nor matter can enter or leave during the experiment. A perfectly insulated thermos flask with an air tight seal would be an example of an isolated system. An open system is one in which both energy and matter can enter or leave. An open glass beaker on a counter top and a polymer reactor are both examples of open systems. Between these two extremes are a variety of gradations including a closed system in which energy can enter or leave but matter cannot. We can create systems that maintain a constant temperature (isothermal), constant volume (isovolumetric), or constant pressure (isobaric), depending on our needs. We can also retain the heat in a system by insulating it to create an adiabatic system. For example, if we want to determine the melting point of a semicrystalline polymer, we are likely to create an isobaric system by placing a sample of the polymer in an open heating chamber and then heating the polymer while monitoring the temperature of the sample. This is the basis for differential scanning calorimetry measurements, which we will discuss in Chapter 7.

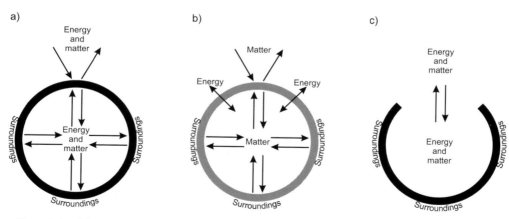

**Figure 3.1**   Schematic representations of thermodynamic systems:
a) isolated system, b) closed system and c) open system

The field of thermodynamics explores energy changes associated with a process. In learning thermodynamics there is one important energetic fact to remember – all systems desire to be at as low an energy level as possible. This statement leads to the critical question: *what exactly is energy?* There are two general forms of energy: kinetic and potential. Kinetic energy is the energy of motion. The beaters on a mixer in your kitchen have kinetic energy as does a rotating screw in an extruder, as we will see in Chapter 11. These moving parts introduce kinetic energy to anything they make contact with, such as cake batter or the molecules of a polymer in the extruder. In a polymer, the individual, long chain molecules have kinetic energy even without mechanical agitation. This arises from the movement of the chain or segments of the chain due to Brownian motion. In fact, the temperature of a system is defined by the average kinetic energy of the molecules which make up the system. A system stores energy as potential energy. Potential energy is often referred to as the energy of position. There is potential energy of attraction between a positively charged object and a negatively charged object that are near one another. An object sitting on a shelf above the floor has potential energy – if the shelf is removed, the object falls to the floor due to gravitational forces.

### 3.1.2   Internal Energy

Systems contain both kinetic and potential energy as internal energy, $U$. The internal energy of the molecules in the system is defined by their translational, vibrational, and rotational motion, the energy associated with their electrons and the energies of their interactions with one another. Translational energy is the energy of movement of molecules or portions of molecules. A simple example of translational motion is that of a baseball as it flies through the air. Vibrational energy arises from the movement of atoms relative to one another across a bond. This motion is often depicted as balls connected by a spring, where the balls represent atoms and the spring represents the bond. Rotational energy is the energy associated with the rotation of a molecule or segments of a polymer chain. The electronic energies define the interaction between the electrons and nuclei in the atoms as well as the strength of the bonds between atoms. There is both a kinetic and potential energy component in this term because of quantum mechanical considerations. Luckily for us, given the complicated nature of polymer molecules, we often ignore the electronic energy as we are more interested in the movement of the whole polymer rather than the movement of the electrons. The final type of internal energy, the energy of interactions, arises from attractive and repulsive forces between atoms in a system.

We cannot directly measure a system's internal energy. Therefore, we have to observe changes in internal energy by monitoring the work ($w$) and heat ($q$) generated or utilized by the system. This fact led to the development of the first law of thermodynamics. This law states that energy cannot be created or destroyed. In our parlance, we say that any change in internal energy ($\Delta U$) in a system has to result in work and/or heat exchanging between the system and the surroundings. Mathematically this can be expressed by Eq. 3.1.

$$\Delta U = w + q \qquad (3.1)$$

Although we all have an intuitive sense of the word heat, thermodynamics provides a rigorous definition. Heat is a form of energy that flows as a result of differences in temperature. For example, when a cup of hot coffee is placed in a cool room, energy moves from the coffee to the room by heat because of the temperature difference. Work is energy that can be harnessed to do something. For example the expansion of a gas, the formation of an electrical current and injection of molten polymer into a mold are all examples of work being done.

The sign of the values on heat and work define the direction in which the energy is flowing. In this text, energy flowing into the system is defined as positive and energy leaving the system is defined as negative. When we extrude a polymer we add internal energy to the polymer by heating it, shearing it, and pumping it through the extruder. The sign of the internal energy change of the polymer would be positive as would the sign of both the work and the heat. When that same polymer exits the die and cools under ambient conditions, its change in internal energy would be negative as it loses heat to the environment.

### 3.1.3   Enthalpy

Internal energy is one way to describe the energy in a system. Another way is to use the concept of enthalpy ($H$). Like internal energy, we cannot measure enthalpy directly. Again we must rely on changes in enthalpy ($\Delta H$) to describe a process experienced by a system. Under isobaric conditions where any work is only due to changes in volumes of gases, changes in enthalpy are equivalent to the changes in heat experienced by a system. This is a rather specific definition based on several assumptions with limited use for our area of study. It does, however, help us make a connection between heat and enthalpy. When enthalpy (or heat) is lost from a system, it is referred to as an exothermic reaction, while any gain in enthalpy (heat) is considered to be endothermic. Equating enthalpy to heat makes the concept easier to understand, but may mislead us when we work on real systems. If we need to explore a system where a gas reacts to create a solid in a closed container, for example the free radical polymerization of ethylene, we must eliminate all assumptions and use the mathematical definition for changes in enthalpy, as defined in Eq. 3.2

$$\Delta H = \Delta U + p \, \Delta V + V \, \Delta p \tag{3.2}$$

Where:

$P$  =  pressure

$V$  =  volume

Enthalpies are often used to describe the energetics of bond formations. For example, when an amide forms through the condensation reaction between an ester and an amine, the new C-N bond, has an enthalpy of formation of –293 kJ/mole. The higher the negative value for the bond enthalpy of formation, the stronger the bond. An even more useful concept is the enthalpy of a reaction. For any reaction, we can use the fact that enthalpy is a state function. A state function is one whose value is independent of the path traveled. So, no matter how we approach a chemical reaction, the enthalpy of the reaction is always the same. The enthalpy of reaction can be determined by knowing the enthalpy of formation, $\Delta H_{form}$, of the reactants and products and subtracting the sum of the enthalpies of the reactants from the sum of the enthalpies of the products, where each term is multiplied by its stoichiometric coefficient $n$ from the balanced equation. For any reaction, the enthalpy change for the reaction, $\Delta H_{rxn}$, is given by Eq. 3.3.

$$\Delta H_{rxn} = \left[ \sum_j n_j \, (\Delta H_{form})_j \right]_{Products} - \left[ \sum_i n_i \, (\Delta H_{form})_i \right]_{Reactants} \tag{3.3}$$

This formula, known as Hess' Law, allows us to determine the magnitude of enthalpy associated with a process and whether that process is endothermic or exothermic.

Enthalpic effects are often used to describe phase transitions. For example, when a low molecular weight material, such as water, undergoes a phase transition, we can monitor the changes in enthalpy through the transition. Figure 3.2 shows a typical enthalpy diagram for the solid to liquid transition of indium at atmospheric pressure. A similar diagram for low density polyethylene can be found in Fig. 3.3. The observed change in enthalpy is associated with the formation of crystallinity in the material. The melting peak in the polyethylene graph is wider than the one seen with indium due to its broad molecular weight distribution, entanglements and range of crystalline domain sizes.

Another valuable use of enthalpy is in the determination of heat capacities. When heat flows into a material, its temperature rises. The temperature increase results from the increased movement of the molecules comprising the material. The composition of the material defines how easily it will undergo a change in its temperature. We quantify the energy required to increase a substance's temperature by determining its heat capacity. The specific heat capacity of a material is defined as the change in enthalpy required to raise the temperature of one mole (or gram) of the material one degree Kelvin. The high heat capacity of polymers is

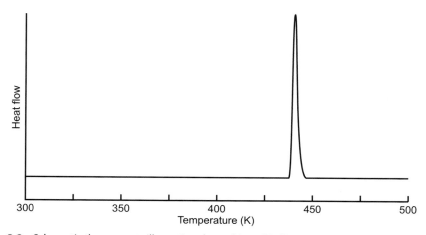

**Figure 3.2**   Schematic thermogram illustrating the melting of indium

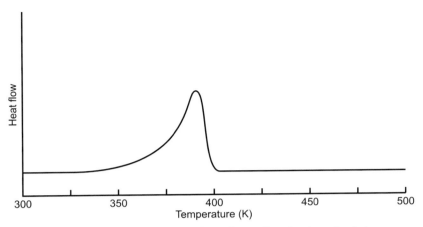

**Figure 3.3**   Schematic thermogram illustrating the melting of low density polyethylene

an important factor in developing a processing system. It will help define how much heat is required to plasticate (make fluid) a polymer so that it can be molded. It will also help us predict how much heat has to be released from the melt to return the polymer to room temperature after processing.

## 3.1.4     Entropy and Spontaneity

Chemists want to know whether or not a reaction will occur spontaneously. A spontaneous process occurs without influence from the outside. For example, ice spontaneously melts at room temperature, rust spontaneously forms on iron nails used in exterior construction, semicrystalline polymers spontaneously crystallize when they cool to their crystallization temperature. Looking again at the formation of nylon, we know that polymerization will occur spontaneously when we combine di-amine and di-carboxylic acid. We also saw earlier that the bond formation has a negative enthalpy of formation (i.e. it is exothermic). This observation makes sense because a process that evolves heat is one where the final state has a lower internal energy than it did initially. Some processes, though, are endothermic but still occur spontaneously. Polymer degradation is typically an endothermic process, but it, unfortunately, occurs spontaneously. Crystalline regions melt spontaneously when heated above their melting temperature. Based on the energy argument that all systems tend towards their lowest energy state, this seems counterintuitive. The apparent paradox was solved when the concept of entropy was developed. Entropy is a measure of the disorder in a system. Systems not only wish to achieve their lowest energy level, they also want to maximize their entropy. A polymer chain is highly ordered relative to the individual monomer molecules from which it is polymerized. Therefore, decomposition into monomer or low molecular weight gaseous products increases the entropy of the system. A solid polymer has lower entropy than a molten one. There is less chain movement and, therefore, less disorder.

So far we have focused on the system under study. The most general criterion for spontaneity is phrased in terms of the entire universe and is summarized in the second law of thermodynamics. This law, that the entropy of the universe must always increase, provides a tool with which we can define the spontaneity of a system. If a process increases the entropy of the universe, that process is spontaneous. The entropy of the universe can increase through any process that a) increases the disorder in the system and/or b) releases energy to the surroundings. The release of energy from the system generates entropy in the surroundings by creating molecular disorder through thermal motions. To quantify this relationship, we utilize the concept of free energy. A negative change in free energy indicates a positive change in entropy of the universe. Most commonly, we use the Gibbs free energy function. The change in Gibbs free energy of a system, $\Delta G$, is defined (at constant pressure and temperature) by Eq. 3.4

$$\Delta G = \Delta H - T\,\Delta S \tag{3.4}$$

Where:

$\Delta H$ = the change in enthalpy of the system

$T$ = the absolute temperature

$\Delta S$ = the change in entropy of the system

When the change in Gibbs free energy is negative, the change in the entropy of the universe is positive, so the process occurs spontaneously. Negative Gibbs free energy changes can be a result of an exothermic process with a corresponding increase in the entropy of the system, for example the burning of gasoline generates both heat and disorder. Another example of negative free energy arises when the exothermic enthalpic effect overpowers the negative entropic effect. This situation occurs when we polymerize monomers. The bond formation is exothermic, but at the same time we are creating order in our system. This means that for a negative Gibbs free energy, the magnitude of the heat released during the reaction must be greater than the product of the absolute temperature and the change in entropy. Whenever an endothermic process occurs spontaneously, the entropic effect dominates the process. An example is when a polymer depolymerizes. Heat must be added, but the high entropy due to the formation of monomers creates a negative Gibbs free energy.

The third law of thermodynamics states that the entropy of a perfect crystal is zero at a temperature of absolute zero. Although this law appears to have limited use for polymer scientists, it is the basis for our understanding of temperature. At absolute zero ($-273.14\,^{\circ}\mathrm{C} = 0\,\mathrm{K}$), there is no disorder or molecular movement in a perfect crystal. One caveat must be introduced for the purist – there is atomic movement at absolute zero due to vibrational motion across the bonds – a situation mandated by quantum mechanical laws. Any disorder creates a temperature higher than absolute zero in the system under consideration. This is why absolute zero is so hard to reach experimentally!

## 3.1.5     Equilibrium

Most chemical reactions do not progress completely from reactants to products. Instead, the net reaction stops in the forward direction when equilibrium is established. Analysis of the contents of the reaction vessel would show a constant concentration of monomers and polymer once equilibrium is reached. This situation is actually a dynamic equilibrium, where the monomers are forming polymers at the same rate as the polymers depolymerize to monomer. Therefore, at equilibrium, the net concentrations of any one species remains constant. The amount of monomer converted into polymer will be defined by the equilibrium constant, $K$. This constant is the ratio of the concentration of the products to the reactants, with each concentration raised to the stoichiometric coefficients in the balanced equation. For Eq. 3.5:

$$a \text{ A} + b \text{ B} \leftrightarrow c \text{ C} + d \text{ D} \tag{3.5}$$

the equilibrium constant is defined according to Eq. 3.6.

$$K_{eq} = \frac{[\text{C}]^c \, [\text{D}]^d}{[\text{A}]^a \, [\text{B}]^b} \tag{3.6}$$

where the brackets indicate concentrations.

Equilibrium equations are often difficult to write out for polymerization processes, because there are several intermediates (dimers, trimers, oligomers, etc.), but, in general, the larger the equilibrium constant, the more polymer will be produced before the forward reaction appears to halt.

The magnitude of the equilibrium constant is defined by the free energy of polymerization. The larger the negative value of the change in the free energy due to polymerization, the more products will form before equilibrium is established. This observation has been summarized (and greatly simplified) in Eq. 3.7.

$$\Delta G^\circ = - R \, T \ln K \tag{3.7}$$

Where:

$\Delta G^\circ$ =  the molar standard state Gibbs free energy (the change in free energy of a reaction when the products and reactants are maintained at standard conditions)

$T$  = the absolute temperature

$K$  = equilibrium constant

$R$  = the ideal gas constant, which is a statement of how much energy is held in a system at a given temperature.

The equilibrium constant is fixed regardless of the concentrations of species or the pressure under which the reaction occurs. Because it is a constant, any change in pressure, or concentrations of species must correspond to the same ratio of concentrations as was observed before the change was imposed on the system. This fact is known as Le Chatelier's principle and is used to optimize many processes. For example, if we knew that the equilibrium constant for a chain growth polymerization reaction describes a 50% conversion of the monomers to polymer, we can increase this yield by increasing the starting concentration of one of the monomers. This addition shifts the ratio of the concentration of products to reactants such that the system does not describe the equilibrium condition. Therefore the reaction must progress to generate more polymer to restore the equilibrium ratio. As we will soon see, the operation of pressurized reactors and thermal stabilizers are predicated on this knowledge.

Equilibrium constants are temperature dependent. In general, the addition of heat to a polymerization reactor will promote a higher yield for an endothermic reaction while reducing the yield for exothermic reactions. Since polymerization is exothermic the temperature of the system increases as the reaction progresses. The heat generated, though, will reduce the yields of the reaction as we just described. For this reason, many polymer reactors employ cooling mechanisms to promote higher yields from the polymerization process.

In summary, the meaning of the different forms of energy, the three laws of thermodynamics and the relationship between equilibrium and free energy are the framework for modern thermodynamics. In the next section we will apply these ideas to a general description of polymerization.

## 3.2    **Energy of Polymerization Reactions**

The reaction of monomers to form a polymer occurs in a multiple-step process. Because thermodynamics depends only on the initial and final state (i.e., a vat of monomers is the initial state and the same vat full of polymer is the final state), we describe polymerization energies as the energy of the polymer relative to the energy of the monomers. The types of energies that we are interested in are enthalpies, entropies, and the combination of these in terms of free energy. The enthalpy of formation of the polymer may be measured directly by adding monomers together in a calorimeter or indirectly through the enthalpy of combustion where the polymer is burned in the presence of oxygen to produce carbon dioxide and water. The enthalpy of polymerization is negative and results from the stability of the polymer relative to the monomer. The negative enthalpy is observed as heat evolved. The magnitude of the enthalpy of polymerization depends on the energetic stability of the polymer relative to its component monomers. Steric hindrance, conjugation, interactions of neighboring atoms, and the reduction of repulsion between groups after polymerization all help define the enthalpy of polymerization.

Because any given polymer sample contains a distribution of different chain lengths and branching, any enthalpy of polymerization values reported will be an average value. Also, no polymerization reaction proceeds fully to completion. Instead, the reaction stops when an equilibrium is established, leaving a mixture of components with a range of concentrations. These variables lead to the approximate nature of enthalpy of polymerization values.

The formation of a polymer from monomers is not entropically favorable. This is because we convert many monomer molecules into a few polymer molecules. This greatly reduces the disorder and motion of the system. The ordering effect observed in polymerization is mitigated somewhat in condensation polymerization processes, by the evolution of low molecular weight species, which contribute to the entropy of the system.

Since the Gibbs free energy of a process is the balance of both enthalpic and entropic considerations, we apply this concept to determine the entropy of polymerization. The simplest method determines the ceiling temperature of polymerization. The ceiling temperature is the temperature at which long chain polymers will not form when the monomer is at a set concentration. This concept is easy to describe in terms of the Gibbs free energy equation. When the change in the Gibbs free energy is zero, there is no net forward or backward reaction. It is at this point that Eq. 3.7 holds true – in other words we are at equilibrium. This means that the change in the enthalpy of polymerization equals the product of the temperature and the change in entropy of polymerization, as shown in Eq. 3.8.

$$\Delta H = T \Delta S \tag{3.8}$$

Knowing this, we can add monomers together at a series of temperatures and determine the point at which no further polymerization occurs, regardless of how long the reaction is observed. We use this temperature and the enthalpy of polymerization to determine the entropy of polymerization.

Once we have determined the entropy and enthalpy of polymerization, we can calculate the free energy of the process at a variety of temperatures. The only time this is problematic is when we are working near the temperatures of transition as there are additional entropic and enthalpic effects due to crystallization. From the free energy of polymerization, we can predict the equilibrium constant of the reaction and then use this and Le Chatelier's principle to design our polymerization vessels to maximize the percent yield of our process.

## 3.3    Energies of Interactions

Atoms interact with one another in a variety of ways. The strongest types of interactions are called bonds. A bond is the strong attraction between two atoms due to the attraction of one atom's nucleus with the other atom's electrons and vice versa. We can understand bonding with a simple potential energy diagram, such as the one shown in Fig. 3.4, which illustrates potential energy as a function of the separation between two atoms. At the right hand side of the curve, the two atoms are far apart and have negligible energy of interaction. As they approach one another, the electrons of each atom are attracted to the positive charge of the nucleus lowering the overall potential energy. The minimum on the curve corresponds to the equilibrium distance due to the minimization of the potential energy. As the atoms continue to get closer, strong repulsive forces begin to dominate. The electrons from the two atoms repel one another as do the two nuclei. The strength of the bond can be determined from the

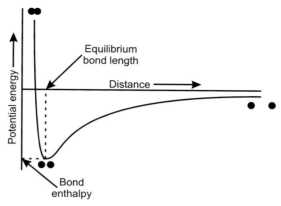

**Figure 3.4**   Potential energy versus separation between two atoms

**Table 3.1**   Lengths and enthalpies of formation for bonds frequently encountered in polymer science

| Bond | Bond Length (pm) | Enthalpy of Formation (kJ/mole) |
|---|---|---|
| C–H | 110 | −413 |
| C–C | 154 | −348 |
| C=C | 134 | −614 |
| C≡C | 120 | −839 |
| C–O | 143 | −358 |
| C=O | 120 | −799 |
| C–Cl | 178 | −328 |
| N–H | 100 | −436 |
| O–H | 97 | −463 |
| O=O | 121 | −495 |

depth of the potential energy well, while the bond length is the equilibrium distance. Some relevant bond lengths and bond energies are shown in Table 3.1.

## 3.3.1   Bonding

Atoms have different abilities to form bonds depending on their valence shell electrons. Most atoms present in polymers strive to meet an octet, i.e. eight electrons in their valence. (Hydrogen atoms are a very important exception, particularly in the case of polymers that contain more hydrogen atoms than carbon atoms.) To reach this goal, atoms form bonds.

Each bond comprises two electrons shared between the atoms. This leads us to the concept of bond order. Bonds exist as single, double, triple, or fractional order, based on how many electrons are associated with the bond. Single, double, and triple bonds are the ones most of us are familiar with, containing two, four or six electrons, respectively. Partial order bonds exist as a result of resonance or the molecular orbital configuration. An example of a polymer that contains delocalized electrons leading to non-integer bond order is polystyrene. The carbon atoms of the aromatic ring are bound to one another through bonds with an order of one and a half.

One of the great misconceptions in chemistry is that all bonds can be categorized as either ionic or covalent. In introductory classes we are taught that a covalent bond is the sharing of electrons between two nonmetal atoms such that the electrons act as internuclear "glue". At the same time, we are told that the simplest ionic bond is the transfer of an electron from a metal atom to a non-metal atom, creating a cation and an anion, respectively. The bond forms because these two species are now attracted to one another by their opposing charges. One could argue that these definitions are true for the most part. The problem arises in that atoms each have their own affinity for electrons as described by electronegativity. For example, a fluorine atom is highly electronegative, meaning it will pull electrons from bound atoms to fill its valence shell. Therefore, when fluorine is bonded to a non-metal, the sharing is not equal. The electrons will spend the majority of their time in the orbital around the fluorine nucleus, creating a polar covalent bond. This creates a partial negative charge on the fluorine. The other atom will have a corresponding partial positive charge. A strongly polar covalent bond is, in essence, almost an ionic bond. We can define a bond's ionic character based on how unevenly the electrons are apportioned among the atoms. By doing this, we see that all bonds fall on a continuum from purely covalent (like the one forming between two nitrogen atoms) and purely ionic (like those in sodium chloride). This concept is illustrated in Fig. 3.5.

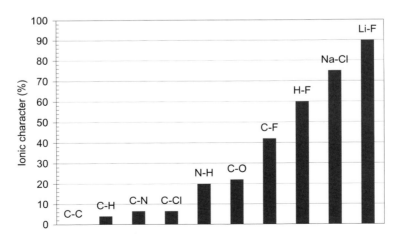

**Figure 3.5**   Ionic character of common bonds found in polymers and those of reference salts

As we have explained in the previous two chapters, polymers are long chains of repeat units connected by bonds. One consequence of these long chains is the high number of intermolecular forces present between polymer molecules. The character of the bonds in a polymer will partially define the strength of the intermolecular interactions between the molecules.

## 3.3.2    Intermolecular Interactions

Intermolecular forces exist between the atoms of molecules as a result of the interactions between the nuclei of one of the atoms and the electrons of the other. Although this sounds very similar to a general description of chemical bonding, there are a number of differences. Chemical bonds are permanent. In this case, "permanent" does not mean that they cannot react; instead, it means that the atoms will remain bonded if they are not disturbed. Intermolecular forces do not share this permanency. The interactions occur very quickly and then, just as quickly, cease when translational and rotational motions separate the interacting species.

### 3.3.2.1   Dispersion Forces

The nucleus at the center of an atom is considered to be a stationary, positively charged center around which electrons orbit. These orbits are not the elliptical paths seen with a satellite traveling around the earth; instead, the electrons are constrained to spend approximately 90% of their time within three-dimensional regions, called orbitals, that lie outside of the nucleus. An electron can be anywhere within this region (and even outside of it the other 10% of the time). This makes for an interesting effect. Every once in a while an atom's electrons all shift to one side of the atom. This creates an instantaneous separation of charge, i.e., a dipole. The negative end will contain the electrons, and the positive end will contain a deshielded nucleus (one that is not evenly covered by the electrons). This situation only lasts for a brief moment, but it is sufficiently long to influence neighboring atoms, inducing corresponding electron shifts in their electron clouds. This effect then continues throughout the material. These instantaneous rearrangements of electrons on their atoms create a type of intermolecular force known as dispersion forces. When we consider that a typical polymer molecule can contain in excess of 50,000 electrons and each polymer molecule is in close proximity to many other polymer molecules we can see that this type of interaction has a huge impact on the behavior of a polymer. In fact, millions of dispersion interactions happen throughout a small polymer sample in any given second. Polyethylene, polypropylene, and polystyrene are examples of polymers whose properties are dominated by dispersion forces.

### 3.3.2.2   Specific Interactions

In addition to the dispersion forces, polymers can interact due to specific interactions when there are polar groups on the polymer chain. Many types of polymer, including polyvinyl chloride, polyesters, and polycarbonates, contain highly polar bonds. In polymers where the polar bonds are not symmetrically placed around the backbone carbon, the polarity of the bonds will affect the properties of the polymer. In general, we see that the polymers are more cohesive and stronger when they contain asymmetric polar bonds.

Another effect can be seen in polymers with highly symmetric polar bonds. Poly(tetrafluoroethylene) is a polymer with very polar bonds. This material, more commonly known by its trade name Teflon®, consists of four carbon-fluorine bonds symmetrically placed in the monomer unit. The fluorine atoms strongly attract the electrons away from the carbon atoms to which they are attached, creating highly polar bonds. This polarity is balanced by equal pulls across the carbon atom making the net polymer non-polar, despite the highly polar carbon-fluorine bond. Since the fluorine has a very high electron density with a net negative charge after borrowing the electron from the carbon, it repels the electrons of other species that come in proximity of the polymer. For this reason, poly(tetrafluoroethylene) is not very cohesive which is part of the reason it is difficult to process. It experiences limited attractions to other materials and is therefore ideal for non-stick coatings on bake ware.

An extremely powerful type of intermolecular forces occurs when hydrogen is bonded to nitrogen, fluorine, or oxygen. The large difference between the electronegativities of hydrogen and these atoms creates highly polar bonds. Such highly polar groups can strongly interact with one another, leading to the formation of "hydrogen bonds". Hydrogen bonds are not true bonds, because electrons are neither shared nor donated and received and they are transient in both liquids and gases.

A final type of intermolecular force, the ion-ion force, is very similar to an ionic bond. In this type of interaction, a charged group, which is part of the polymer, molecularly interacts with an ion in the surrounding matrix. Ionomers rely heavily on these interactions for their properties. For example, a typical ionomer chain contains many carboxylic acid groups that can be deprotonated by reacting with a base. The deprotonated acid groups then have a net negative charge. When the positive cation from the base, such as the calcium or sodium ion, is added to the polymer, it acts as a tie between the chains by attracting the deprotonated groups from different chains.

All of these intermolecular forces influence several properties of polymers. Dispersion forces contribute to the factors that result in increased viscosity as molecular weight increases. Crystalline domains arise in polyethylene because of dispersion forces. As you will learn later in the text, there are other things that influence both viscosity and crystallization, but intermolecular forces play an important role. In polar polymers, such as polymethylmethacrylate, polyethylene terephthalate and nylon 6, the presence of the polar groups influences crystallization. The polar groups increase the intensity of the interactions, thereby increasing the rate at which crystalline domains form and their thermal stability. Polar interactions increase the viscosity of such polymers compared to polymers of similar length and molecular weight that exhibit low levels of interaction.

## 3.4     **Statistical Thermodynamics**

Thus far we have explored the field of classical thermodynamics. As mentioned previously, this field describes large systems consisting of billions of molecules. The understanding that we gain from thermodynamics allows us to predict whether or not a reaction will occur, the amount of heat that will be generated, the equilibrium position of the reaction, and ways to drive a reaction to produce higher yields. This otherwise powerful tool does not allow us to accurately describe events at a molecular scale. It is at the molecular scale that we can explore mechanisms and reaction rates. Events at the molecular scale are defined by what occurs at the atomic and subatomic scale. What we need is a way to connect these different scales into a cohesive picture so that we can describe everything about a system. The field that connects the atomic and molecular descriptions of matter with thermodynamics is known as statistical thermodynamics.

Statistical thermodynamics is based on a statistical interpretation of how atoms and molecules behave. This statistical nature arises because we have so many atoms and molecules in systems and because matter is intrinsically defined based on probabilities, which is the crux of all quantum mechanics. Rather than delve into the great details of statistical thermodynamics, which would far exceed the scope of this text, we will present its foundations only.

First, any system contains many molecules and each molecule has energy. In introducing thermodynamics, we discussed internal energy – that is the type of energy we start with in statistical thermodanymics also. The difference is that we will start with the internal energy of the molecules or atoms rather than the system. The internal energy results from electronic interactions in atoms and the vibrations, rotations, and translations of the atoms and molecules themselves. The total internal energy of the system is the sum of the internal energy of each of the molecules. Since we cannot isolate a single molecule to measure its internal energy, we instead have to describe the probability that it will have a specific internal energy, based on statistical considerations.

We know that the molecules all have their own energy and that the total energy in the system will be distributed among all the molecules based on a function called the partition function. One analogy for this is to imagine a stadium full of people that entered by a single door where they were given some portion of one million dollars randomly. Some people would have enough money to buy a nice car, some might not get any, but most would have about enough to buy a pennant and a soda. The total money in the stadium would be constant. If everyone passed their money around at random, the total currency would not change, but its distribution would. This idea could be mathematically captured in the partition function of money. The same is true of molecules and their energies. A system has some total amount of internal energy. There is a small fraction of molecules with very high energy, a much larger percentage with an intermediate amount of energy and another small fraction with low energies. What we observe when we look at a system is the average of all the molecular energies.

The understanding of the distribution of energies in a system provides an important tool in describing many processes important in polymer science. For example, the rates of reactions, crystallization and degradation rely on energy distributions.

## 3.5     Heat Transfer

The field of heat transfer explores the rate at which heat flows from a region of high temperature to one of low temperature. Heat flow occurs as molecules transfer their thermal energy, in the form of molecular motion, to nearby, lower energy molecules or by fast moving molecules moving to another region of the system. The mechanisms of heat transfer can be categorized as occurring by convection, conduction, or radiative processes. Convection occurs in a liquid or a gas as high-energy, fast moving molecules create an area of low density media that rises relative to the slower moving, denser regions. This molecular movement redistributes the energy in a system. This type of heat transfer is how heated air from a register moves around a room to reach a pleasant 22 °C (72 °F) during the winter. Conduction occurs as high energy molecules collide with lower energy molecules thereby transferring some of their kinetic energy to their collision partner. Conduction transfers the thermal energy around more evenly, allowing heat to travel from warmer to cooler regions. Conductive heating is used in electric stoves to heat pans for cooking. Radiative heat transfer occurs when a warm object emits electromagnetic radiation. The radiation can be used to heat an object at a distance from the heat source. An example of this type of heating is the broiling element in an electric oven.

The mechanisms described above tell us how heat travels in systems, but we are also interested in its rate of transfer. The most common way to describe the heat transfer rate is through the use of thermal conductivity coefficients, which define how quickly heat will travel per unit length (or area for convection processes). Every material has a characteristic thermal conductivity coefficient. Metals have high thermal conductivities, while polymers generally exhibit low thermal conductivities. One interesting application of thermal conductivity is the utilization of calcium carbonate in blown film processing. Calcium carbonate is added to a polyethylene resin to increase the heat transfer rate from the melt to the air surrounding the bubble. Without the calcium carbonate, the resin cools much more slowly and production rates are decreased.

## 3.6     Summary

This chapter surveys the roles that energy and heat play in polymer science. Energy considerations are important in determining the probability of a specific reaction occurring, the viability of a processing design, the role of molecules in large system properties and the mechanism of heat transfer. The goal of this whirlwind chapter is to introduce the connections between the physical concepts allowing a better understanding of polymer processes.

# Review Questions for Chapter 3

1. Explain how the molecules in a solid polymer exhibit both kinetic and potential energy.

2. Why can we not measure the absolute internal energy of a system?

3. When cooling a polymer after plasticating it, we have to remove the excess energy from the part allowing it to reach room temperature. What molecular properties of the polymer will define how much energy must be removed for the part to cool down?

4. Why do the enthalpy diagrams for the melting of indium and low density polyethylene differ from one another? What does the breadth of the melting peak tell us about the polymer under study?

5. Why is the polymerization of diamines with dicarboxylic acids spontaneous, despite the fact that the process dramatically increases the order of the system as the bonds form?

6. How can we determine the entropy of polymerization for step growth polymers?

7. Why is polyethylene a solid at room temperature when the only intermolecular forces it experiences are the weak dispersion forces?

8. Why does poly(tetrafluoroethylene) effectively create a non-stick surface when each bond between the carbon and fluorine is so polar?

9. What is statistical thermodynamics? How does this approach to understanding the thermodynamic behavior differ from the classical approach?

10. How does heat transfer occur in a polymeric part as it cools to room temperature following its high temperature manufacture?

# 4  Kinetics

## 4.1  Introduction

So far, we have discussed the types of reactions that produce polymers, the energy of these reactions, and the role of equilibrium in the polymerization process. Until now we have ignored the length of time required for these processes to occur. This fundamental time-dependent rate information is an essential component of a complete understanding of any reaction.

Rate considerations have enormous practical implications to anyone working with polymers. As an example, it may be possible to make an incredible new polymer, but would we be able to profitably commercialize this super new polymer if its polymerization took weeks, months, or even years to occur? Rather obviously, the answer is "no". Therefore, we must study the rates of reactions in an effort to understand how to produce materials in the time scales we have at our disposal. The study of kinetics provides us with the tools and the knowledge necessary to understand the rates of the polymerization reactions that are important to us. Kinetic studies allow us to understand the energetic considerations necessary for a reaction to progress. We also gain the tools to propose mechanisms that describe how a reaction actually occurs at the molecular level.

### 4.1.1  Defining Rates

One of the challenges in kinetic studies is how to define the rate of a reaction. Typically, we have two options: we can define the rate in terms of how quickly reactant molecules are depleted from the reaction vessel, or we can monitor the rate at which product molecules are formed during the reaction. Either way, we are attempting to define the rate for the same reaction, so we ought to get the same answer. But, often, the stoichiometric coefficients for the individual reactant and product species are different. For example, for the general reaction shown in Eq. 4.1

$$2\,A + B \rightarrow C + 2\,D \tag{4.1}$$

the rate of formation of species D would be twice that of the formation of species C. A graphical way to represent this process can be seen in Fig. 4.1, where the rate at any time, $t$, is the instantaneous slope of any one curve. As seen in the graph, this implies that the rate of a reaction depends on which species we are monitoring. Since the possibility of different rates for the same reaction makes no sense, we must define the rate of a reaction independently of the stoichiometric coefficients of any of the species associated with the reaction. Therefore, we define the rate according to Eq. 4.2:

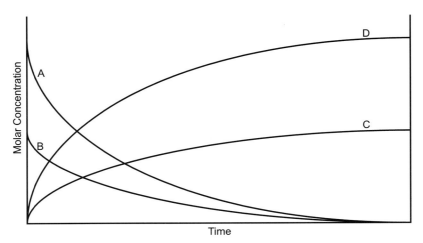

**Figure 4.1**   Concentration of reactants A and B, and products C and D as a function of elapsed time since the start of the reaction

$$\text{Rate} = -\frac{1}{n_{\text{reactant}}}\frac{d[\text{reactant}]}{dt} = +\frac{1}{n_{\text{product}}}\frac{d[\text{product}]}{dt} \tag{4.2}$$

where [reactant], and [product] represent the concentrations of reactant or product species in the reaction. The fraction preceding the rate of change of concentration of each species normalizes the rate for the effect of the stoichiometric coefficient, $n$, from the balanced equation. The sign indicates whether the species is being consumed (a negative sign) or produced (a positive sign). For the example in Eqn. 4.1, the rate is defined as shown in Eq. 4.3.

$$\text{Rate} = -\frac{1}{2}\frac{d[A]}{dt} = -\frac{d[B]}{dt} = +\frac{d[C]}{dt} = +\frac{1}{2}\frac{d[D]}{dt} \tag{4.3}$$

The definition of rate given in Eq. 4.2 allows us to define a rate for any given reaction. The rate calculated will be specific for that reaction, independent of which species we monitor.

## 4.2      The Rate Law: Orders, Rate Constants, and Their Use

Now that we have a definition of rate, we need to develop a basic model to describe the rates of chemical reactions. The basic model takes the form of a rate law. A general rate law allows us to relate the rate of a reaction to the concentration of the reactants present. It makes sense, intuitively, that at low reactant concentrations the rate of the reaction will be quite slow. Also, at high reactant concentrations in the same reaction we would expect a faster rate. The general rate law reflects this intuition by stating that, in simple reactions, such as the one in Eq 4.1, we define the rate according to Eq. 4.4.

$$\text{rate} = k[A]^m [B]^n \tag{4.4}$$

Where:

[A]  =  the concentration of reactant species A
[B]  =  the concentration of reactant species B
$k$    =  the rate constant for the reaction at a specific temperature
$m$   =  order of reactant species A
$n$   =  order of reactant species B

The orders of each reactant are empirically determined values. They provide us with a way to describe the effect of concentration of each species on the overall rate. In simple reactions, orders are zero or integer values, ranging from one to three, though they can be non-integer values. The higher the order for any one species, the more the reaction rate depends on the concentration of this species. The rate constant is also an empirically determined value that provides us information about how "easily" a reaction occurs.

Once we have a defined rate law, we can predict the concentration of any species in the reaction at any time in the reaction by integrating the rate law equation with respect to time. By doing this, the concentration of a species at some time, $t$, relative to its initial concentration at time zero can be determined.

## 4.2.1    The Mechanistic Meaning of Order

The order of an individual reactant or of the overall reaction can provide us with mechanistic clues as to how the reaction takes place. Specifically, the order can provide us with an idea of the molecularity of a reaction. If we think about the ways in which a reaction can occur, there are normally two options. The first is that a molecule simply decomposes. Simple decomposition is represented by an order of one, and is called unimolecular in nature. The second option is that two reactant molecules collide to create products. If this is the case, the reaction has an overall order of two; this type of reaction is considered bimolecular. There are even some reactions that are termolecular, meaning that three reactant molecules must collide simultaneously for a reaction to occur. As you can imagine, the probability of this type of collision is low. The molecularity of a reaction can describe any of the fundamental mechanistic steps and provide us with a step-by-step picture of how a reaction progresses. Most chemical reactions occur through several different mechanistic steps. Often this leads to an overall rate equation which is described with a non-integer value for the order of the reactants. This is only indicative of several intermediate steps, which forces us to recognize that the way the reaction occurs is more complex than the simple, stoichiometric balanced equation describing the overall reaction. The complex nature of most reactions and our dependence on assumptions to simplify the problem means that these mechanisms are hypotheses based on the kinetic data we can obtain for our reaction. This statement may leave you feeling that the field has a lot of grey area left in it. Most of the time, these assumptions are adequate for the task at hand so kineticists, in general, accept the limitations of their field.

## 4.2.2     The Meaning of the Rate Constant

As we explained previously, the rate constant provides us with information about the ease with which a reaction occurs at a given temperature. A high value of the rate constant, $k$, indicates that the reaction progresses quickly and that it is relatively easy to get from reactants to products. A low value of $k$ tells us that the rate is slow; therefore the formation of products from reactants can be difficult. Rather than think in terms of ease or difficulty of a reaction process, we return to a description of what is actually occurring during a reaction at the molecular level. Firstly, we have said that the reactant molecules have to meet. In addition to having to meet, they have to meet under specific conditions. If they simply collide at very low speeds, they are likely to bounce off each other. The attraction between the two species will be insufficient to permit the reactants' bonds to break and then reform to create the product. If, on the other hand, they collide head-on at high speeds, like two freight trains on the same track approaching rapidly from opposite directions, there is plenty of energy and there will be a reaction. This indicates that a reaction requires some threshold energy of impact between the reactant species before the reaction occurs. The higher the threshold energy, the less likely it is that any given collision will have enough energy to result in a reaction, resulting in a low reaction rate. A graphical representation of the activation energy as a function of the reaction coordinate (a way to express how far the reaction has progressed from the perspective of the reactant molecules themselves) with respect to the energy of the reactants and products is shown in Fig. 4.2. Figure 4.2 a) shows that the energy, $E$, of the products is lower than that of the reactants, making the reaction a thermodynamically exothermic reaction. Figure 4.2 b) shows an endothermic reaction, where the energy of the products is higher than that of the reactants.

The simplest model describing the meaning of the rate constant is the Arrhenius equation, as shown in Eq. 4.5.

$$k = A\, e^{\frac{-E_a}{RT}}$$

(4.5)

Where:

$k$ = the rate constant
$E_a$ = the activation energy
$A$ = the pre-exponential factor
$R$ = the ideal gas constant
$T$ = the absolute temperature

Equation 4.5 shows that the rate constant, $k$, is related to the activation energy, $E_a$, of the reaction by an inverse exponential operation. This means that the greater the activation energy, the smaller the rate constant, i.e., it is difficult to get the reactants to meet at high enough energies for the reaction to progress. The pre-exponential factor is a constant that includes information about how orientation of the reactant species to one another and the size of the reactant molecules affect the rate constant. From this equation, we see that the rate constant increases with temperature. This temperature dependence arises from the statistical distribution of the energy of molecules described in Chapter 3. As the temperature rises, more of the reactant molecules collide with sufficient energy to overcome the activation energy barrier. A general rule is that for every 10 °C increase in temperature, we observe a doubling in the rate of the reaction.

a)

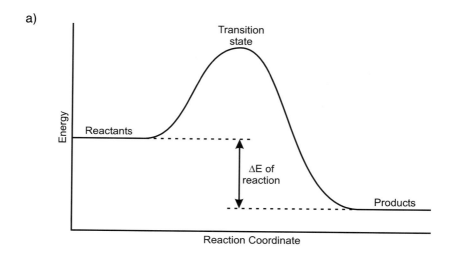

b)

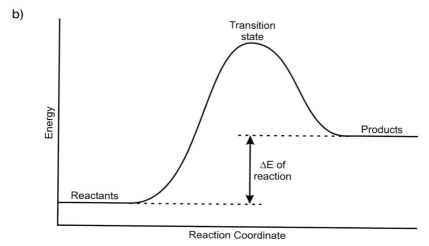

**Figure 4.2**   Reaction profiles for an:
a) exothermic chemical reaction and b) endothermic chemical reaction

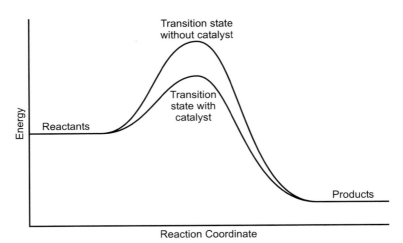

**Figure 4.3**   Effect of catalyst on the energy of the transition state

## 4.2.3    The Role of Catalysts

Most industrial processes rely on catalysts to speed up and control the mechanism of chemical reactions. The manufacturing of high polymers is no exception as described in Chapter 2. There are two general classes of catalysts, homogeneous and heterogeneous catalysts. A homogeneous catalyst is present in the same phase as the reactants, while heterogeneous catalysts exist in a separate phase. Catalysts increase the rate of a reaction by reducing the activation energy of the reaction. Once the activation energy drops, the rate constant increases, allowing the reaction to proceed at a much higher rate. Although catalysts participate in a reaction, their concentration does not change over time because they are regenerated after each use. Despite this, some processes require continuous addition of the catalysts. Impurities can inactivate catalysts by poisoning the active site. Catalysts can also be incorporated into the growing polymer, creating an impurity in the polymer chain. Heterogeneous catalysts also act to concentrate the reactant species in localized areas on the surface of the catalyst through adsorption processes. By gathering the reactants in one area, the probability of the reactant molecules meeting to react increases thereby increasing the rate. Figure 4.3 shows the reaction profile for a general reaction with and without a catalyst.

### 4.2.3.1    Homogeneous Catalysts

We encounter homogeneous catalysts in both step-growth and chain-growth polymerization processes. We saw several examples of these types of reactions in Chapter 2. For example, the acid catalyzed polymerization of polyesters occurs via a homogeneous process as do some metallocene catalyzed polymerization of polyolefins.

Homogeneous catalysts are also often used in cationic and anionic polymerization processes. Lewis acid catalysts, such as boron trifluoride and stannic chloride, accept protons from cocatalysts, such as water and methanol, and donate these protons to the monomer to initiate cationic polymerization. The proton can then be transferred back to the catalytic species, thereby regenerating the catalyst and terminating polymerization on that specific chain. Polyester polymerization can be catalyzed this way, increasing the reaction rate by a factor of 100 to 200 relative to the uncatalyzed reaction.

The term "catalyst" is often misused in anionic polymerization. These mechanisms require the use of initiators that differ from catalysts in that they are not regenerated at the end of the reaction. The similarity between initiators and catalysts is that they both create a situation that permits polymerization via a reduction in the activation energy of the process.

A final example of homogeneous catalysis is the use of metallocene catalyst systems in chain growth polymerization processes. The metallocene, which consists of a metal ion sandwiched between two unsaturated ring systems, is activated by a cocatalyst. The activated catalyst complexes with the monomer thereby reducing the reaction's energy of activation. This increases the rate of the reaction by up to three orders of magnitude.

### 4.2.3.2 Heterogeneous Catalysts

Heterogeneous catalysts are often used in the free radical polymerization of polyethylene. Ziegler-Natta catalysts supported on a silica substrate are an excellent example of this type of catalyst. The complexed transition metal halide and group I–III halide alkyl species bind the ethylene monomer, as described in Chapter 2, holding it in place and advantageously lowering the energy of activation by destabilizing the ethylene bond. This process promotes polymerization with the next ethylene monomer that approaches. In addition to speeding up the reaction, heterogeneous catalysts can direct the polymerization process in such a way as to create linear polymers. This is a result of the catalyst directing the bonding so that the free radicals cannot react randomly in the solution but, instead, specifically react with free monomers making a linear addition to the growing end of the chain. Anionic polymerization processes can also use heterogeneous catalyst systems. For example, metal oxides coated on inert species can be used to create stereospecific polymers of dienes, such as the polymerization of 1,3 butadiene to create polybutadiene with a controlled vinyl content.

## 4.3    Kinetics of Step Growth Polymerization Processes

During step growth polymerization, chain lengths increase as reactive groups meet and react. For example, the polymerization of polyamides results when the amine group and carboxylic acid group meet, creating an amide bond and water as shown in Fig. 2.11a). This process can be thought of solely in terms of the functional group concentration and mobility of those functional groups relative to one another. This means that if we wished to describe the polyamidization process, the rate of the reaction could be defined in the general rate law shown in Eq. 4.6

$$\text{Rate} = k \, [\text{COOH}]^m \, [\text{NH}_2]^n \tag{4.6}$$

As can be see in Eq. 4.6, we need not include any information about the chain length of the reacting species, only the concentration of the reactive groups. This assumption has been shown to be valid as long as the chain length exceeds 3 or 4 residues. Below this point, the rate constant varies due to the effects of the mobility of the reacting groups. The observations of the validity of the model and the limitations of it have been explained via the liquid cage theory, developed in the 1930s by Rabinowitch and Wood. According to this theory, the reaction is described in terms of the time that the reacting groups remain in proximity of one another before diffusing away out of the range in which a reaction could occur. At the onset of polymerization, when the chain length affects the rate of polymerization, the diffusion of reactive groups both towards and away from one another is very fast. Therefore, the reactive groups spend very little time in close proximity to one another. Because typical condensation reactions are rather slow, the reaction kinetics will depend on how often the reactive groups meet. Once the viscosity of the solution within the reaction vessel has increased, the movement of the functional groups is hindered and diffusion is slower. This means that the reactive functional groups stay together for longer times period and therefore experience multiple collisions with one another, thereby increasing the likelihood of a successful reaction. At very high viscosities, this still holds true because, even though there are fewer interactions experienced by any one individual group, the length of interaction is long, making the likelihood of a reaction the same as that seen in lower viscosity media.

## 4.3.1     Kinetic Experiments

To determine the rate behavior of chain growth polymerization reactions, we rely on standard chemical techniques. We can choose to follow the change in concentration of the reactive groups, such as the carboxylic acid or amine groups above, with spectroscopic or wet lab techniques. We may also choose to monitor the average molecular weight of the sample as a function of time. From these data it is possible to calculate the reaction rate, the rate constant, and the order of the reacting species.

## 4.4     Kinetics of Chain Growth Polymerization

Chain growth polymerization processes are more difficult to study than step growth polymerization processes. This is because chain growth polymerization is not the result of a simple organic chemical reaction, such as polyamidization or polyesterification. Chain growth polymerization involves several different processes. First, the process must be initiated. Once initiated, the chain lengths increase due to propagation reactions. During the time that propagation occurs, we also observe chain transfer reactions. The final step is termination. The rate constant describing each of these steps differs because the activation energy differs based on the type of reaction. In order to study the kinetics of chain growth, we need to look at each of these steps and make as many simplifications as reasonable to make a rather intractable problem solvable.

## 4.4.1    Initiation

The initiation step of chain growth creates a reactive site that can react with other monomers, starting the polymerization process. Before the monomer forms the reactive site, the initiator ($I$) (which may be either a radical generator or an ionic species) first creates the polymerization activator ($A$) at a rate defined by the rate constant $k_1$. This process can be represented as shown in Eq. 4.7.

$$I \xrightarrow{\ k_1\ } A \tag{4.7}$$

The activator reacts with the monomer ($M$) present in the reaction vessel to create the propagating species, $MA$, according to Eq. 4.8, with a rate constant $k_2$:

$$M + R \xrightarrow{\ k_2\ } MA \tag{4.8}$$

The rates of the formation of both the activator and the propagating species define the rate of the initiation step. Since the order of any reactant in a mechanistic step refers to the molecularity of the reactant in that step, we can describe the order of the initiator in the reaction. Initiators act by decomposition (either into radicals or ions). Most decomposition reactions are unimolecular and are therefore first order (the exceptions are those that require a collision with a partner to gain the energy to overcome the activation barrier). An additional consideration associated with the formation of the activator is the efficiency, $\varepsilon$, of the interaction of the generated radicals with the monomer to form the $MA$ species. This stems from the fact that not all activators will be successful in finding a monomer before finding another species to react with. This can be represented as a fraction of effective particles and can be used as a multiplicative factor in the general rate equation. For the initiation process, the rate can be approximated by the rate of the formation of the activator, as this is typically the rate limiting step. The preceding considerations lead us to a general rate law for initiation shown in Eq. 4.9.

$$\text{Rate} = k_1\, \varepsilon\, [I] \tag{4.9}$$

where the rate can be described in terms of the extent of the formation of the activated species, $MA$, over time.

## 4.4.2    Propagation and Chain Branching

Propagation and chain branching both maintain the number of radicals generated by the initiation steps, but affect the polymerization kinetics differently. Propagation increases the chain length of the growing polymer by incorporating monomer. Chain branching occurs across two polymer chains and does not incorporate additional monomer.

### 4.4.2.1  Propagation

Propagation steps lead to the incorporation of additional monomers to the polymer at the growing end of the chain. We make the assumption that the rate of addition is constant, regardless of the chain length, because the reaction itself is the same. This is the same assumption we made for the overall polymerization process in step growth polymerization. The reaction can be represented as shown in Eq. 4.10.

$$M + MA \xrightarrow{\ k_3\ } R'MA \tag{4.10}$$

where $R'$ represents the chain attached to the activated group from the prior addition of monomers. The general rate law is represented by Eq. 4.11.

$$\text{Rate} = k_3\,[M][MA] \tag{4.11}$$

As discussed previously, we are able to infer the order of each of the reactants from the mechanisms presented in Chapter 2. In this case, the process by which monomers add to the reactive group is bimolecular, making the order of each of the reactants equal to one.

### 4.4.2.2  Chain Branching

The kinetic description of chain branching is complex, because the probability of a chain branching event depends on many things that we cannot simplify for the model we are developing. Suffice it to say that chain branching slows down the polymerization process. This is because any reaction occurring between chains does not incorporate the free monomer, leading to a reduced rate of monomer consumption.

### 4.4.3  Termination

Termination results in the removal of the activated species from the reaction. It involves the bimolecular reaction between the $MA$ and a specific reactive species, $D$. Depending on the type of polymerization reaction, the reactive species may be a radical or an ion acceptor.

The reaction, then, can be defined as Eq. 4.12.

$$MA + D \xrightarrow{\ k_4\ } M + AD \tag{4.12}$$

In Eq. 4.12, $AD$ is the deactivated product of the reaction. This allows us to write the rate of the reaction for termination as Eq. 4.13.

$$\text{Rate} = k_4\,[MA][D] \tag{4.13}$$

## 4.4.4    The Steady State Approximation

For most chain growth polymerizations, we can assume that the concentration of the activated species, $MA$, remains constant throughout most of the reaction. One caveat in this discussion is that we are assuming that the catalyst's activity remains constant. This assumption is reasonable because the formation of the activated complex occurs simultaneously with termination and at the same rate. Deviations from this simplification arise when we attempt to include the very early and very late stages of the polymerization process. This generally observed behavior allows us to invoke the steady-state approximation, treating the change in the activated species concentration as zero. The value of this approximation is that it allows us to determine how much activated species will be present in a reaction vessel once we know the rate constants, $k_2$ and $k_4$. For the process discussed thus far, the concentration of the activated complex is defined in Eq. 4.14 by the concentration of the initiator and the ratio of the rate constants describing initiation and termination:

$$[MA] = \left( \frac{\varepsilon \, k_2 [I]}{k_4} \right)^{\frac{1}{2}} \tag{4.14}$$

By substituting the definition of the concentration of $MA$ in Eq. 4.14 into Eq. 4.11 we can determine the overall rate of polymerization shown in Eq. 4.15.

$$\text{Rate} = k_3 \left( \frac{\varepsilon \, k_1 [I]}{k_4} \right)^{\frac{1}{2}} [M] \tag{4.15}$$

Here, we see that the rate of the reaction depends on the square root of the concentration of the initiator and linearly on the concentration of the monomer. The steady state approximation fails when the concentration of the monomer is so low that the initiation reaction cannot occur at the same rate as the termination reaction. Under these conditions, the termination reaction dominates the observed kinetics.

## 4.4.5    Measuring the Kinetics of Chain Growth Polymerization

The typical means of measuring the kinetics of chain growth polymerization requires determining the change in molecular weight of the polymer over time. This can be accomplished by analyzing aliquots of the polymerization products at specified times or by using a dilatometer. In a dilatometer, the reactants are mixed together and the volume noted. As polymerization progresses and the average molecular weight of the polymer system increases, the total volume of the materials in the system increases. The polymerized chains hinder one another's motion increasing the free volume of the system. The use of a dilatometer has a major advantage over methods requiring the removal of aliquots; by never touching or removing a portion of the sample, there is no need to alter the system to take a measurement.

## 4.5     Other Kinetic Processes Studied in Polymeric Systems

In addition to polymerization, we often study other processes in polymeric systems via kinetic methods. Most often, we are interested in the kinetics of a process as it relates to the properties of a polymeric material over its lifetime. For example, we are interested in how long it takes for a polymer to develop crystallinity after it solidifies from the melt, as a function of temperature, strain, or additives. The process of crystallization will be discussed in Chapter 5. We are also interested in the time it takes for polymer additives to migrate through a material. Migration kinetics are especially important to manufacturers who are interested in the surface properties of their product because they are concerned about the interaction of the migrating species with their decorating inks or adhesives. Some examples of migratory additives include antifogging agents, slip agents, and antistatic additives, all of which need to diffuse to an interface to produce the desired effect. As we will discuss in Chapter 9, there are also several degradative processes that influence a polymer's properties and its functional lifetime which require an understanding of their kinetics. These processes include thermal, ultra-violet, mechanical, chemical, and biological degradation, as well as oxidation.

## 4.6     Conclusions

The study of polymerization kinetics allows us to understand how quickly a reaction progresses and the role of temperature on the rate of a reaction. It also provides tools for elucidating the mechanisms by which polymerization occurs. In addition, we are able to study the effect of catalysts on the rates of polymerization reactions, allowing us to develop new and better catalysts based on the measured performance.

## Review Questions for Chapter 4

1.   Define rate, order, rate constant and activation energy.

2.   How does a catalyst alter the rate of a reaction?

3.   During step growth polymerization, the rate of reaction slows down as the reaction progresses. What factors are responsible for this observation?

4.   Explain how a metallocene complex reduces the activation energy of the polymerization reaction.

5.  Why are the kinetics of chain growth polymerization more difficult to study than those of step growth polymerization? What simplification do we use to treat the kinetics of the chain growth process? How does this simplification reduce the complexity of the problem and what are the limitations of this method?

6.  Why must we include an efficiency term in the rate equation describing the effect of the initiator on the kinetics of polymerization? What does this term account for?

7.  When describing the kinetics of a polymerization reaction involving a heterogeneous catalyst, we must assume that the activity of the catalyst remains constant. Why does this assumption fail?

8.  Why does chain branching reduce the rate of polymerization in chain growth polymers?

9.  How does a dilatometer work to provide a means of measuring the rate of chain growth reactions? From a practical perspective, what issues are likely to arise during this method of measurement?

10. Why would a manufacturer of polymeric items be interested in the rate of crystallization within a semicrystalline polymer?

# 5　Molecular Characterization of Polymers

## 5.1　Introduction

Polymers differ from most chemicals in that they do not consist of identical molecules. All polymers comprise a distribution of molecules that share certain chemical characteristics, but may vary widely in terms of their lengths and precise compositions. Polymer scientists typically describe polymers in terms of their average molecular characteristics or some readily measurable value that represents the average composition of the ensemble of molecules. Thus, we commonly describe the lengths of polymers in terms of one or more averages that reflect moments of their overall molecular weight distribution. Similarly, we describe chemical composition in terms of the average comonomer content, which does not completely describe the distribution of comonomers within a given chain. Other commonly measured values that are related to polymer composition include melt flow rate, density, and melting temperature.

In order to take advantage of a polymer's properties we do not need to know its precise chemical composition, i.e., the exact chain length, number, and location of branches and comonomers, etc. in every molecule. Under most circumstances it is sufficient for a polymer engineer to know and understand the significance of a relatively small number of key characteristics, such as average molecular weight, molecular weight distribution, average chemical composition, branching type and concentration, crosslinking density, and average tacticity. Not all of these characteristics are applicable to all polymers, so an engineer may need to compare only three or four key variables in order to select an appropriate polymer grade for a given application. Often it is not necessary to know the absolute value of a given molecular characteristic, providing that there is some readily measurable physical property that can be directly correlated to the characteristic in question. For instance, the stiffness of an injected molded polyethylene item is directly related to its density, which can be accurately and quickly measured.

Polymers exhibit various degrees of complexity. At the simplest level are linear homopolymers, which are made from a single type of monomer that is polymerized to form unbranched chains with a statistical distribution of chain lengths. Complexity increases as comonomers are introduced, either at random intervals along the chain, or as blocks of various lengths. Chain branching adds another level of complexity, as does crosslinking. Even linear homopolymers can be made more complex if steric centers are present, which can lead to different types and degrees of tacticity.

## 5.2    Molecular Weight

The average molecular weight and molecular weight distribution of a polymer are extremely important because these characteristics, in combination with chemical composition, largely control a polymer's melt flow characteristics. Other intrinsic properties that are directly influenced by molecular weight include tensile strength, extensibility, and toughness. Molecular weight distribution also strongly influences molecular orientation, which in turn affects many other physical properties, including elastic modulus, tensile strength, tear strength, coefficient of friction, extensibility, refractive index, melting temperature, and solvent resistance. Oriented polymers exhibit anisotropy; that is, their measured properties depend on the direction of testing. We commonly see this in packaging films. As everyone has doubtless experienced, it is easier to tear open a snack package in one direction than another. In this case, the tear propagates much more readily in a direction parallel to the chain orientation than in does in the perpendicular direction.

### 5.2.1    Molecular Weight versus Molecular Length and Molecular Volume

Many polymer properties are related to the size of the constituent molecules. In order to understand these properties it is important that we make clear the relationship between molecular weight, molecular length, and molecular volume. Molecular weight is not a unique descriptor of a polymer molecule. For instance, molecules of polystyrene and high density polyethylene may share the same molecular weight, but polyethylene will have a much longer chain, as illustrated schematically in Fig. 5.1 a) and b). In the absence of long chain branching, the high density polyethylene molecule contains approximately 3.7 times as many backbone carbon atoms as the polystyrene. The difference reflects the ratio between the molecular weights of the styrene and ethylene monomers, of 104 and 28 g/mole respectively.

Chain branching also plays a significant role in producing a more compact molecule as illustrated schematically in Fig. 5.1 b) and c), in which high density and low density polyethylene are compared. Both the molecules contain the same number of carbon and hydrogen atoms, but the low density polyethylene incorporates significant numbers of short and long chain branches. The presence of branches (particularly long chain branches) results in a much more compact molecule than that of a linear molecule of similar molecular weight.

When polymer chains are allowed to reach a fully relaxed state, they adopt a configuration known as a "three-dimensional random walk". This concept was introduced in Chapter 1 to help us visualize a polymer's configuration (see Fig. 1.15). The random angular configurations adopted by the bonds in the backbone of the molecule define the random walk. The volume encompassed by a molecule's random walk depends on its molecular weight, monomer type(s), short and long chain branching, chain flexibility, temperature, and the solvating power of the surrounding medium. Other factors remaining constant, the volume encompassed by a polymer molecule in a random walk configuration increases as molecular weight increases, chain flexibility decreases, branching decreases, and temperature decreases. (The inverse

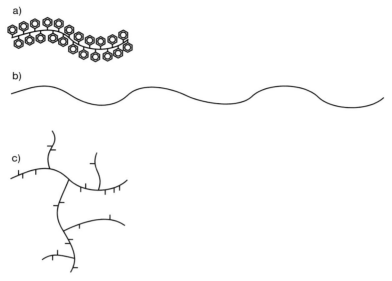

**Figure 5.1**    Schematic representation of relative chain lengths of molecules of:
  a)  polystyrene,
  b)  high density polyethylene, and
  c)  low density polyethylene, sharing the same molecular weight
      (for simplicity, the carbon and hydrogen atoms are not shown)

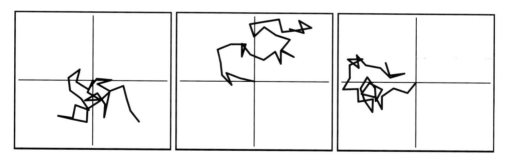

**Figure 5.2**    Planar random walks for chains comprising 30 freely jointed segments

relationship between temperature and volume may seem to be counterintuitive, but is readily explained in terms of chain flexibility. Higher temperatures enhance molecular motion, which increases chain flexibility and thus reduces the molecular volume.) Random walk configurations vary widely from molecule to molecule; in fact, no two are alike. Figure 5.2 illustrates this point for random walks comprising 30 freely jointed segments. Each of the three configurations shown is distinctly different from the other two.

The three-dimensional random walk of an isolated polymer molecule encompasses a very large fraction of volume that is not occupied by the polymer. In practice, polymer molecules are not found in isolation and the random walks of numerous chains overlap each other. Overlap between polymer molecules results in entanglements between neighboring chains. The degree of entanglement becomes more complex as molecular weight increases. Entanglements between molecules play an important role in determining polymer properties in both the molten and solid states.

## 5.2.2    Molecular Weight Distribution

Polymer scientists normally represent molecular weight distribution as a plot of fractional mass versus the logarithm of molecular weight. Figure 1.14 shows a generalized example of such a plot. These plots, while being accurate representations of a polymer's molecular weight distribution, do not provide a ready means for comparing the molecular weight of polymers. As a more practical method, we commonly resort to the use of a few key numerical values that reflect one or more moments of the molecular weight distribution. These average values are increasingly weighted to the longer chains of the distribution.

We calculate the number average molecular weight ($M_n$) from Eq. 5.1.

$$M_n = \frac{\sum M_i N_i}{\sum N_i} = \frac{\sum W_i}{\sum N_i} \tag{5.1}$$

Where:

$M_i$ =  Molecular weight of chains in fraction i

$N_i$ =  Number of chains in fraction i

$W_i$ =  Weight of chains in fraction i

Similarly, we calculate the weight average molecular weight ($M_w$), z average molecular weight ($M_z$), and z + 1 average molecular weight ($M_{z+1}$) from Eqs. 5.2, 5.3, and 5.4, respectively.

$$M_w = \frac{\sum M_i^2 N_i}{\sum M_i N_i} = \frac{\sum M_i W_i}{\sum W_i} \tag{5.2}$$

$$M_z = \frac{\sum M_i^3 N_i}{\sum M_i^2 N_i} = \frac{\sum M_i^2 W_i}{\sum M_i W_i} \tag{5.3}$$

$$M_{z+1} = \frac{\sum M_i^4 N_i}{\sum M_i^3 N_i} = \frac{\sum M_i^3 W_i}{\sum M_i^2 W_i} \tag{5.4}$$

Each of the molecular weight averages is sensitive to a different portion of the overall molecular weight distribution. The shorter chains in a distribution have the greatest effect on the number average molecular weight. The central portion of the distribution has the largest effect on the weight average molecular weight. The z and z + 1 averages are increasingly sensitive to the higher molecular chains, particularly any high molecular weight tails that may be present in the distribution.

It may seem redundant to employ several different molecular weight averages, but each one has its use, reflecting different aspects of a polymer's molecular structure. Thus, the viscosity of a molten polymer at low shear rates is directly related to the weight average molecular weight, whereas the flow rate at higher shear rates is more sensitive to the z average molecular weight.

In order to summarize the breadth of a molecular weight distribution we commonly refer to the ratio of two of the molecular weight averages, such as $M_w/M_n$, $M_z/M_w$, or $M_z/M_n$. Generally, the larger the ratio, the broader the molecular weight distribution. For example, we frequently refer to molecular weight ratios when comparing the flow properties of different polymers. Thus, a polymer with a larger $M_z/M_n$ ratio would be expected to have a lower viscosity at high shear rates than a polymer with a similar weight average molecular weight but a smaller $M_z/M_n$ ratio.

We need to be careful when referring to a polymer by a single molecular weight average without reference to its actual molecular weight distribution. Figure 5.3 illustrates this point, showing four very different molecular weight distributions. Each of these polymers shares a common weight average molecular weight, but their other averages vary widely.

The molecular weight distribution in Fig. 5.3 a) exhibits a "most probable molecular weight distribution", which is characteristic of polymers produced by metallocene catalysts. This distribution contains relatively few molecules with either extremely high or low molecular weights. Products made with this type of distribution are relatively difficult to process in the molten state, exhibit modest orientation, and have good impact resistance.

Resins with a bimodal molecular weight distribution, as illustrated in Fig. 5.3 b), are more readily processed than those with a most probable molecular weight distribution, but are more likely to be oriented in the solid state and have a somewhat reduced impact resistance.

Resins with a significant low molecular weight tail, as shown in Fig. 5.3 c), exhibit similar melt flow characteristics to the most probable molecular weight distribution, but may be more flexible in the solid state due to the plasticizing effect of the shorter chains.

Resins that have a high molecular weight tail, as illustrated in Fig. 5.3 d), often display good melt processing characteristics. The long chains take longer to relax than the shorter ones and are responsible for the high orientation often associated with these products. Thin films made from such resins are likely to display a high degree of anisotropy; they may be readily torn parallel with the preferential chain orientation, but show good tear resistance in the perpendicular direction.

In practice, product developers often blend two or more resins together in order to obtain a product that has the required melt flow and solid-state characteristics. Thus, we frequently combine metallocene catalyzed linear low density polyethylene, having a most probable molecular weight distribution, with low density polyethylene, having a broad molecular weight distribution. The linear low density polyethylene provides good impact resistance, while the low density polyethylene improves melt flow characteristics.

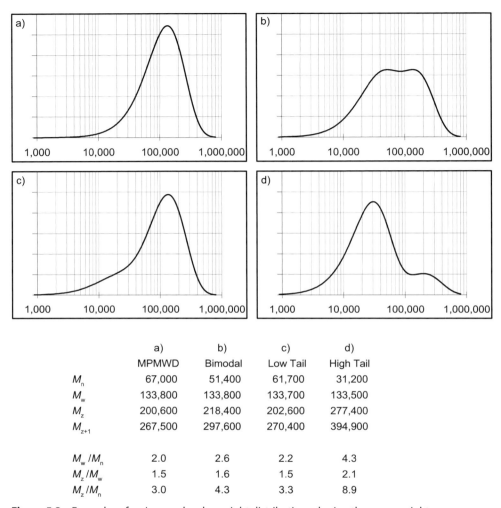

| | a) | b) | c) | d) |
|---|---|---|---|---|
| | MPMWD | Bimodal | Low Tail | High Tail |
| $M_n$ | 67,000 | 51,400 | 61,700 | 31,200 |
| $M_w$ | 133,800 | 133,800 | 133,700 | 133,500 |
| $M_z$ | 200,600 | 218,400 | 202,600 | 277,400 |
| $M_{z+1}$ | 267,500 | 297,600 | 270,400 | 394,900 |
| | | | | |
| $M_w/M_n$ | 2.0 | 2.6 | 2.2 | 4.3 |
| $M_z/M_w$ | 1.5 | 1.6 | 1.5 | 2.1 |
| $M_z/M_n$ | 3.0 | 4.3 | 3.3 | 8.9 |

**Figure 5.3**   Examples of various molecular weight distributions sharing the same weight average molecular weight

### 5.2.3    Molecular Weight Determination

We can measure a polymer's molecular weight by one of two general methods. In the first, we fractionate the polymer to generate a plot of fractional mass versus molecular weight. The most commonly used method of this type is size exclusion chromatography. The second family of molecular weight determination methods yields a single value that represents a complex function of the overall molecular weight distribution. This category includes light scattering, viscometry, and melt flow rate measurements.

### 5.2.3.1    Size Exclusion Chromatography

Size exclusion chromatography (which is also known as gel permeation chromatography) is based on the premise that a polymer molecule in solution adopts a random coil configuration, which encompasses a volume (known as its hydrodynamic volume) that is proportional to its molecular weight. We fractionate polymers according to their hydrodynamic volumes to generate a molecular weight distribution plot.

Figure 5.4 illustrates the principles of size exclusion chromatography. The first step is to inject a dilute polymer solution into a column packed with crosslinked polymer microspheres, the surfaces of which are honeycombed with pores of various sizes. The more pores that a polymer molecule can enter, the longer it takes to elute from the column. Thus, the largest molecules, which can only enter the largest pores, elute first. Smaller molecules can enter a larger fraction of the pores and thus their passage through the column is retarded. We record the concentration of the polymer solution exiting the column as a function of time. The molecular weight of each fraction is calculated from a calibration curve that relates molecular weight to elution time. We can then plot the molecular weight distribution based on the mass and molecular weight of each fraction. Finally, we calculate the various molecular averages and their ratios. Naturally, all the calculations and graphing is carried out by computer.

Long chain branches complicate the measurement of molecular weight. This is because the hydrodynamic volume of a molecule is not simply related to its molecular weight. For a given molecular weight, branched molecules have smaller hydrodynamic volumes than linear ones. In order to correct for this complication, we employ two or more detectors to analyze the solution as it elutes from the column. One of the detectors (typically a refractive index detector) measures the polymer concentration in the eluent, while the other (normally a light scattering or viscometric detector) measures the hydrodynamic volume of the eluting polymer molecules. The variance between the observed relationship between the two detectors and a theoretical value for linear molecules provides a measure of the branching. The greater the variance, the more branched are the polymer molecules.

### 5.2.3.2    Intrinsic Viscosity

The viscosity of a dilute solution of polymer depends on the molecular weight of the polymer. This gives us a simple method for measuring molecular weight based on viscosity, which is readily measured.

To perform this analysis, we first prepare a dilute solution of polymer with an accurately known concentration. We then inject an aliquot of this solution into a viscometer that is maintained at a precisely controlled temperature, typically well above room temperature. We calculate the solution's viscosity from the time that it takes a given volume of the solution to flow through a capillary. Replicate measurements are made for several different concentrations, from which the viscosity at infinite dilution is obtained by extrapolation. We calculate the "viscosity average molecular weight" from the Mark-Houwink-Sakurada equation (Eq. 5.5).

$$[\eta] = K\, M_v^{\alpha} \tag{5.5}$$

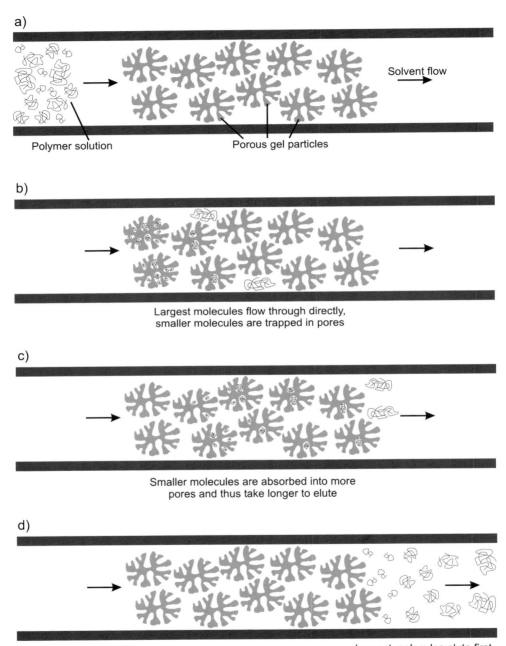

a)

Solvent flow

Polymer solution

Porous gel particles

b)

Largest molecules flow through directly,
smaller molecules are trapped in pores

c)

Smaller molecules are absorbed into more
pores and thus take longer to elute

d)

Largest molecules elute first,
followed by increasingly
smaller molecules

**Figure 5.4**   Principle of size exclusion chromatography

where:

$[\eta]$     = viscosity at infinite dilution

$K$ and $\alpha$   = Mark-Houwink constants (which reflect polymer/solvent interactions)

$M_v$     = viscosity average molecular weight

The viscosity average molecular weight typically falls somewhere between $M_n$ and $M_w$.

### 5.2.3.3   Melt Flow Rate

The rate at which a molten polymer flows through a capillary under standard conditions is known as its "melt flow rate" (MFR). The melt flow rate of a polymer is inversely related to its molecular weight. Theoretically, a melt flow rate can be converted to a molecular weight, but this is rarely done in practice. Polymer engineers commonly refer to melt flow rates when comparing material specifications and make comparisons directly based on these values.

The melt flow rate of a polymer is the weight of polymer in grams that extrudes from a standard capillary die under a standard load, at a standard temperature, over a ten minute period. The term "melt index" is used exclusively for polyethylene; melt flow rate is the preferred term for all other polymers. We measure melt flow rates using a piece of equipment called a "melt indexer". The capillary dimensions, testing temperature, and load are specified for a given polymer by the National Institute for Standards and Testing.

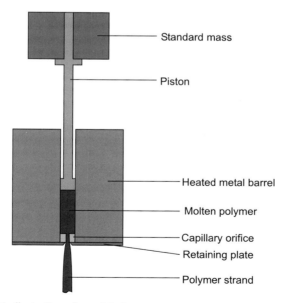

**Figure 5.5**   Schematic illustration of a melt indexer

Figure 5.5 shows a schematic diagram of a melt indexer (which is also sometimes referred to as an extrusion plastometer). To determine the melt flow rate of a polymer resin, we place a suitable mass of it into the barrel, which is pre-heated to a standard temperature appropriate to the polymer. We then place a weighted piston on top of the sample. After allowing the polymer to reach the temperature of the barrel we allow it to extrude from the capillary orifice. The melt flow rate is the mass of polymer in grams that extrudes in ten minutes.

## 5.3     Homopolymer Chemical Composition

Homopolymers can exhibit two types of chemical variation along their lengths. Regiodefects occur when asymmetric monomers are incorporated into the polymer in some combination of "head-to-tail", "head-to-head" and "tail-to-tail" configurations. Homopolymers display tacticity when their monomer residues contain a steric center that is incorporated into the backbone. Stereodefects occur when a regular sequence of tacticity is interrupted by a steric center with the opposite configuration from one or both of its neighbors. Certain polymers, notably polypropylene, can exhibit both regiodefects and stereodefects

### 5.3.1     Regiodefects

Regiodefects occur in addition polymers when a monomer is added to the growing chain in a configuration that is reversed relative to those around it. Figure 5.6 shows an example of a misinsertion in a polypropylene chain. "Head-to-tail" polymerization is followed by a misinserted monomer that introduces a "head-to-head" and "tail-to-tail" configuration, followed by a return to head-to-tail polymerization. Condensation polymers typically do not contain regiodefects, because the polymerization process inevitably results in a head-to-tail configuration.

Regiodefects are less readily incorporated into crystallites than defect-free chain sequences. In semicrystalline polymers, increasing levels of misinsertion result in reduced crystallinity. This can affect numerous physical properties, resulting in reduced modulus, lower heat distortion temperature, and decreased tensile strength.

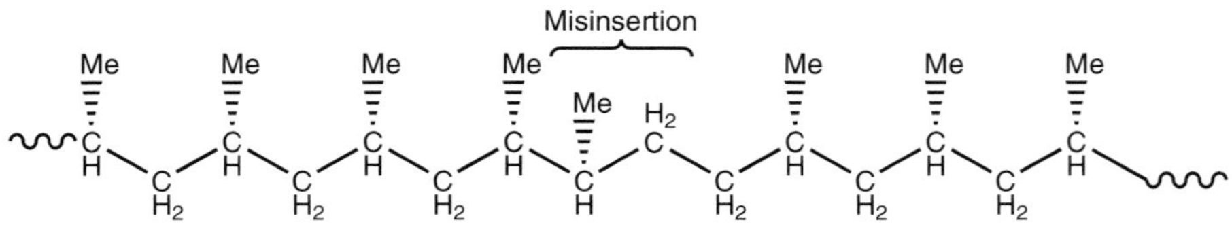

**Figure 5.6**   Regiodefect in a polypropylene chain

## 5.3.2 Tacticity

Polymers that incorporate steric centers into their backbones can display various types of tacticity. The three principal types of tacticity are isotactic, syndiotactic, and atactic, as illustrated in Fig. 1.8 for polypropylene. Other polymers that display tacticity include polystyrene and poly α-olefins.

When adjacent monomers in a backbone share the same stereoconfiguration, the placement is known as a meso diad. When adjacent monomers have opposing stereoconfigurations, the placement is known as a racemic diad. Thus, a purely isotactic polymer comprises all meso placements, and a syndiotactic polymer consists of all racemic placements.

In practice, the monomers comprising a polymer are never arranged in a purely isotactic or syndiotactic fashion. All stereoregular polymers contain some misinsertions. The polymerization catalyst and polymerization conditions control the level of stereoregularity.

### 5.3.2.1 Isotacticity

In isotactic polymers the configuration of the steric centers on the backbone is identical. The net result is that all side groups are positioned on the same side of the chain, as illustrated schematically in Fig. 1.8 a).

The presence of all the side groups on the same side of the polymer chain has a significant effect on chain configuration. The side groups tend to interfere with each other, impeding free rotation of the backbone bonds and rendering certain configurations energetically unfavorable. The chain configuration shown in Fig. 1.8 a), in which the backbone carbon atoms are arranged to form a planar zig-zag, is unstable. As drawn, the pendant methyl groups would sterically interfere with their nearest neighbors. In practice, the closest that an isotactic chain can come to a linear configuration is a helix with the pendant groups extending outwards from the principal chain axis. The larger the pendant group, the greater the number of monomers per turn of the helix. Thus, isotactic polypropylene, with its small methyl side group, forms a helix with three monomers per turn, while poly-1-butene, which has an ethyl side group, adopts a helix with four monomers per turn. Due to their regular nature, isotactic polymers will crystallize from the molten state when thermodynamic conditions are favorable.

Stereodefects are the result of one or more racemic diads interrupting a sequence of meso diads. Figure 5.7 illustrates the two principal types of stereodefect. In Fig. 5.7 a), a single racemic placement results in the subsequent methyl groups being placed on the opposite side of the chain from those of the preceding sequence. In Fig 5.7 b), a pair of racemic placements interrupts the meso sequence. In this case, both the meso sequences have their methyl groups on the same side of the chains.

Stereodefects reduce the overall regularity of an isotactic polymer chain and hinder its ability to crystallize. As the concentration of defects increases, the degree of crystallinity falls, resulting in reduced density, reduced melting temperatures, lower heat distortion temperatures, reduced modulus, and reduced yield stress.

a)

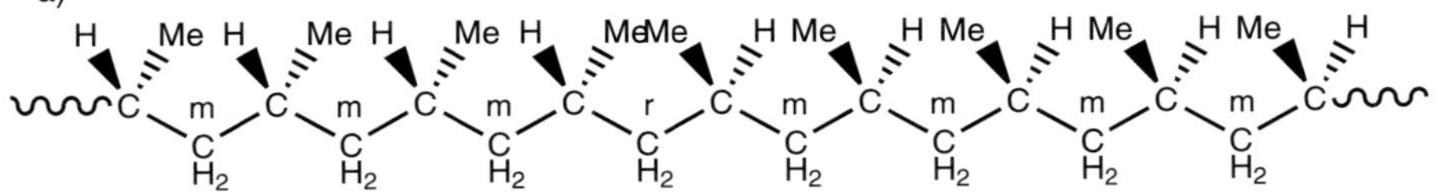

b)

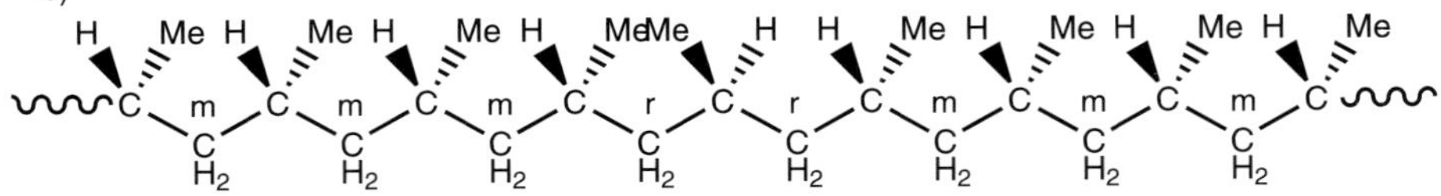

**Figure 5.7**  Principal types of stereodefect found in polypropylene:
a)  single racemic placement,
b)  paired racemic placements (m = meso placement, r = racemic placement)

### 5.3.2.2  Syndiotacticity

In syndiotactic polymers the configurations of the steric centers on the backbone alternate. The net result is that side groups are positioned on the opposite side of the chain from their nearest neighbors, as illustrated schematically in Fig. 1.8 b).

The syndiotactic configuration is less sterically hindered than the corresponding isotactic configuration. However, there is still significant steric interference between pendant groups and the backbone, which hampers bond rotation. Due to the alternating steric configuration in syndiotactic polymers, it takes two monomer residues to form a single repeat unit. Syndiotactic polymers can crystallize, but tend to do so fairly slowly due to steric interference and the fact that repeat units comprising two monomers are required to organize into repeating arrays.

Stereodefects in syndiotactic polymers have similar effects to those in their isotactic counterparts, reducing crystallinity levels and changing the associated physical characteristics.

### 5.3.2.3  Atacticity

In atactic polymers, side groups are irregularly positioned on either side of the chain, as illustrated schematically in Fig. 1.8 c). A truly atactic polymer would comprise a random distribution of steric centers. In practice, atactic polymers typically show some preference for either meso or racemic placement. The tendency towards stereoregularity is due to the fact that polymerization catalysts often contain steric centers, which tend to direct the incoming monomers and the growing chain into preferred configurations.

The approximately random placement of side groups in atactic polymers prevents them from developing regular structures. For this reason, atactic polymers are non-crystalline and behave as rubbers or glasses, depending on whether they are above or below their glass transition temperature.

### 5.3.3    Composition Analysis

We use carbon-13 nuclear magnetic resonance (NMR) spectrometry to determine the regio- and stereoregularity of homopolymers in dilute solution. NMR spectrometry measures the characteristic electromagnetic frequencies of certain types of atomic nuclei as they relax from a high energy state to a low energy state after being excited in the core of a powerful magnet. Each carbon-13 nucleus exhibits a resonance frequency that is characteristic of its chemical configuration and other carbon atoms in close proximity, with which it interacts. The area of each peak in the spectrum is directly proportional to the concentration of carbon atoms in a particular configuration. By comparing the area of individual peaks with the integrated area under all the peaks, we can accurately determine the concentrations of each configuration present. We normally describe isotactic and syndiotactic polymers in terms of the percentage of placements that are either meso or racemic.

## 5.4    Copolymer Chemical Composition

When two or more monomers are polymerized into the same molecular chain they produce a copolymer. The distribution of monomers, in terms of their relative concentrations and placements, is responsible for controlling a copolymer's properties. Figure 5.8 illustrates five possible comonomer distributions for a copolymer comprising equal numbers of two types of monomer. The relative concentrations of the different monomers and the lengths of the various blocks can be varied widely. Relatively small changes in comonomer concentration and placement can result in significant changes in physical and chemical properties. Properties that can be modified include such diverse characteristics as extensibility, elastic recovery, modulus, heat resistance, printability, and solvent resistance.

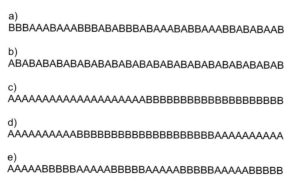

**Figure 5.8**   Various types of copolymer, where A and B are different monomers: a) random, b) alternating, c) diblock, d) triblock, e) short blocks

## 5.4.1     Comonomer Distribution

The monomers that comprise copolymers can be introduced in one of three general ways, randomly, as regularly alternating series, or as blocks of identical monomers. The polymerization catalyst and the reaction conditions control the type of comonomer distribution produced.

### 5.4.1.1     Random Copolymers

Random copolymers having equal numbers of two different comonomers, as illustrated in Fig. 5.8 a), exhibit no regularity and thus do not crystallize. However, if one of the monomers outnumbers the other by an order of magnitude or more, the situation is quite different. When one comonomer dominates, the other comonomer can be considered as a source of defects that separate homopolymer sequences of the major comonomer. If the homopolymer sequences are sufficiently long, they may be capable of crystallizing. Such copolymers are typically made by addition polymerization. Examples include linear low density polyethylenes, which comprise a majority of ethylene, copolymerized with small numbers of $\alpha$-olefins, and propylene-ethylene copolymers in which propylene is the major component. The concentration of the minor comonomer strongly influences the properties of such copolymers. When the minor comonomer is present at a negligible concentration, the properties of the copolymer approach those of a homopolymer of the major component. As the concentration of the minor comonomer increases, the regular sequences of the major comonomer become shorter, which can affect properties drastically. Thus, a copolymer comprising 95 wt% isotactic polypropylene and 5 wt% ethylene has a peak melting temperature of approximately 130 °C, compared to an isotactic homopolypropylene that has a peak melting temperature of approximately 160 °C.

### 5.4.1.2     Alternating Copolymers

Alternating copolymers, as illustrated in Fig. 5.8 b), are generally made by condensation polymerization of two different monomers. Such copolymers display regularity and are capable of crystallizing under the appropriate conditions. Examples of such copolymers include nylons 66 and 610, and various types of polyurethane.

### 5.4.1.3     Block Copolymers

Diblock copolymers, as illustrated in Fig. 5.8 c), comprise homopolymer sequences of the two monomers linked together. The homopolymer blocks may be either compatible or incompatible, depending on their chemical structure. If the sequences are compatible, they will mix to form a material with characteristics similar to those of a blend of the two homopolymers. On the other hand, if the blocks are incompatible, they will tend to segregate from one another to form distinct phases. Each phase will display properties characteristic of the homopolymer, modified by the constraints placed on them by having one end attached

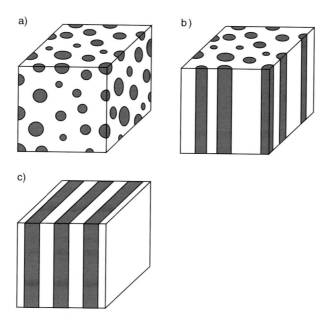

**Figure 5.9**  Various types of diblock phase structures:
a) spherical minor phase domains, b) rodlike minor phase domains, c) planar phases

to the other phase. The blocks can organize themselves in various ways, depending on the relative lengths of the blocks, as shown schematically in Fig 5.9. When the length of one component is much shorter than the other, it typically forms spherical domains within a continuous matrix of the other block, as shown in Fig. 5.9 a). As the length of the minor component increases, rods and planes can form, as shown schematically in Fig. 5.9 b) and c), respectively. Other more complex forms involving interpenetrating networks can also form under specific conditions.

Triblock copolymers, as shown in Fig. 5.8 d), comprise a central homopolymer block of one type, the ends of which are attached to homopolymer chains of another type. As with other block copolymers, the components of triblocks may be compatible or incompatible, which will strongly influence their properties. Of particular interest are triblocks with incompatible sequences, the middle block of which is rubbery, and the end blocks of which are glassy and form the minor phase. When such polymers phase-segregate, it is possible for the end blocks of a single molecule to be incorporated into separate domains. Thus, a number of rubbery mid-block chains connect the glassy phases to one another. These materials display rubber-like properties, with the glassy domains acting as physical crosslinks. Examples of such materials are polystyrene/isoprene/polystyrene and polystyrene/polybutadiene/polystyrene triblock copolymers.

Multiblock copolymers, as shown in Fig. 5.8 e), with incompatible components form similar structures to those found in diblocks and triblocks.

## 5.4.2     Spectroscopic Analysis

Spectroscopic analyses are widely used to identify the components of copolymers. Infrared (IR) spectroscopy is often sufficient to identify the comonomers present and their general concentration. Nuclear magnetic resonance (NMR) spectrometry is a much more sensitive tool for analysis of copolymers that can be used to accurately quantify copolymer compositions and provide some information regarding monomer placement.

### 5.4.2.1     Infrared Spectroscopy

Many characteristic molecular vibrations occur at frequencies in the infrared portion of the electromagnetic spectrum. We routinely analyze polymers by measuring the infrared frequencies that are absorbed by these molecular vibrations. Given a suitable calibration method we can obtain both qualitative and quantitative information regarding copolymer composition from an infrared spectrum. We can often identify unknown polymers by comparing their infrared spectra with electronic libraries containing spectra of known materials.

### 5.4.2.2     Nuclear Magnetic Resonance Spectrometry

NMR spectrometry measures the frequency generated by certain types of atomic nuclei as they decay from a high energy state after excitation by a powerful radio frequency pulse. The precise resonance frequency of an atom is influenced in a predictable fashion by the local magnetic fields induced by electrons in the orbitals surrounding nearby atoms and in adjacent bonds. Hydrogen and carbon-13 atoms (both of which are found in the vast majority of synthetic polymers) can be excited by magnetic fields, which makes this technique very useful in determining copolymer composition.

We use hydrogen NMR spectrometry to measure the relative concentrations of the hydrogen atoms that are part of the different monomer residues making up copolymers. We can measure monomer residue concentrations directly by comparing the relative areas of the various peaks with which they are associated.

We use carbon-13 NMR spectrometry to identify the monomer units present in copolymers, their absolute concentrations, the probability that two or more monomer units occur in proximity, and long chain branching concentrations. For instance, in the case of polyethylene, we can not only distinguish and quantify ethyl, butyl, and hexyl branches, but we can also determine whether branches are present on carbon backbone atoms separated by up to four bonds. We can compare the observed adjacency of branches to a theoretical value calculated for random comonomer incorporation. By this method, we can determine whether comonomers are incorporated at random, as blocks, or in some intermediate fashion.

# 5.5    Branching

Branching in polymers can take one of two basic forms, either short chain branches or long chain branches. The distinction between the two types of branching is largely pragmatic, based on the limits of detection methods. When we can accurately determine the length of a branch it is a short chain branch. Branches of a longer, but indeterminate length are known as long chain branches.

## 5.5.1    Short Chain Branches

We can introduce short chain branching into polymers by three methods; copolymerization, "backbiting", and chemical modification. The first two occur during polymerization, while the last requires a secondary chemical reaction. Short chain branches have well defined chemical structures, the nature of which we can accurately determine via analytical methods or know from the structure of the reactants.

### 5.5.1.1    Copolymerization

We can incorporate short chain branches into polymers by copolymerizing two or more comonomers. When we apply this method to addition copolymers, the branch is derived from a monomer that contains a terminal vinyl group that can be incorporated into the growing chain. The most common family of this type is the linear low density polyethylenes, which incorporate 1-butene, 1-hexene, or 1-octene to yield ethyl, butyl, or hexyl branches, respectively. Other common examples include ethylene-vinyl acetate and ethylene-acrylic acid copolymers. Figure 5.10 shows examples of these branches.

### 5.5.1.2    Backbiting Reactions

Backbiting reactions occur during the high pressure free radical polymerization of ethylene to produce low density polyethylene. Figure 5.11 shows an example of one type of backbiting. In this example, the growing chain end bearing the radical turns back on itself and abstracts a hydrogen atom from a methylene group four carbons back along the chain. In the process, the radical is transferred to the carbon atom from which the hydrogen was removed. Chain growth proceeds from the carbon bearing the free radical, leaving a pendant butyl group. Other, more complex, backbiting mechanisms exist, which create paired ethyl branches, 2-ethylhexyl branches, and various other branches.

### 5.5.1.3    Chemical Modification

We can introduce short chain branches by means of many classic organic chemistry reactions. The variety of branches that can be introduced covers the full range that is accessible via standard reactions.

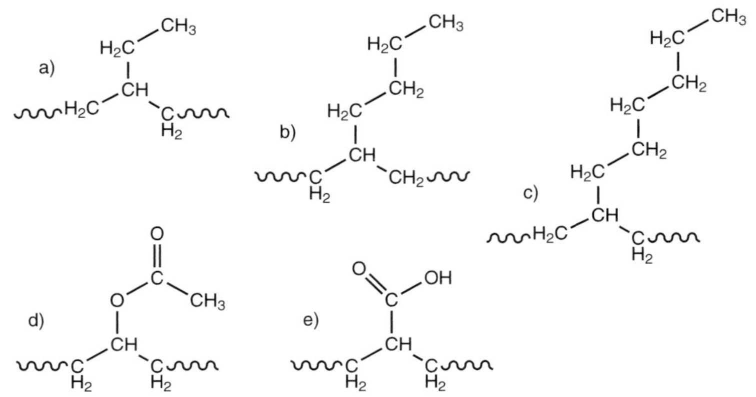

**Figure 5.10**   Examples of short chain branches found in ethylene copolymers:
a)  ethyl branch derived from 1-butene,
b)  butyl branch derived from 1-hexene,
c)  hexyl branch derived from 1-octene,
d)  acetate branch derived from vinyl acetate, and
e)  carboxylic acid branch derived from acrylic acid

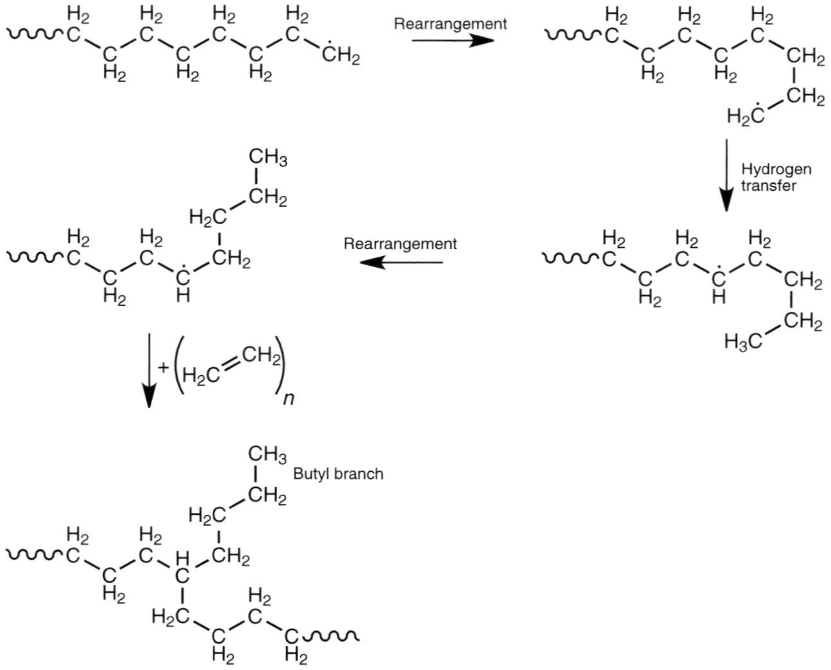

**Figure 5.11**   Example of backbiting reaction to form a butyl branch during the high pressure polymerization of polyethylene

## 5.5.2    Long Chain Branches

We can introduce long chain branching into polymers in four ways; copolymerization of macromers with monomers, backbiting, chemical grafting, and incorporation of tri-functional monomers into condensation polymers. The first three methods are analogous to those used to introduce short chain branches. Crosslinking, which proceeds via the formation of numerous long chain branches to generate a network of interconnected branches, is addressed separately in Section 5.6.

### 5.5.2.1    Copolymerization with Macromers

Macromers are polymer chains that are terminated by a reactive group, such as a vinyl group, which can be incorporated into a growing polymer chain. The copolymerization of monomers with macromers to form long chain branches is analogous to copolymerization to form short chain branches. Macromers can either be produced *in situ* by chain growth termination that generates a reactive group, or in a separate reaction, the product of which is added to the polymerization vessel. In the first case, the branches have the same chemical composition as the backbone. In the latter case, the branches may or may not share the same chemical composition as the main chain. If the branches are chemically distinct, we call the product a "branched block" copolymer.

When a single macromer is incorporated into a polymer chain, the result is a three armed "star", as shown in Fig. 5.12 a). When numerous macromers are incorporated into a single backbone, the result is a "comb", as shown in Fig. 5.12 b). In the case, where macromers are produced *in situ*, the macromers themselves can be branched and the resulting polymers can be branched in a complex fashion, as shown in Fig. 5.12 c).

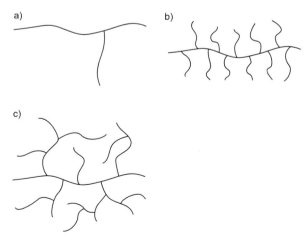

**Figure 5.12**    Examples of long chain branched molecules produced by macromer incorporation: a) three-armed "star", b) "comb" molecule, and c) complexly branched molecule

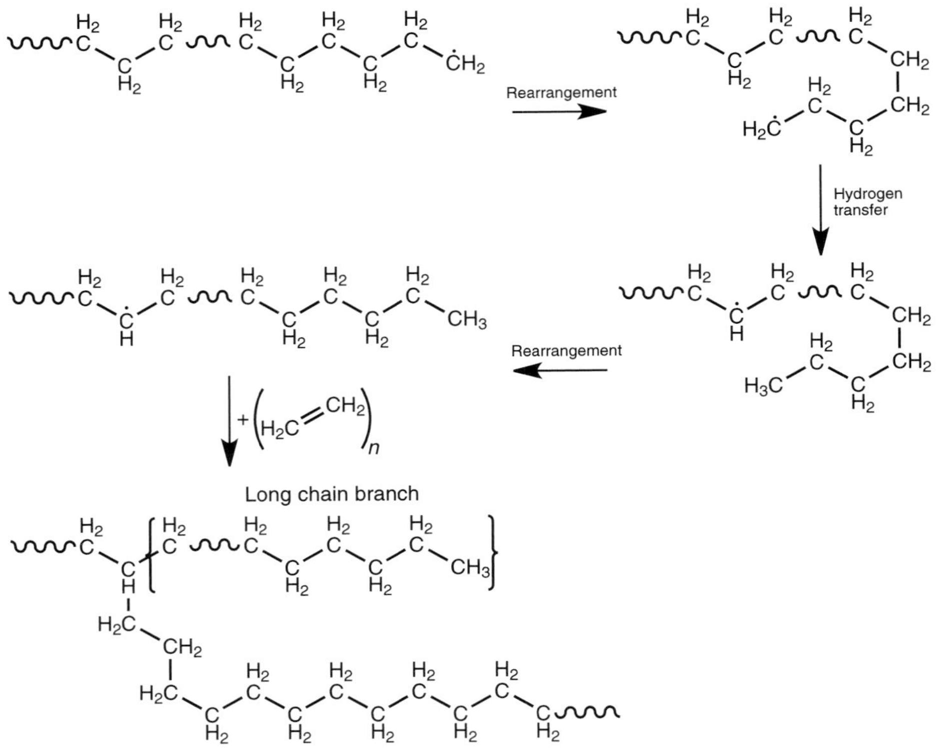

**Figure 5.13**   Example of backbiting reaction to form a long chain branch during the high pressure polymerization of polyethylene

## 5.5.2.2   Backbiting Reactions

The backbiting reactions that produce long chain branches during the high pressure free radical polymerization of polyethylene are similar to those that yield short chain branches. In order to create a long chain branch, the growing end of a chain abstracts a hydrogen atom from a point remote from the chain end, as illustrated in Fig. 5.13. Chain growth proceeds from the newly created free radical, creating a pendant long chain branch. Such reactions can occur either intra- or intermolecularly. The long chain branches created by backbiting are of indeterminate length.

## 5.5.2.3   Grafting

Grafting to create long chain branches employs well-known organic reactions to incorporate polymer chains that have reactive end groups. When there is no suitable reactive group on the backbone that can be attacked directly (as is the case with polymers such as polypropylene,

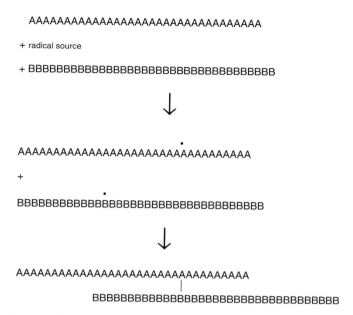

**Figure 5.14**   Schematic illustration of radical grafting to create long chain branches

polyethylene, polystyrene, etc.), we commonly employ a free radical approach. In the first step we blend a free radical source, such as an organic peroxide or an azide, with the polymer, under conditions that do not cause the radical source to degrade. Once intimately mixed, we heat the blend to decompose the radical source. The free radicals generated abstract hydrogen atoms from the polymer chains, leaving unpaired electrons that are available for further reaction. The free radical approach can be used to create grafts of two or more polymers to create long chain branches of one polymer attached to the other, as illustrated schematically in Fig. 5.14. We can view the products of such free radical grafting as four-armed stars, two arms of which are made up of one polymer while the other two arms consist of a second polymer. Byproducts include grafts consisting of two polymer chains of the same type. When high concentrations of the radical source are used, this grafting yields complex multiple branched structures.

### 5.5.2.4   Incorporation of Tri-Functional Monomers

We can readily introduce long chain branches into condensation polymers by copolymerizing a small number of tri-functional monomers with a majority of bi-functional monomers. Figure 1.12 illustrates the general principle of creating long chain branches using tri-functional monomers. The incorporation of one tri-functional monomer into a polymer molecule generates a single long chain branch, which yields a three-armed star molecule. As the concentration of tri-functional monomers increases, branching becomes more complex.

### 5.5.3    Long Chain Branching Analysis

We cannot accurately determine the precise nature of a long chain branch. Analytical methods are not sensitive enough to distinguish between monomer residues in a long chain branch remote from its end, and those in the backbone of the main chain of the same overall chemical composition. We can use spectroscopic techniques, such as infrared spectroscopy and nuclear magnetic resonance, to identify and quantify the branch points and terminal groups, but they do not provide information regarding the placement or lengths of the branches.

Two analytical methods, size exclusion chromatography and rheology, provide a long chain branching index. The application of size exclusion chromatography to long chain branching analysis was described in Section 5.2.3.1.

The viscosity of a long chain branched polymer is lower than that of an unbranched polymer of equivalent molecular weight and chemical composition. Highly branched polymers exhibit a greater viscosity discrepancy from their linear equivalents than do lightly branched polymers. Additionally, the viscosity discrepancy increases as the shear rate, at which the viscosity is measured, rises. Thus, we can obtain a long chain branching index by comparing the viscosity versus shear rate relationship for a branched polymer to that of a linear polymer. As with the long chain branching index derived from size exclusion chromatography, the viscosity branching index is a single value that reveals nothing about the distribution of long chain branch lengths or their placement on the backbone.

## 5.6    Crosslinking

In a fully crosslinked polymer, each chain segment is chemically linked, directly or indirectly, to every other chain segment, as shown schematically in Figs. 1.4 and 1.12 d). The interconnected chain segments form a complex three-dimensional network, which has distinct property differences from its un-crosslinked analogs. The best known class of crosslinked polymers are the materials that we commonly call rubbers. Other classes of polymers that are routinely crosslinked are epoxy and phenolic resins, polyester, and polyethylene. Crosslinked polymers are typically tougher than their linear counterparts, and exhibit dimensional stability and elastic recovery at temperatures above their crystalline melting temperature or glass transition temperature. We take advantage of these properties in such diverse applications as tires, pleasure boat hulls, adhesives, and electrical cable insulation.

### 5.6.1    Types of Crosslink

Figure 5.15 illustrates the two general types of crosslink. In Fig. 5.15 a), a short crosslink connects adjacent chains to form a "four-armed star", the arms of which are normally unequal in length. In Fig. 5.15 b), a longer crosslink connects two adjacent molecules. When

a)                                    b)

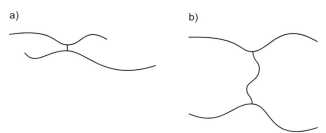

**Figure 5.15**   Principal classes of crosslinks:
   a)  short linkages between adjacent chains, and
   b)  extended chain segments linking chains

the crosslink is of sufficient length, it behaves as a polymer segment. In Fig. 5.15 we show only the simplest examples of the two types of crosslink. In practice, crosslinked polymers include much more complex molecules, with multiple crosslinks connecting two or more adjacent chains. The distribution of crosslinks and the relative lengths of the various arms and connectors control the properties of a crosslinked polymer.

## 5.6.2   Statistics of Network Formation

Theoretically, if each molecule in a polymer sample were to be linked to two of its neighbors, a single highly branched molecule would form that would encompass the whole sample. In practice, due to the statistical distribution of chain lengths and the random incorporation of crosslinks, the situation is far more complex.

If we assume a random distribution of crosslinking events, the average number of crosslinks along the length of a chain is proportional to its length. Thus, we would expect a chain with 10,000 atoms in the backbone to have twice the number of crosslinks as a chain with only 5,000 atoms in the backbone. The actual number of crosslinks on a chain is controlled by the probability of a crosslinking event occurring at any given site on the chain. The net result is that for a given population of molecules sharing the same molecular weight, the number of molecules with 0, 1, 2, ..., $n$ crosslinks will fall into a statistically predictable distribution. Shorter molecules will be more likely to experience zero crosslinking events. These molecules are not incorporated into the network. Molecules that experience a single crosslinking event may either be incorporated into a network, or, if they are linked to another molecule that also experiences only one crosslinking event, a four-armed star will be created (as shown in Fig. 5. 15 a)), which is independent of the network. The more crosslinking events per chain, the more likely it is to be incorporated into a network. Thus, longer chains will be more likely to form a network than shorter chains.

For a given initial molecular weight distribution, the weight fraction of chains incorporated into a network increases as the number of crosslinking events increases. Similarly, the number of crosslinking events required to create a network decreases as the average molecular weight increases. These relationships are illustrated schematically in Fig. 5.16.

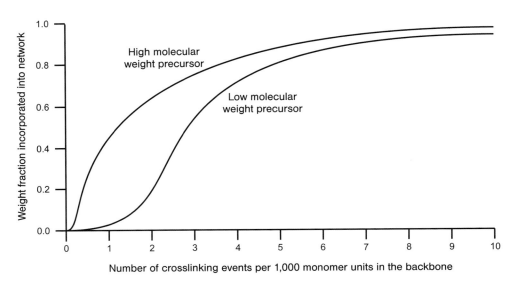

**Figure 5.16**   Schematic representation of the effect of number of crosslinks and initial average molecular weight on network formation

### 5.6.3      Crosslinking Analyses

As with many other polymer molecular characteristics, we cannot precisely determine the molecular structure of crosslinked polymers. In practice, we can measure a crosslinked polymer's gel content and its average crosslink density. Each of these analyses provides a single value that represents a complex situation.

#### 5.6.3.1      Gel Content Analysis

We use gel content analysis to determine the weight fraction of a crosslinked polymer that is bound into an insoluble network. We immerse a stainless steel mesh basket containing a known weight of the crosslinked polymer in a suitable solvent (which may be heated to facilitate dissolution). If necessary, we can slice or grind the sample to increase its surface area. After 24 hours or more, we remove the basket from the solvent and dry it to constant weight. We calculate the gel content from Eq. 5.6.

$$\text{Gel Content (\%)} = \left( \frac{\text{final mass}}{\text{original mass}} \right) \times 100 \tag{5.6}$$

### 5.6.3.2 Crosslink Density Analysis

We can determine a sample's crosslink density from the degree to which it swells when immersed in a suitable solvent. We use this method to calculate the average molecular weight of the "effective chains", that is, those chain segments terminated at each end by a crosslink. We can only apply this analysis to samples that contain no soluble material.

We immerse a weighed sample of a crosslinked polymer in a suitable solvent and allow it to swell for up to 24 hours. We calculate the molecular weight of effective chains from Eq. 5.7.

$$M_c = \frac{\rho_p}{V} \tag{5.7}$$

where:

$M_c$ = Molecular weight of effective chains
$\rho_p$ = Density of polymer $(g/cm^3)$
$V$ = Concentration of effective chains $(mol/cm^3)$

We calculate $V$ from Eq. 5.8.

$$V = -\frac{V_r + \mu\, V_r^2 + \ln(1 - V_r)}{V_o\, (V_r^{1/3} - v_r/2)} \tag{5.8}$$

where:

$V_r$ = Volume fraction of polymer in swollen gel
$\mu$ = Huggins solvent/polymer interaction parameter
$V_o$ = Molar volume of solvent $(cm^3)$

We calculate $V_r$ from Eq. 5.9.

$$V_r = \frac{1}{\dfrac{M_s\, \rho_p}{M_p\, \rho_s} + 1} \tag{5.9}$$

where:

$M_s$ = Mass of solvent in gel (g)
$M_p$ = Mass of polymer in gel (g)
$\rho_s$ = Density of solvent $(g/cm^3)$

## 5.7     Conclusions

In this chapter we have explored the various methods by which polymer scientists characterize the molecular structure of polymers. Given the complex molecular distribution found in most polymers, the best that we can do in many cases is to measure some average value or distribution of values that represents the polymer. Armed with these values polymer scientists and engineers can design or select resins suitable for a myriad of practical applications.

## Review Questions for Chapter 5

1.   Name three factors that affect a polymer's random walk configuration. How does each factor affect molecular volume?

2.   In words, what are weight average molecular weight and number average molecular weight? What does the ratio of $M_w$ to $M_n$ tell us about the molecular characteristics of a polymer?

3.   When performing size exclusion chromatography measurements, do the large polymer molecules or small polymer molecules elute from the column first? Why?

4.   What is meant by the term "melt flow rate"? How is it related to molecular weight?

5.   Name the three forms of tacticity. For polypropylene, which form crystallizes fastest and why?

6.   Why does the addition of 5% ethylene into an isotactic polypropylene drastically reduce the crystallinity of the polypropylene relative to the isotactic homopolyer?

7.   What can IR spectroscopic analyses tell you about a polymeric material? What can NMR spectrometry analyses tell you?

8.   How does backbiting create branches during free radical polymerization?

9.   Define grafting. How is free radical attack employed during a grafting process?

10.  Describe two methods of analyzing crosslinked polymers? What is learned from each?

# 6 Rheological Properties of Polymeric Materials

## 6.1 Introduction

Rheological studies explore the flow of a material as an external force acts upon it. This flow depends not only on the magnitude and directionality of the external force, but also on the molecular composition and structure of the material that experiences the force. In this chapter, we will focus on the flow behavior of molten polymers, as it relates to their molecular structure. It is important to note that the molecular characteristics that determine a molten polymer's behavior also define the polymer's solid state behavior. Therefore, many of the concepts introduced in this chapter will reappear in Chapter 8, Solid State Properties of Polymers.

## 6.2 Flow, Stresses, Strains and Deformations

When describing the effect of an external force, we must first define the force itself. A lay person's definition of a force is the amount of effort to get the desired effect. As scientists, we need a more precise definition of force. With a precise definition we can understand and quantify the effect of an applied force on a polymeric material. The mathematical definition of force is the work (which is a form of energy) required to move an object over some distance. Another way to define a force is in terms of the acceleration it creates when applied to some object of a mass $m$. In our everyday experiences, the first explanation is a simple idea to relate to. When we push a stalled car we exert a force on it. We could easily quantify the force from the weight of the car, the slope of the hill it is sitting on, and how far we must push it. Once we begin to talk about forces in polymer systems, the ideas become a bit more complicated. For example, the force required to open a bag of candy is defined by the work required to deform the bag until it ruptures by overcoming the intermolecular forces which hold the plastic together.

### 6.2.1 Normal Stresses and Strains

Rather than talk in terms of force, polymer scientists use the concept of stress. This concept allows us to talk about the applied force independent of the size of the object. A stress is the force applied to a system over some area. The area in question is the cross-sectional area over which the deformation occurs. This concept is illustrated in Fig. 6.1. Here we see a piece of plastic being deformed by a force and the cross sectional area of deformation, which is

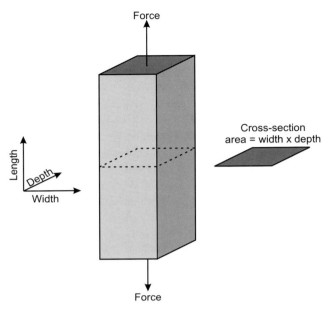

**Figure 6.1**   Rectangular sample subjected to tensile force

perpendicular to the applied force. We define the stress as the ratio of the applied force to the area over which the force is applied. The imposed stress induces a strain within the specimen experiencing the applied force.

The strain is the magnitude of the deformation of a material in the direction of the applied stress as related to its length in that direction, as in Eq. 6.1 and Fig. 6.2.

$$\text{Strain} = \frac{\text{Change in length}}{\text{Original length}} = \frac{\Delta l}{L} \tag{6.1}$$

In Fig. 6.2, the direction of the stress, as well as the directionality of the deformation, is indicated. It is important to note that the directionality of the deformation is perpendicular to the area over which the stress is applied, making this an example of the effect of a normal stress on a material.

## 6.2.2    Shear Stresses

In addition to the normal stress, we can also apply a shear stress to a material to cause a deformation. A shear stress, δ, induces a deformation parallel to the applied stress as shown in Fig. 6.3. This type of deformation occurs when the applied stress pulls the material along with it in the direction in which it is applied. For example, when you make a peanut butter sandwich you apply a shear stress to spread the peanut butter over the bread.

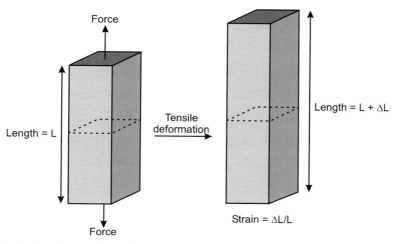

**Figure 6.2**   Rectangular sample undergoing tensile strain

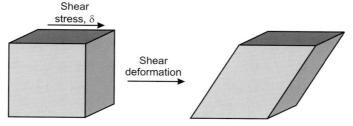

**Figure 6.3**   Shear strain induced by shear stress

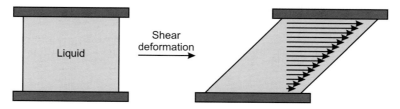

**Figure 6.4**   Effect of shear stress on a liquid sandwiched between metal plates

To better understand the way that shear stresses affect a system, we can look at an idealized system. In the example shown in Fig. 6.4 we will sandwich a layer of liquid between two metal plates. We hold the bottom plate stationary, while the top one can slide parallel to the bottom plate at some velocity, $v$, while maintaining continual contact with the liquid. Between the plates, the top "layer" of the fluid that is in direct contact with the top plate moves with it. The top layer in turn drags the adjacent layer with it, but the force is reduced incrementally since it has already acted on the top layer, so that the second layer cannot travel as far. This reduced displacement of layers continues throughout the fluid, all the way to the bottom layer. The fluid layer directly in contact with the bottom plate is stationary with no displacement. This situation is represented in Fig. 6.4, where the arrows represent the displacement of the fluid.

The rate at which this layer-by-layer deformation occurs is called the strain rate, $\gamma$. Since each layer of the fluid is deformed to a different degree by the applied stress, we have to define the strain rate as the displacement of one layer relative to the next layer, separated by some vertical distance, in some unit time. The equation used to describe this is:

$$\gamma = \frac{d}{dt}\left(\frac{dx}{dy}\right) \qquad (6.2)$$

where the ratio $\left(\dfrac{dx}{dy}\right)$ represents how far a fluid layer is displaced from its starting position, $dx$, relative to its distance, $dy$, from the sliding plate. The change of this ratio with time provides us with the strain rate. The actual mathematics used to describe the effect of shear stresses on fluid systems is quite complex. Complications arise from the fact that the shear stress is only one of the stresses that a polymer melt will experience. The true stresses that the melt experiences are three dimensional and therefore require the use of general constitutive equations and tensor treatments.

### 6.2.3    Viscosity

We now have the tools necessary to describe how a polymeric material will respond to applied stresses. The next step is to add a method to characterize individual polymers in terms of the ease by which they deform. Imagine that we impose a shear stress on two different materials for the same length of time. In the first material we observe a great deal of deformation; in the second there is very little. What is the reason for this? The answer lies in the fact that there are fundamental differences in the response of each of the materials to the imposed stress. We define these differences by taking the ratio of the applied stress to the strain rate and calling it the material's viscosity, $\eta$, which is defined in Eq. 6.3.

$$\eta = \frac{\delta}{\gamma} \qquad (6.3)$$

The more resistant a material is to an applied stress, the higher is its viscosity. Molecular differences account for differences in viscosity.

For many materials, the application of a stress creates a strain rate in a linear fashion, i.e., the rate of strain is proportional to the applied stress. This linear relationship, which defines a Newtonian fluid, does not hold true for polymers. Most molten polymers respond to stresses in a non-linear fashion, such that the greater the applied stress the more effective the stress is at inducing a strain rate. This non-Newtonian behavior is referred to as "shear thinning"

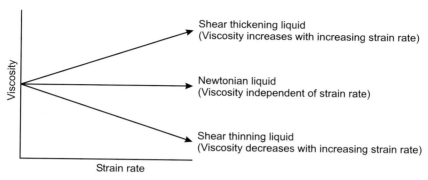

**Figure 6.5**   Effect of strain rate on various types of liquid

and is a consequence of the molecular structure of polymer melts, which we shall discuss in the next section.

Certain polymeric systems can become more viscous on shearing ("shear thickening") due to shear-introduced organization. These systems become more resistant to flow as the crystals form so that the introduction of the shear increases their viscosity. Figure 6.5 shows the viscosity versus strain rate relationship for Newtonian and non-Newtonian fluids, highlighting the differences in their behaviors.

## 6.3    The Molecular Origin of Rheological Properties

The large size of polymer molecules creates vast numbers of intermolecular interactions due to the dispersive, polar, and sometimes ionic components of the polymer molecule. High molecular weight polymers also intertwine with their neighbors creating entanglements between the chains. These molecular characteristics create a bulk material that resists flow, despite the intrinsic flexibility of the polymer chains themselves. In general, molecular characteristics that increase the number of interactions or entanglements increase a material's resistance to flow. Therefore, longer chain lengths, more and stronger interactions, more entanglements, the presence of crosslinks, long chain branches, and polar interactions all contribute to high melt viscosity. Any factor that increases the stiffness of a polymer chain also increases the polymer's viscosity. Short chain branches, therefore, reduce interactions and permit fewer entanglements because the interfering branches lower the polymer's melt viscosity. The free volume in a polymeric material also plays a role in defining the rheological properties. Free volume arises from the fact that the atoms in polymer molecules cannot pack tightly together. Steric interference holds the chains apart. This means that the material encompasses some free volume that is devoid of atoms. The more free volume present in a material, the more room that chain segments have to move around in. This makes the system less rigid and more able to flow.

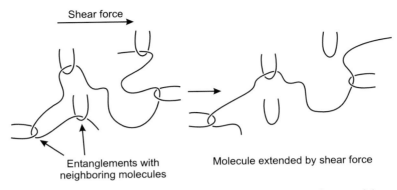

Shear force

Entanglements with
neighboring molecules

Molecule extended by shear force

**Figure 6.6**  Polymer molecule disentangling from its neighbors under the influence of shear

It might seem counter-intuitive that molten polymers are typically shear thinning. Why should they not behave as simple Newtonian fluids? We can answer this question by returning to our understanding of the chemical composition of a polymer. The long chains of interconnected monomers experience a variety of intermolecular interactions, such as dispersion forces and other intermolecular forces, which depend on the chemical constituents. In addition, the long chains often intertwine among themselves. If we expose the polymeric material to a strong, constant shear force as shown in Fig. 6.6, the polymeric material responds by breaking some of the intermolecular interactions and by disentangling to relieve the stresses induced by the shear force. In this example, the force is constant, therefore many of the broken associations are prevented from reforming. Since we now have fewer interactions, the material is weaker, so it is more susceptible to shear and more associations break, making it even more susceptible. This continual weakening is the reason for shear thinning. The shear thinning behavior of molten polymers enables us to more successfully convert them to useful items, as we shall see in Chapters 11 to 16. Keep in mind that the large size and bulk of polymer molecules means that when we are describing polymer flow we are referring to a material that behaves more like silly putty than water.

### 6.3.1    Molecular Structure Effects on Melt Viscosity

Since the chemical composition of a polymer defines its flow behavior, we will now explore the effects of molecular architecture, such as chain length, the presence of branching, and polarity on these properties.

### 6.3.1.1    Chain Length Effects

Although it is generally true that, as a polymer's chain length increases, its melt viscosity also increases, there are some nuances to this observation. By increasing the chain length, the number of repeat units increases. This creates more entanglements and raises the viscosity. One

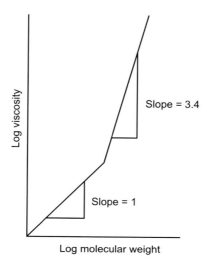

**Figure 6.7**  General relationship between viscosity and molecular weight

way to explore the relationship between chain lengths and melt viscosity is by plotting the log of the viscosity as a function of the log of the molecular weight of linear polymers. This plot, as shown in Fig. 6.7, consists of two distinct, linear regions. The viscosity of the melt in the first region exhibits a slope of one, indicating a linear relationship between the viscosity and chain length at low chain lengths. In the second region, the slope has a value of approximately 3.4, indicating that the melt viscosity is more strongly related to the chain length relative to that seen in the first region. The differences between these two regions can be understood in terms of two competing properties of the polymer: free volume and entanglements. At low degrees of polymerization, the polymer has a great deal of free movement since it has few entanglements. Therefore its viscosity will be defined by its free volume. At high degrees of polymerization, a polymeric material has many entanglements with limited motion. This means that the resulting viscosity will be more determined by the entanglements than by free volume effects. What is so powerful about this finding is that the same behavior is observed for a wide range of polymers.

### 6.3.1.2    Branching Effects

The number and types of branches in a polymer can have a marked effect on its rheological behavior. This is because the branches affect the entanglement density of the polymer, either by enhancing or reducing it. The entanglement density describes how many entanglements exist in any given volume. Another component of how branches affect rheological properties arises from their influence on the free volume of the polymer. Anything that increases the free volume increases the freedom of motion of the polymer chain, as there is empty space to accommodate configurational changes.

In general, short chain branches reduce the viscosity of the polymer by creating a molecular scaffolding that forces the polymer chains apart. Therefore the short chain branches reduce the entanglement density of the material and increase the free volume of the material. On the other hand, long chain branches, greater than some polymer-specific critical length, create the opposite effect. They increase the viscosity of the polymer by increasing the entanglement density. They can also decrease the free volume because of the ability to form associations without steric hindrance.

Fundamental research on laboratory produced polypropylene, polycarbonate, nylon 6, and with the commercially available polyethylenes support the general trends stated for the effects of long and short chain branching on rheology. Deviations from the general trends for branching have been observed with polystyrene. In polystyrene, long chain branching can decrease the viscosity of the material relative to its unbranched analog. These results have been explained, again, by the effect of the branches on the entanglement density: the branches reduce the entanglements in the amorphous polymer.

### 6.3.1.3   Molecular Weight Distribution Effects

It should be no surprise that rheological behavior, like all polymer properties, is a function of the distribution of the chain lengths present in the sample. In polymers there is no unique chain length; instead they comprise a distribution of lengths. This fact is critical to understanding rheology. If we compare two samples of the same polymer with the same average molecular weight, one with a broad molecular weight distribution, the other with a narrow one, we see dramatically different melt behaviors. The narrow molecular weight distribution material will exhibit a much higher melt viscosity than the material with the broad distribution. This is because the low molecular weight species in the broad distribution material lubricate molecular motion within the polymer. Molecular weight distribution effects have had major repercussions in the past twenty years, primarily due to the widespread use of metallocene catalyzed polyolefin resins in a wide variety of processes. These polymers, with their relatively narrow molecular weight distributions, are more difficult to process, since their viscosities are higher than their broad molecular weight counterparts. Therefore, considerable effort has gone into educating processors on the different behavior of the metallocene catalyzed polymers compared to their Ziegler-Natta catalyzed counterparts. In addition, equipment manufacturers have redesigned their machines to address the challenges of processing narrow molecular weight materials.

## 6.4     Temperature Effects on Rheological Properties

Returning to our definitions in Chapter 3, temperature is a measure of the average kinetic energy of the molecules in a system. This means that the higher the temperature, the more movement in the polymer chains. Taking this one step further, the greater movement reduces the opportunities for the chains to form intermolecular associations. Beyond this, there is the even greater effect of expansion, which increases a polymer's free volume. Increased free volume reduces the strength of intermolecular interactions. The net result is that there are fewer and weaker interactions between polymer chains as the temperature increases, which reduce the viscosity of the polymer. Since the viscosity of a particular resin can be altered by temperature changes, the specific needs of any polymer converting process need to be considered when choosing the processing temperature.

# 6.5    Measurement Techniques

We have different methods by which we can measure a polymer's rheological behavior. The method that we choose will be defined by the information that we require. The two main classes of rheological testing are constant and variable strain rate measurements. A constant strain rate method imposes a strain rate, usually by applying a constant force, to the polymer and measures its flow under those conditions. The disadvantage of this type of measurement is that it provides a single data point, which might not represent how the polymer will behave under the range of strain rates that it experiences during processing. The advantage of constant strain measurements is that they are quick and easy to perform, and are understood throughout the polymer industry. The method by which melt flow rates are measured, described in Chapter 5 section 5.2.2.3, is an example of a constant force rheology measurement method.

Variable strain rate tests apply a range of shear stresses to the material and record the viscosity of the material under each strain rate. These tests are non-trivial and, though typically automated, are often rather complicated to interpret. They provide a more complete picture of the behavior of the material. Within this group of methods, there are several different types of tests, including capillary, oscillatory disk and cone and plate rheometry, which predominate. In capillary rheometry, molten polymer is forced through a narrow bore tube at various rates. In oscillatory disk rheometry (also known as parallel plate rheometry), a sample is placed between two metal plates that oscillate through a range of angles at a range of rates. These configurations are shown in Fig. 6.8. The capillary rheometer measures the pressure of the molten polymer at several positions along the wall of the barrel under a specified load to determine the pressure gradient. This information is then converted into a wall shear stress for that specific polymer. The strain rate is determined by the rate at which the polymer exits the capillary. The oscillatory disk rheometer measures the instantaneous torque experienced by the upper plate as transmitted by the molten polymer from the oscillating lower plate. Another option is a bowl mixer with intermeshing mixing elements that rotate at a constant speed, while measuring the torque necessary to maintain that constant speed. Ideally, our choice would be defined by what we are trying to learn, but often (due to the high cost of rheological testing equipment) we use whatever equipment is available to us.

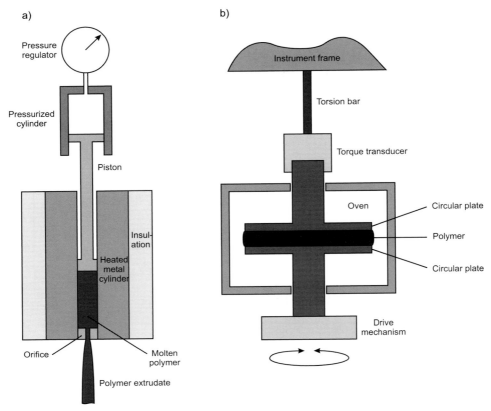

**Figure 6.8**  Schematic diagrams of a:
a) capillary rheometer and b) parallel plate rheometer

## 6.6    Implications of Rheology and Conclusions

Although the field of rheology is, in and of itself, fascinating, most people are interested in it chiefly because they are concerned about its implications. Specifically, we wish to know how a polymer will behave in a particular piece of processing equipment. These same considerations arise whenever we process polymers, so we will address them as we describe the individual polymer converting processes later in the book.

# Review Questions for Chapter 6

1. Define stress and strain. Explain how these two concepts are interrelated and provide three every day examples of stresses and their corresponding strains.

2. What is a shear stress? What is a normal stress?

3. Why does the effect of shear stress decrease as the distance from the surface at which the shear stress is imposed increases?

4. How does the shear stress travel from the contact layer of a fluid into the bulk material?

5. Rank the following fluids in order of increasing viscosity at room temperature:
   a) Water
   b) Modeling clay
   c) Liquid honey
   d) Axle grease

6. Define viscosity and predict what effect each of the following variables would have on the viscosity of a polymer:
   a) Weight average molecular weight
   b) Presence of molecular dipoles in the polymer chain
   c) Presence of hydrogen bonding between polymer chains
   d) Degree of long chain branching

7. Why do thermoplastic polymers exhibit shear thinning behavior?

8. Some polymers undergo degradation that reduces the average chain length during manufacturing – polyesters are notorious for this. Others, polyethylene for example, crosslink during degradation at high temperatures. What would the manufacturer processing these materials observe in each of these situations?

9. It is much more difficult to incorporate additives into polymers than it is into low viscosity liquids. Why is this? What must occur, at the molecular scale, to incorporate a solid additive into a polymeric melt?

10. How does the concept of free volume help explain the observed viscosity of a polymer?

# 7 Development and Characterization of Solid State Molecular and Supermolecular Structure

## 7.1 Introduction

Polymer molecules can organize themselves to form molecular and supermolecular structures at various levels of size and perfection. As we shall see, thermodynamic and kinetic factors control the resulting states of organization. We can categorize the various solid states of polymers according to their freedom of movement and level of organization. As the level of organization increases, the freedom of motion of the constituent chain segments decreases. Preferential molecular orientation improves organization at all levels.

Molten polymers, which are in a high state of thermal agitation, exhibit the lowest level of organization. Molecules in this state are surrounded by a relatively large free volume, which permits a great deal of molecular motion. Short range motion in the molten state manifests itself as low modulus, i.e., high flexibility. Longer range motions result in viscous flow. Despite their high degree of thermal agitation, molecular segments from adjacent chains tend to be aligned with one another on a local scale of less than a nanometer ($1 \times 10^{-9}$ m).

As polymers solidify from the molten state, their free volume decreases and their organization increases. Solid polymers fall into one of three classes: rubbery amorphous, glassy amorphous, and semicrystalline, which we introduced in Chapter 1.

The properties of a rubbery amorphous polymer form a continuum with those of the polymer in its molten state. Rubbery amorphous polymers exhibit the same range of motions as molten polymers, but they happen much slower, due to reduced thermal motion and the associated decrease in free volume.

When the temperature of an amorphous polymer falls below a characteristic temperature, known as its glass transition temperature ($T_g$), it transforms from a rubbery to a glassy condition. Molecules in the glassy state exhibit a limited range of motions, each of which involves only a few atoms. Molecular segments from adjacent chains tend to be aligned with one another, which reduces free volume and increases the material's density. Large scale molecular motions are impeded by the limited free volume surrounding the chains.

As a polymer cools, molecular segments from adjacent chains can organize to form crystallites. Two criteria must be satisfied for this to happen: the crystalline state must be thermodynamically favored and there must be sufficient time for chain segments to reorganize. Crystallites are composed of chain segments from numerous adjacent molecules that are arranged to form well defined arrays, which may have dimensions from 5 to 10,000 nanometers ($5 \times 10^{-9}$ to $1 \times 10^{-5}$ m). Crystallites in turn tend to be aligned relative to one another, forming supermolecular structures that can have dimensions from 1 to 1,000 micrometers ($1 \times 10^{-6}$ to $1 \times 10^{-3}$ m).

We can investigate a polymer's molecular and supermolecular organization in various ways, which provide information on either a direct or indirect basis. For instance, microscopy of various types provides us with images of small regions of polymers, allowing us to directly visualize certain types of structure. Other methods, such as X-ray diffraction and differential scanning calorimetry, sample larger volumes of material and provide information representative of the material as a whole.

## 7.2    Solid State Morphology

Solid polymers can adopt a wide variety of structures, all of which are derived from the three basic states; rubbery amorphous, glassy amorphous, and crystalline. Either of the amorphous states can exist in a pure form. However, crystallinity only occurs in conjunction with one of the amorphous states, to form a "semicrystalline" structure.

### 7.2.1    Amorphous States

In the amorphous states, polymer molecules exhibit no well defined ordering. On a local scale, chain segments from adjacent molecules tend to be arranged parallel with each other, which minimizes the free volume surrounding them. This local ordering extends over only a few Ångstroms ($< 1 \times 10^{-9}$ m) and does not translate to longer range organization.

#### 7.2.1.1    Rubbery Amorphous

In the molten state polymers are viscoelastic; that is they exhibit properties that are a combination of viscous and elastic components. The viscoelastic properties of molten polymers are non-Newtonian, i.e., their measured properties change as a function of the rate at which they are probed. (We discussed the non-Newtonian behavior of molten polymers in Chapter 6.) Thus, if we wait long enough, a lump of molten polyethylene will spread out under its own weight, i.e., it behaves as a viscous liquid under conditions of slow flow. However, if we take the same lump of molten polymer and throw it against a solid surface it will bounce, i.e., it behaves as an elastic solid under conditions of high speed deformation. As a molten polymer cools, the thermal agitation of its molecules decreases, which reduces its free volume. The net result is an increase in its viscosity, while the elastic component of its behavior becomes more prominent. At some temperature it ceases to behave primarily as a viscous liquid and takes on the properties of a rubbery amorphous solid. There is no well defined demarcation between a polymer in its molten and rubbery amorphous states.

The density of the rubbery amorphous state is only slightly higher than that of the molten state, the difference being attributable to reduced thermal motion of its chains. In this loosely packed condition, the polymer incorporates a substantial amount of molecular scale void

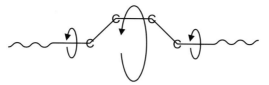

**Figure 7.1**   Crankshaft rotation involving three carbon-carbon backbone bonds

space, known as free volume, which permits the molecules a high degree of short range motion. Chain segments, comprising several backbone atoms and their associated atoms, can undergo concerted motion within the general confines of the surrounding chain segments. "Crankshaft rotation" is a characteristic segmental motion of this type, which is illustrated schematically in Fig. 7.1. (For the sake of clarity, only the backbone carbon atoms are shown in the figure.) Rotations of this type require abundant local void space. The local freedom of molecular motion of rubbery amorphous polymers translates into long range flexibility, which is characteristic of this state.

If we subject a rubbery amorphous polymer to strain for an extended period of time, its molecules will begin to slide past one another and disentangle. This results in permanent deformation, which is generally undesirable. For this reason, rubbery amorphous polymers have few practical applications, unless they are crosslinked to form a network, as shown in Figs. 1.4 and 1.12 d). The covalent crosslinks between neighboring chains prevent long range motion, thus providing dimensional stability. Such crosslinks can take the form of short bridging chains or direct bonds between backbone carbon atoms of neighboring chains, as illustrated in Fig. 5.15. Crosslinks are responsible for the elastic properties of the materials that we commonly call rubbers. As crosslink density increases, molecular motion is increasingly restricted, resulting in stiffer and more elastic materials. We discussed the statistics of network formation in Chapter 5.

### 7.2.1.2   Glassy Amorphous

As a rubbery amorphous polymer cools down, its molecular motions becomes less energetic, allowing its chain segments to approach more closely to one another. This results in a corresponding increase in the polymer's density. When its temperature falls to a characteristic value, known as its glass transition temperature, it enters the glassy amorphous state. A step change increase in density accompanies this transition, while molecular motion undergoes a step change downwards. In the glassy state, molecular vibrations are limited to those involving atoms that are directly linked to one another. Limited molecular motion is responsible for high stiffness, good dimensional stability, and brittle failure, which are characteristic of this state. We can only stretch a glassy amorphous polymer a few percent before it cracks or breaks. Examples of polymers that exist in the glassy state at room temperature include atactic polystyrene, polyethylene terephthalate, and polycarbonate. We take advantage of the properties of glassy polymers in applications such as soda bottles, jewel cases for CDs, motorcycle windshields, and bus shelters.

## 7.2.2    Polymer Crystallites

Polymer crystals consist of well defined arrays of aligned chain segments from adjacent molecules. Polymer crystals (with only a few exceptions) are too small for us to see with the naked eye, or even with an optical microscope. Given their small size, we normally refer to polymer crystals as "crystallites".

We call the building blocks that make up crystallites "unit cells". Each unit cell within a crystallite is identical, consisting of short segments from two or more neighboring chain segments. A crystallite can contain many thousands of unit cells arranged to form a three dimensional matrix. Figure 7.2 shows two views of an orthorhombic polyethylene unit cell. In this case, the unit cell consists of a central chain segment, surrounded by parts of four others. In Fig. 7.2 a) we see an orthogonal projection showing parts of five chains, with a cuboid outlining the unit cell superimposed upon it. In Fig. 7.2 b) we see the same unit cell viewed down its chain axes; the top of the unit cell is lightly shaded. Figure 7.3 illustrates how polyethylene unit cells pack together to form a crystallite.

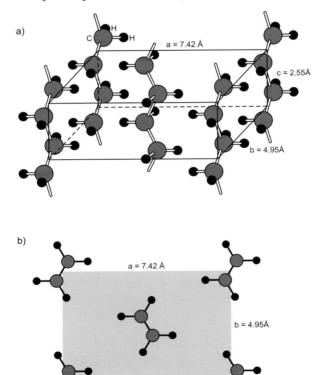

**Figure 7.2**   Orthorhombic unit cell of polyethylene viewed:
a) orthogonally and b) parallel to the chain axes

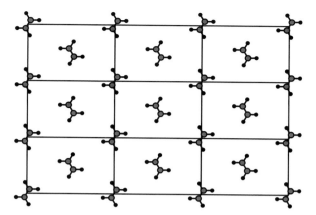

**Figure 7.3**  Nine polyethylene unit cells arranged to show how they are packed in a crystallite

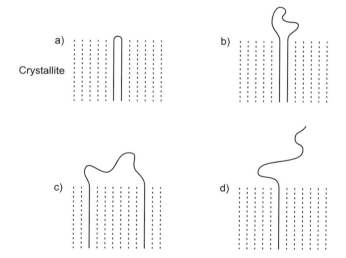

**Figure 7.4**  Polymer chain configurations at crystallite surfaces:
a)  tight fold with adjacent re-entry,
b)  loose loop with adjacent re-entry,
c)  loose loop with non-adjacent re-entry,
d)  departure from the crystallite surface

The dimensions of crystallites vary widely; some measure only a few nanometers in any direction, while others, known as "lamellae", are platelets with lateral dimensions of several tens of nanometers and thicknesses of a few nanometers. The chain axes in lamellae typically span the thickness of the crystallite. With reference to the unit cell illustrated in Fig. 7.2 a), the "c" direction corresponds to the thickness of the crystallite.

At the surface of crystallites, polymer chains can adopt one of four configurations, as shown in Fig. 7.4. They can:

a) Form a tight fold and re-enter the crystallite immediately adjacent to the point from which they emerged,

b) Form a loose loop and re-enter the crystallite adjacent to their point of emergence,

c) Form a loose loop and re-enter the crystallite at a point remote from their emergence, or

d) Depart the immediate environs of the crystallite.

### 7.2.2.1 Semicrystalline State

Due to the extreme length of polymer molecules, their distribution of chain lengths and the constraints applied to them by entanglements, they cannot crystallize completely. A matrix of amorphous chains – either glassy or rubbery – surrounds the crystallites that form from the molten state. We call this the semicrystalline state, which is illustrated schematically in Fig. 7.5. The degree of crystallinity varies widely depending on the polymer type and crystallization conditions. Ultra low density polyethylene may be less than 10% crystalline, whereas polytetrafluoroethylene (Teflon®) can be as much as 90% crystalline. Polymer fibers generally have a higher degree of crystallinity than films, which in turn are more crystalline than molded items.

The length of a polymer molecule is typically at least an order of magnitude greater than the thickness of a crystallite. Similarly, the average size of a polymer's random walk is many times greater than the lamellar thickness. Given the large size of a polymer's random walk relative to its crystallite dimensions, it is clear that a single polymer molecule can contribute to more than one crystallite and that segments of the same molecule will be found in both the crystalline and amorphous regions. Figure 1.13 illustrates a polymer chain contributing to several crystallites. "Tie chains" spanning the amorphous region link adjacent crystallites and also connect crystallites to the amorphous zones.

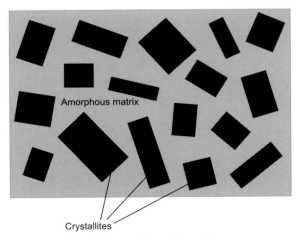

**Figure 7.5**   Schematic representation of the semicrystalline state

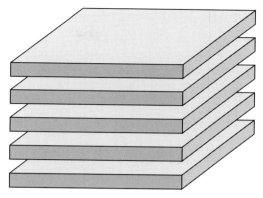

**Figure 7.6**  Stacking of lamellae

The lamellae within a semicrystalline polymer are often aligned with one another. On a local scale, neighboring lamellae tend to be stacked, so that their lateral planes are parallel, as shown in Fig. 7.6. On a longer scale, lamellae can arrange themselves into extended stacks, known as "cylindrites", or radially in three dimensions, to form "spherulites", as illustrated schematically in Fig. 7.7 a) and b) respectively.

The properties of a semicrystalline polymer are controlled by its degree of crystallinity, the alignment of crystallites relative to one another, the number and type of links between the crystallites and amorphous regions, and the overall orientation of molecules within the material.

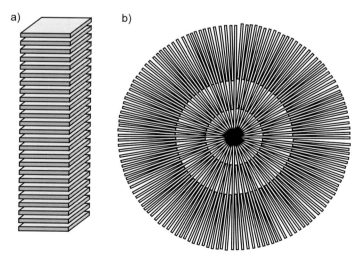

**Figure 7.7**  Schematic representation of long range organization of lamellae:
a) cylindrite b) spherulite

## 7.3    Crystallization

Polymers crystallize when the crystalline state is more thermodynamically stable than the molten one. The system progresses towards a low energy state, but full crystallization is hindered by chain entanglements, branches, side group interactions, and viscous drag on chains.

The change of Gibbs free energy governs the direction of the process according to Eq. 7.1, which we encountered previously in Chapter 3.

$$\Delta G = \Delta H - T\,\Delta S \tag{7.1}$$

Where:

$\Delta G$  =  change of Gibbs free energy

$\Delta H$  =  change of enthalpy

$T$     =  absolute temperature

$\Delta S$  =  change of entropy

When $\Delta G$ is negative, crystallization is energetically favorable. As the temperature of the system falls, the $\Delta G$ becomes more negative making crystallization more favorable, but the viscosity of the molten polymer increases; this slows molecular motion and impedes chain alignment. The magnitude of the change of Gibbs free energy and the mechanisms of molecular reorganization control the rate of crystallization.

Each crystallizable polymer exhibits a characteristic equilibrium melting temperature, at which the crystalline and amorphous states are in equilibrium. Above this temperature crystallites melt. Below this temperature a molten polymer begins to crystallize.

Polymers crystallize from the molten state by the two-step process of nucleation and crystal growth. Nucleation initiates crystallization, followed by the addition of linear chain segments to the crystal nucleus.

### 7.3.1    Nucleation

Nucleation can occur either homogeneously or heterogeneously. Homogeneous nucleation occurs when random molecular motion in the molten state results in the alignment of a sufficient number of chain segments to form a stable ordered phase, known as a nucleus. The minimum number of unit cells required to form a stable nucleus decreases as the temperature falls. Thus, the rate of nucleation increases as the temperature of the polymer decreases. The rate of homogeneous nucleation also increases as molecular orientation in the molten polymer increases. This is because the entropy difference between the molten and crystalline states diminishes as molecular alignment in the molten state increases.

Certain types of contaminants within a molten polymer act as heterogeneous nuclei. Polymer chains that solidify against the pre-existing surface of a contaminant create less new solid/liquid interface than the same volume of polymer chains forming a homogeneous nucleus. Thus, the presence of foreign particles reduces the energy barrier to the formation of a stable nucleus. The most effective heterogeneous nucleating agents are those having surface periodicities that match the length of the "a" or "b" dimensions of the polymer's unit cell. Resin manufacturers often add heterogeneous nucleating agents to polymers in order to raise their crystallization rates and degrees of crystallinity.

## 7.3.2    Primary Crystallization

Primary crystallization occurs when chain segments from a molten polymer that is below its equilibrium melting temperature deposit themselves on the growing face of a crystallite or a nucleus. Primary crystal growth takes place in the "a" and "b" directions, relative to the unit cell, as shown schematically in Fig. 7.8. Inevitably, either the "a" or "b" direction of growth is thermodynamically favored and lamellae tend to grow faster in one direction than the other. The crystallite thickness, i.e., the "c" dimension of the crystallite, remains constant for a given crystallization temperature. Crystallite thickness is proportional to the crystallization temperature.

Conceptually, chain segments can add to the growing crystallite face by tight fold adjacent re-entry or independent deposition of chain segments, as shown schematically in Fig. 7.9 a) and b). These are two extreme cases, involving quite different modes of molecular re-arrangement. For a chain to be laid down in a series of tight folds, as shown in Fig. 7.9 a), the chain has to be "reeled in" from the molten material ahead of the growing crystallite face. Tight fold adjacent re-entry requires that molecular motion along a chain's contour length occur several orders of magnitude faster than the rate of crystal growth. In contrast, for chain segments to organize themselves independently on a crystallite growth face, as shown in Fig. 7.9 b), only local chain motion is required. Each chain segment only needs to reorganize sufficiently to align itself with the chain segments making up the crystallite face. When this mechanism dominates, the crystal growth rate is of a similar magnitude to that of molecular reorganization. In practice, some combination of the two competing mechanisms takes place. At higher crystallization temperatures, adjacent re-entry assumes greater importance. As the crystallization temperature falls, independent deposition of chain segments will dominate.

During commercial processing of polymers, primary crystallization is normally complete within a matter of a few seconds or, at most, a minute or two.

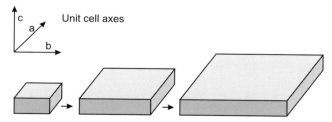

**Figure 7.8**    Growth of a lamella in the "a" and "b" directions

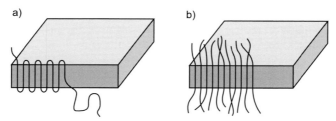

**Figure 7.9**   Mechanisms of laying down chain segments on the growing face of a crystallite: a) tight folds with adjacent re-entry, b) independent deposition of chain segments

### 7.3.3    Secondary Crystallization

Primary crystallization leaves the polymer in a thermodynamically metastable state. Further crystallization is energetically favored, but is limited by restricted molecular motion in the amorphous regions surrounding the crystallites. Over an extended period of time, re-arrangement of molecular segments within the amorphous regions may permit further crystal growth, by a process known as secondary crystallization. Secondary crystallization can take two forms, thickening of pre-existing crystallites, and the creation of new crystallites. These processes may continue for hours, days, or even weeks after primary crystallization has ended.

Re-organization of amorphous chain segments adjacent to crystallite surfaces leads to crystallite thickening. These segments align themselves to form additional crystalline unit cells that increase the "c" dimensions of the crystallites.

Re-organization of amorphous chains in the interstitial regions between pre-existing crystallites leads to the formation of secondary crystallites. The new crystallites have smaller "a", "b", and "c" dimensions than the primary crystallites.

Secondary crystallization occurs most readily in polymers that have been quench-cooled. Quenched samples have low degrees of crystallinity and thus have relatively large volumes of amorphous material. A pre-requisite for secondary crystallization is that the amorphous regions must be in the rubbery amorphous state. Increased temperature accelerates the rate of secondary crystallization. The new volumes of crystallinity that form during secondary crystallization are generally quite small, amounting to less than 10% of the crystalline volume created during primary crystallization.

### 7.3.4    Crystallization Rate

In the broadest terms, the rate of polymer crystallization is increased by factors that increase the free energy difference between the amorphous and crystalline states and factors that favor the re-organization of amorphous chain segments. The factors that influence the crystallization rate fall into two categories: molecular characteristics and external conditions. Molecular characteristics include chemical composition, stereoregularity, copolymer composition, molecular weight, molecular weight distribution, and long chain branching. External factors include temperature, orientation, and heterogeneous nucleation.

In general, the enthalpic difference between the amorphous and crystalline states is greatest for polymers with a simple chemical composition. A large enthalpic difference favors crystallization, because it makes the change of Gibbs free energy more negative, which increases the driving force towards crystallization. Thus, polyethylene, with its simple chemical structure and unit cell, crystallizes more readily than isotactic polypropylene. Certain stereoregular forms are more favored than others; isotactic polypropylene, which has a relatively simple unit cell, crystallizes much faster than syndiotactic polypropylene, which has a much more complex unit cell.

Copolymers generally crystallize slower than homopolymers for two reasons. Comonomers reduce the enthalpic difference between the molten and crystalline states and they limit the length of uninterrupted identical monomer sequences. As the length of the crystallizable monomer sequences decreases, the crystallites in which they can participate become smaller and are thus stable only at lower temperatures, where viscosity is higher. Stereodefects have a similar effect to that of comonomers, reducing the average crystallizable sequence length and thus reducing the crystallization temperature. Additionally, the non-crystallizable sequences can physically block the alignment of crystallizable sequences with crystal growth faces. If the minor comonomers are physically larger than the major comonomer, they can act as anchors, which hinder the slippage of chains along their length. Similarly, long chain branching increases the viscosity of molten polymers and thus impedes crystallization. Other molecular factors that increase melt viscosity are elevated molecular weight and a narrow molecular weight distribution. Polymer chain flexibility favors molecular re-organization and thus increases the crystallization rate. Examples of flexible polymer chains include polyethylene and polyoxymethylene, which do not have bulky chemical groups or large atoms attached to their backbones. Factors such as double bonds and ring structures within the backbone reduce molecular flexibility and thus reduce the crystallization rate.

Temperature has a complex effect on crystallization rate. Initially, as the temperature falls below the equilibrium melting temperature, the crystallization rate increases because nucleation is favored. However, as the temperature continues to fall, the polymer's viscosity increases, which hampers crystallization. As a rule of thumb, a polymer crystallizes fastest at a temperature approximately mid-way between its glass transition temperature and its equilibrium melting temperature.

Molecular orientation increases the crystallization rate for two reasons: the statistical probability of chain segments becoming suitably aligned to form a nucleus is increased and the entropic difference between the molten and crystalline states is reduced, which increases the driving force towards crystallization.

## 7.4    Characterization of Solid State Molecular and Supermolecular Structure

Polymer scientists refer to the solid state structure of a polymer as its morphology. This includes both the arrangement of the crystalline and amorphous phases, and the orientation of molecules and crystallites. We can analyze a polymer's solid state structure in many ways, which provide us with information on various levels of morphology. We use scattering measurements, such as X-ray diffraction and neutron diffraction, to determine electron density and mass fluctuations, which yield information regarding ordering. Microscopy provides us with images of structures at scales down to a few nanometers. Other techniques measure physical characteristics that can be related to some aspect of morphology. For example, the density of a polyethylene sample is largely governed by its degree of crystallinity. Differential thermal techniques raise the temperature of samples while measuring one or more physical properties related to solid state structure or transitions.

### 7.4.1    Scattering Measurements

We use scattering measurements to characterize the regularity of various morphological features. The size of the features of interest determines our choice of incident energy; the scale of ordering probed is of a similar magnitude to the wavelength of the incident beam. Thus, we use X-ray diffraction to characterize organization at the atomic scale, and light scattering to probe much larger structures, such as spherulites and cylindrites.

Bragg's Law (Eq. 7.2) governs all scattering measurements. This relationship defines the scattering angle in terms of the wavelength of the incident radiation and the regular spacing being probed.

$$n \lambda = 2\, d \sin\theta \tag{7.2}$$

where:

$n$  =  an integer

$\lambda$  =  wavelength of incident radiation

$d$  =  periodicity of scattering features

$\theta$  =  scattering angle

Figure 7.10 illustrates the general configuration of a scattering experiment. Radiation from an electromagnetic source is collimated (i.e., non-parallel rays are eliminated) prior to striking the sample. A position-sensitive detector, which measures radiation intensity as a function of position, intercepts the scattered radiation. Computer analysis converts the raw signal into a two- or three-dimensional plot of intensity as a function of scattering angle.

X-rays are the most common type of radiation used to analyze polymers. We also employ visible light and neutron beams for specialized analytical purposes. We use small-angle laser

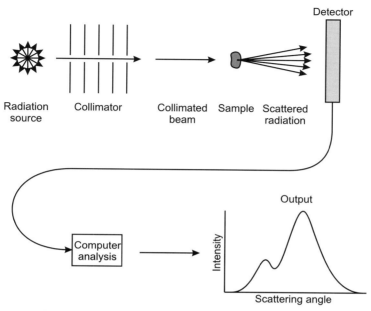

**Figure 7.10**   General configuration of an energy scattering experiment

light scattering to determine the size and perfection of spherulites and other supermolecular structures. Small-angle neutron scattering can be used to determine the average molecular dimensions of deuterated polymer molecules within a matrix of chemically similar protonated molecules.

### 7.4.1.1   Wide-Angle X-Ray Diffraction

We use wide-angle X-ray diffraction to investigate the interatomic spacings in polymers. We typically use it to analyze regularly arrayed atoms, such as those found in crystallites, but it can also provide information regarding interatomic distances in non-crystalline polymers and in the amorphous regions of semicrystalline polymers. Wide-angle X-ray diffraction measures scattering angles from approximately 2 to 90°. The wide angle diffraction from a single crystal, in which the atoms are perfectly arrayed, consists of a two-dimensional pattern of high intensity spots, as shown in Fig. 7.11 a). The precise positions of the reflections are determined by the interatomic spacings in the crystallites and the orientation of the sample relative to the beam and the detector. We can use wide-angle X-ray diffraction to determine interatomic spacings to a precision of one thousandth of an Ångstrom ($1 \times 10^{-13}$ m). In practice, semicrystalline polymers contain a multitude of crystallites, each of which has a slightly different orientation relative to the X-ray beam. The unoriented crystallites scatter X-rays as a series of cones, giving rise to concentric rings on the scattering diagram, as shown in Fig. 7.11 b). Non-crystalline regions create a diffuse ring, known as an amorphous halo, on the diffraction pattern.

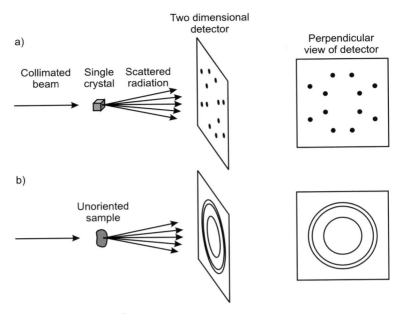

**Figure 7.11**    Scattering patterns from:
a) single crystal and b) unoriented semicrystalline sample

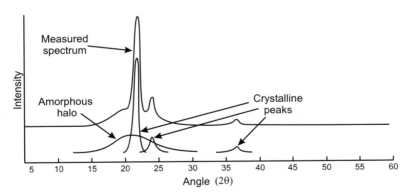

**Figure 7.12**    Wide-angle X-ray diffraction spectrum from high density polyethylene

X-ray patterns can be quantified by plotting the scattering intensity as a function of diffraction angle (which by convention is reported as the $2\theta$ angle, i.e., twice the actual scattering angle), as shown in Fig. 7.12. The measured spectrum can be deconvoluted into its component crystalline peaks and an amorphous halo. The angular position of the peaks reveals the interatomic spacings in the unit cell according to well established crystallographic principles. We can calculate the sample's degree of crystallinity from the relative areas of the crystalline peaks and the amorphous halo. The widths of the crystalline peaks can be analyzed to obtain a measure of the perfection of the atomic arrays within crystallites. The broader the peak, the less well ordered and smaller are the crystallites within a sample. The breadth of the amorphous halo is an indication of the range of interatomic distances found in the non-crystalline regions of a sample.

In practice, commercially fabricated polymer items are generally oriented to some degree. The scattering patterns from such materials comprise arcs, which are parts of the full circles obtained from unoriented samples. The lengths and positions of these arcs reveal much about the orientation of crystallites within a sample. The shorter the arcs, the more oriented the sample. In cases of extreme orientation, as found in highly oriented fibers such as Kevlar™, the scattering pattern can approach that of a single crystal.

### 7.4.1.2 Small-Angle X-Ray Diffraction

We use small-angle X-ray diffraction to analyze the periodic spacings of stacked lamellae, such as those illustrated in Figs. 7.6 and 7.7. The scattering angles measured range from less than 0.1 to 2°. The accurate measurement of such small angles requires the use of sophisticated equipment. The most precise measurements of small-angle X-ray diffraction patterns involve the use of high intensity, highly collimated X-rays produced by a cyclotron. Small-angle X-ray diffraction can be performed on polymers as they crystallize, capturing a series of images as a function of time. Such work provides insight into the mechanisms of crystallite growth.

## 7.4.2 Microscopy

Polymer microscopy covers a variety of techniques that provides images of samples at scales ranging from a few nanometers up to a few millimeters. Microscopy is not limited to optical methods; polymer analysts frequently use electron microscopy and atomic force microscopy to obtain images of polymers. The various microscopic techniques tend to be complementary; thus it is common to apply more than one technique to the analysis of any given sample. All microscopy techniques, especially at high magnifications, suffer from a common drawback: micrographs capture an image of a small portion of the sample, which does not necessarily represent the material as a whole. We must always keep this in mind when examining micrographs. It is good microscopic practice to examine many images from different parts of a sample before drawing any conclusions. Given the wide range of methods encompassed by microscopy, it is not our intention to provide anything more than an abbreviated review of some of the more important techniques.

### 7.4.2.1 Optical Microscopy

We use optical microscopy to examine samples at magnifications from about 5× up to approximately 1,000×. Samples may be examined using either transmitted or reflected light, depending on the nature of the sample and the information that we are seeking.

*Transmitted Light Optical Microscopy*

We use transmitted light optical microscopy to examine transparent or lightly pigmented samples. It is ideally suited to looking at defects within thin films and fibers. When samples are too opaque or thick for direct viewing, we use a microtome to cut thin slices that light can penetrate. To enhance contrast in samples that have regions with different refractive indices, we illuminate the sample with polarized light and view it through a polarizing filter. This principle is exploited in the examination of oriented samples, where the refractive index is a function of the angle at which it is measured. Thus, adjacent regions with different levels of orientation exhibit different colors when examined using a polarizing light microscope. (We can readily demonstrate this principle by viewing the clear plastic of a CD jewel case held between lenses removed from a pair of polarizing sunglasses.) If the refractive index and thickness of a sample are accurately known, it is possible to quantify the degree and direction of orientation within a sample. Material analysts routinely use transmission microscopy to examine sections cut from multilayer polymer items, such as blow molded tomato ketchup bottles and snack food bags. Such seemingly simple samples typically comprise five or more layers, each of which provides a specific property, such as tear strength, resistance to water, or resistance to oxygen permeation. Optical microscopy provides us with information regarding the number, type, and thickness of the various layers.

*Reflected Light Optical Microscopy*

Analysts use reflected light microscopy to examine the surface of polymers. By changing the angle of illumination they can accentuate surface texture and other features of interest. Reflected light microscopy is well suited to the examination of opaque and pigmented samples. Polymer scientists make extensive use of reflected light microscopy when examining the fracture surfaces of failed samples.

### 7.4.2.2   Infra-Red Microscopic Spectroscopy

Optical microscopy is often coupled with infra-red spectroscopy. We use the optical portion of the instrument to identify regions of interest, onto which we direct a highly focused infra-red beam. We obtain an infra-red spectrum from the radiation that penetrates the sample. The region of interest may be as small as 250 μm ($250 \times 10^{-6}$ m) in diameter. We can compare the spectrum with a library of reference samples in order to identify the chemical structure of the area of interest. Polymer scientists make extensive use of this technique when examining multi-layer samples or when performing contaminant analyses.

### 7.4.2.3   Electron Microscopy

Polymer scientists use electron microscopy to examine polymer samples at magnifications ranging from approximately 15× up to 250,000×. Samples may be viewed using either transmission or scanning configurations. In the former, a position-sensitive detector measures the electrons that pass through a thin section. Electron dense regions permit fewer electrons to pass through and appear darker in the image. We use transmission electron microscopy to examine crystallite arrangement in semicrystalline samples. In scanning electron microscopy, electrons reflected from the surface of a sample are detected. Polymer scientists make extensive use of scanning electron microscopy when analyzing fracture surfaces and other textured samples, such as those of breathable films and polymer fabrics.

*Transmission Electron Microscopy*

Transmission electron microscopy – which we normally refer to as TEM – is used to examine thin sections sliced from samples using an ultra-microtome. Sections are typically only a few tens of nanometers in thickness, which allows the electron beam to penetrate readily. Transmission electron microscopy relies on differences in electron density within the sample to provide contrast. Regions of high electron density scatter the incident beam more than low electron density regions and thus appear darker on the electron micrograph. Material analysts commonly stain microtomed sections with a solution of a heavy metal oxide, such as ruthenium tetroxide. Amorphous regions preferentially absorb the stain because they have a higher free volume than crystallites. Transmission electron microscopy is used to examine the relationships between the amorphous and crystalline regions, either in semicrystalline samples or in multiphase blends. In order to obtain information regarding the three-dimensional arrangement of crystallites it is necessary to cut sections in two – or preferably three – mutually perpendicular planes. Under ideal conditions we can obtain images at magnifications as high as 250,000×, which permit us to resolve features down to approximately two nanometers.

*Scanning Electron Microscopy*

Scanning electron microscopy – which we normally abbreviate to SEM – is used to examine the surfaces of polymers. We use it to investigate the surfaces of molded samples, fracture surfaces, or the interior of samples revealed by microtomy. Scanning electron microscopy relies on differences in surface topography to create the image. Smooth samples may be etched to create surface texture representative of the sample, using various forms of radiation, solvents, acids, or bases. Etching tends to remove amorphous regions, leaving crystallites or other more resistant regions standing slightly proud of the surface. We can obtain scanning electron micrographs at magnifications from approximately 15× to 50,000×. A distinct advantage of scanning electron microscopy is its great depth of field. This permits us to visualize surfaces that have relatively high levels of topographic variation.

## 7.4.2.4    Atomic Force Microscopy

We use atomic force microscopy – commonly known as AFM – to examine the surfaces of molded polymer items, fracture surfaces, or etched surfaces. It can yield images that depict surface topography or local hardness. Images are created as a series of pixels by tapping the surface of a sample with an ultra sharp probe that is scanned in lines across the sample to build up the image (much as a television image is built up from a series of lines). A computer determines the height of the tip when it contacts the surface in order to provide topographic information. The computer can also measure the depth to which the tip penetrates the sample to yield hardness images, which are referred to as "phase" images that provide information regarding crystallite organization. The level of magnification is determined by the tapping frequency. Using atomic force microscopy we can resolve features ranging from approximately two nanometers up to several tens of microns.

### 7.4.3    Differential Scanning Calorimetry

We use differential scanning calorimetry – which we invariably shorten to DSC – to analyze the thermal properties of polymer samples as a function of temperature. We encapsulate a small sample of polymer, typically weighing a few milligrams, in an aluminum pan that we place on top of a small heater within an insulated cell. We place an empty sample pan atop the heater of an identical reference cell. The temperature of the two cells is ramped at a precise rate and the difference in heat required to maintain the two cells at the same temperature is recorded. A computer provides the results as a thermogram, in which heat flow is plotted as a function of temperature, a schematic example of which is shown in Fig. 7.13.

Thermograms provide information regarding the transitions that polymers pass through in the temperature range of interest. Glass transitions are evident as step changes. Melting is represented by one or more peaks, corresponding to the crystallites within a sample. Unlike small organic molecules, which melt at a sharply defined temperature, polymers exhibit a broad range of melting temperatures, characteristic of the distribution of crystallite thicknesses. If the distribution of crystallite thicknesses is bimodal, it will be reflected in the thermogram. The peaks in the thermogram of a polymer represent the temperature of maximum heat flow; they should never be referred to as melting points. The area under the melting curve is proportional to the amount of heat required to melt the sample, i.e., its heat of fusion. If we know the theoretical heat of fusion of the pure polymer, we can calculate a sample's degree of crystallinity from its measured heat of fusion using Eq. 7.3.

$$\text{Degree of crystallinity (\%)} = \frac{\text{Measured heat of fusion}}{\text{Theoretical heat of fusion}} \times 100 \qquad (7.3)$$

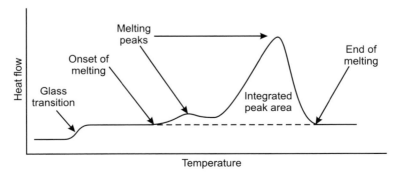

**Figure 7.13**    Schematic representation of a differential scanning calorimetry thermogram

## 7.4.4    Density

The density of a polymer sample depends upon the relative densities and masses of its components according to Eq. 7.4.

$$\rho = \frac{(m_1 + m_2)}{\dfrac{m_1}{\rho_1} + \dfrac{m_2}{\rho_2}}$$

(7.4)

where:

$\rho$ = polymer sample density
$m_1$ = mass of component 1
$m_2$ = mass of component 2
$\rho_1$ = density of component 1
$\rho_2$ = density of component 2

In the case of a semicrystalline polymer, the two components are the crystalline and amorphous regions. If we know the densities of the crystalline and the amorphous regions, we can calculate a sample's degree of crystallinity from Eq. 7.5.

$$\text{Degree of crystallinity (\%)} = \frac{\dfrac{1}{\rho} - \dfrac{1}{\rho_a}}{\dfrac{1}{\rho_c} - \dfrac{1}{\rho_a}} \times 100$$

(7.5)

where:

$\rho$ = sample density
$\rho_c$ = density of crystalline component
$\rho_a$ = density of amorphous component

We can measure a polymer's density by one of two methods: density gradient column analysis and densimetry. Each of these methods can measure the density of a sample to a precision of four significant figures.

### 7.4.4.1    Density Gradient Column Analysis

A density gradient column consists of a glass tube, an inch or two in diameter, filled with a pair of liquids in such a way that it possesses a smooth density gradient; low at the top and high at the bottom. We calibrate the gradient with hollow glass beads that have known densities, which float at various heights within the column. We prepare a calibration curve by plotting the height of each bead against its known density. When we drop a small polymer sample into the column, it sinks until it reaches a point of neutral buoyancy. We measure the sample's height and read its density directly from the calibration curve.

We can use density gradient columns to measure the density of all manner of polymer samples, from fibers and films to specimens cut from molded parts.

### 7.4.4.2  Densimetry

We apply Archimides' Principle to measure a sample's density by a technique known as densimetry. We weigh the sample twice, once when suspended in air and again while suspended in a liquid of known density. We calculate its density from Eq. 7.6.

$$\rho = \frac{(\rho_1 W_a - \rho_a W_1)}{(W_a - W_1)} \tag{7.6}$$

where:

$\rho$  =  sample density
$\rho_a$  =  density of air
$\rho_1$  =  density of liquid
$W_a$ =  weight of sample in air
$W_1$ =  weight of sample in liquid

Densimetry is best suited to the density measurement of specimens with a low surface-to-volume ratio, such as molded parts or pieces cut from them.

## 7.5    Conclusions

In this chapter we discussed the three basic types of solid state structure that we find in polymers and how they form from the molten state. We went on to describe the techniques that polymer scientists use to characterize polymer structures at scales ranging from less than one nanometer ($1 \times 10^{-9}$ m) up to a few millimeters ($> 1 \times 10^{-3}$ m). The wide range of structures that we can generate from polymers contributes to their wide range of properties and corresponding breadth of finished items that we can create.

## Review Questions for Chapter 7

1.   What is meant by the term "supermolecular structure" when describing polymers?

2.   Why is it necessary to consider both the thermodynamic and kinetic behavior of polymer chains when describing the formation of structure in the polymer during cooling from the melt?

3.   Define the terms rubbery amorphous, glassy amorphous, crystalline and semicrystalline.

4.   Why is the density of an amorphous polymer greater in its glassy state than in its rubbery state?

5.   Explain the hierarchy of crystalline regions using the words unit cell, crystallite, lamellae, cylindrites and spherulites.

6.   Which component of Gibbs free energy (entropy or enthalpy) drives crystal formation in a polymer? How?

7.   Describe the mechanism by which crystallites form, starting with nucleation and ending with secondary crystallization.

8.   Sometimes polymer processors quench a polymer by quickly "freezing" the melt. What happens on the molecular scale during quenching? What benefits do the processors gain by quenching their polymer? What material properties are reduced by this process?

9.   How is X-ray diffraction used to explore crystallinity in polymers?

10.  How can we utilize density analysis to characterize crystallinity?

# 8     Solid State Properties of Polymers

## 8.1      Introduction

It is through the solid state characteristics of polymers that we – as users – primarily interact with them. For convenience, we can divide the principal properties of polymers into five categories: mechanical, optical, surface contact, barrier, and electrical. Weather resistance is a sixth category that can influence each of the other five categories. In order to understand these properties we must be able to quantify them. In this chapter we shall concentrate on measurement techniques, since it is through these methods that we learn how a polymer will behave during use.

The mechanical properties of a polymer describe its response to external forces. This family of properties includes tensile, flexural, compressive, and toughness characteristics. Other factors that we must consider when discussing mechanical properties are the temperature and rate of testing. Due to the viscoelastic nature of polymers, their properties change as a function of temperature. Another consequence of viscoelasticity is that polymers tend to deform progressively when we subject them to a fixed load for an extended period of time. We call this type of long term deformation "creep". Optical measurements characterize the way that polymers transmit or reflect light under a variety of conditions. We can divide the surface contact properties of polymers into two categories, those in which the surface remains intact when it comes into contact with another surface, and those that describe the polymer's resistance to surface damage. Barrier properties largely involve either the containment or exclusion of liquids or gases. Where electricity is concerned, we can use polymers either as insulators or as part of a charge storage device. In the first case, we are interested in a polymer's resistive properties, in the latter we need to know its capacitive properties. Many polymer applications involve exposure to the elements for extended periods of time. We commonly measure specific aspects of the five main categories of solid state properties before and after exposure to electromagnetic radiation, heat, and moisture. In commercial and industrial laboratories it is common to test material properties according to standard testing procedures sanctioned by national or international testing standards organizations.

## 8.2      Mechanical Properties

The mechanical properties of a polymer describe how it responds to deforming forces of various types, including tensile, compressive, flexural, and torsional forces. Given the wide range of polymer structures, it should be no surprise that there is a correspondingly wide range of methods to test their properties. Each application requires a different set of tests to define the polymer properties that match the intended end use. Thus we would perform different batteries of tests on nylon gear wheels, polyester yarns, and high density polyethylene milk jugs. Tests largely fall into two types: in the first we apply a deforming force and measure the material's response, in the second we apply a known force and observe whether or not the sample fails.

We must always take into account temperature and testing rate when measuring mechanical properties. Elevated temperatures increase the rate of chain segmental motion, which promotes chain disentanglement and deformation of crystallites, both of which can have profound effects on mechanical properties. This is especially true if the testing temperature approaches the polymer's melting temperature. Conversely, if the testing temperature drops below the sample's glass transition, its failure mechanism will change from ductile to brittle. When evaluating polymers for a specific application it is important that we match the testing temperature with the end use temperature. For example, it is advisable to test the polymers used in car bumpers at temperatures ranging from –50 to 50 °C, to make sure that they can survive Alaskan winters and Arizonan summers.

In many applications, we exert forces on polymer products for extended periods of time. In such cases it is important that they do not deform beyond acceptable limits. We address this concern through creep testing, which may take days, weeks, or even months to perform. We can also change the rate at which we perform certain tests. Thus we might test the polymer resin to be used in a crash helmet at high speed to mimic an impact. In contrast, the film used in a plastic bag would be better tested at slower speeds, which are more representative of what it would experience during use.

## 8.2.1    Force versus Deformation Properties

The classic way that we perform force versus deformation measurements is to deform a sample at a constant rate, while we record the force induced within it. We normally carry out such tests in one of three configurations: tensile, compressive, or flexural, which are illustrated in Fig. 8.1. We can also test samples in torsion or in a combination of two or more loading configurations. For the sake of simplicity, most tests are uni-axial in nature, but we can employ bi-axial or multi-axial modes when needed.

We perform most tests uni-directionally, that is, we increase the deformation in only one direction, as shown if Fig. 8.2 a). Alternatively, we can perform a dynamic test in which the direction of the deforming force is reversed one or more times. In dynamic tests, the waveform of the applied deformation is often sinusoidal, as shown in Fig. 8.2 b), but many other modes are possible, including a sawtooth pattern, or a square wave, as shown in Fig. 8.2 c) and d), respectively.

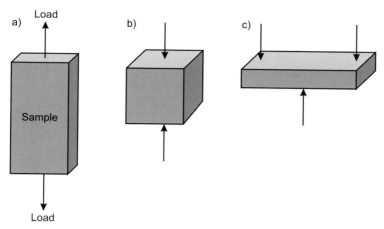

**Figure 8.1**  Schematic illustration of principal testing configurations:
a) tensile, b) compressive, and c) flexural

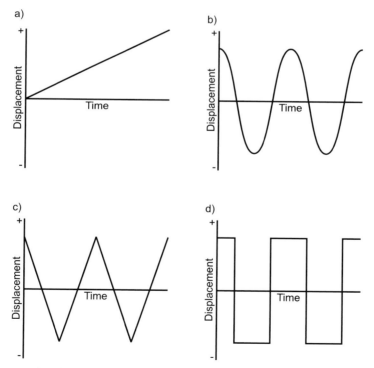

**Figure 8.2**  Examples of uni-directional and dynamic loadings:
a) constant rate uni-directional, b) sinusoidal, c) sawtooth, and d) square wave

### 8.2.1.1   Tensile Testing

Figure 8.3 illustrates the principal features of a tensile tester. A specimen is clamped between a pair of jaws, the lower of which is fixed. The upper jaw is attached to a moveable crosshead that is raised in order to stretch the sample. We can test specimens in many forms; strips are normally cut from films, "dumbbells" or "dogbones" are stamped from thicker samples, or we may injection mold dogbones. Before starting the test, we measure the sample's dimensions then clamp it into the jaws of the instrument. When the crosshead is raised, the load cell records the force induced within the specimen as it stretches. Once the specimen breaks, or when the desired extension level is reached, the test is halted. It is conventional to plot the results as engineering stress versus engineering strain. The engineering stress is calculated from Eq. 8.1, and the engineering strain from Eq. 8.2.

$$\text{Engineering stress} = \frac{\text{Measured load}}{\text{Original cross-sectional area of specimen}} \tag{8.1}$$

$$\text{Engineering strain (\%)} = \frac{\text{Crosshead travel}}{\text{Original gauge length}} \times 100 \tag{8.2}$$

When the specimen is in the form of a strip, the gauge length is the distance between the upper and lower jaws. When dogbones are tested, the gauge length is the length of the parallel portion of the specimen.

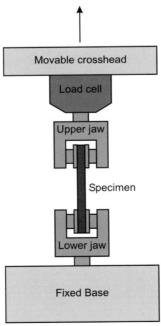

**Figure 8.3**   Schematic diagram of a tensile testing apparatus

Figure 8.4 shows generic load versus elongation curves for rubbery amorphous, glassy amorphous, and semicrystalline polymers. In each case, the effect of extension on a dogbone specimen is shown at various points along the curve.

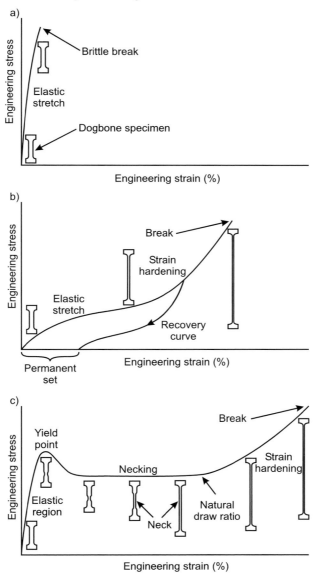

**Figure 8.4**  Generic force versus elongation curves for:
    a)  glassy amorphous polymer,
    b)  rubbery amorphous polymer, and
    c)  semicrystalline polymer

In Fig. 8.4 a), the glassy amorphous polymer extends only a few percent before it breaks abruptly. The extension in the sample up to the point of failure is largely reversible, that is, the material behaves elastically. Polystyrene and polycarbonate, which are used to make CD jewel cases, exhibit this type of behavior.

In Fig. 8.4 b), the rubbery amorphous polymer extends several hundred percent before it breaks. Rubbery polymers commonly exhibit a phenomenon that we call "strain hardening", that is, the force required to deform the specimen increases rapidly prior to break. If we reverse the crosshead travel before the sample breaks, it will elastically recover to some extent, as shown by the lower curve in the figure. Fully crosslinked rubbers may recover completely, returning to their original length. A common example of such a material is a rubber band. Lightly crosslinked samples or thermoplastic elastomers may only partially recover after high extension, taking on a permanent set. Examples of these materials include the rubber covers on the handles of power tools and surgical rubber tubing.

In Fig. 8.4 c) we show a generic force versus elongation curve for a semicrystalline polymer, which deforms in a ductile manner. In this case, the initial force required to stretch the sample increases rapidly, up to a point where it yields. If we remove the load prior to the sample's yield point and allow it to relax, it will recover in a largely elastic fashion. Yielding is normally discernible on the sample as the point at which it starts to neck, i.e., the onset of inhomogeneous deformation, where different parts of the specimen exhibit different strain levels. As we continue to stretch the specimen the neck extends. The force required to extend the neck is essentially constant, resulting in a plateau on the force versus elongation curve. We call the extension at which the neck encompasses the whole of the gauge length the natural draw ratio. Further elongation is normally homogenous up to the point of break.

The examples that we have shown here represent only a small fraction of all the variations possible. There is no such thing as a typical force versus elongation curve for polymers. Samples can break at extensions of only a fraction of a percent up to several thousand percent, with engineering stresses at break ranging from only slightly above zero up to more than 10 GPa.

## Elastic Modulus

In classic terms, the elastic modulus of a material is the stress divided by the strain (i.e., the slope) of the linear portion of its force versus elongation curve at low strain. In this region, the material is assumed to behave in a Hookean fashion, i.e., stress and strain are linearly proportional, as illustrated in Fig. 8.5 a). Most polymers do not behave in this manner.

As we have seen in Fig. 8.4, the force versus elongation curves of polymers are not linear at low strains. To work around the problem of non-linearity, polymer scientists have come up with various methods of calculating the modulus of a polymer. Strategies include measuring the slope of the steepest portion of the force versus elongation curve, measuring the instantaneous slope of the curve at a given strain, measuring the slope of a line drawn between the curve at two specified strain levels, and measuring the slope of a line drawn between the origin and the curve at a specific strain. These methods are illustrated in Fig. 8.5 b) to e), respectively. As can be seen, each of these calculation strategies yields a different value of the modulus.

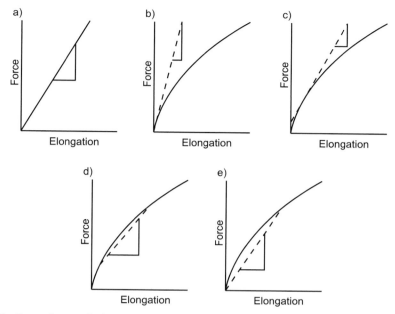

**Figure 8.5** Strategies to calculate modulus:
a) Hookean behavior,
b) steepest slope,
c) tangent at a specific strain,
d) slope between two specified strain values, and
e) slope between zero strain and a specific strain

Molecular orientation and the testing direction strongly influence the observed modulus of a polymer sample. Fibers are typically highly oriented and exhibit much higher modulus values than non-oriented samples prepared from the same polymer. In the case of films, we typically observe anisotropy: a film exhibits a range of modulus values depending upon the testing direction.

As a practical manner, we typically think of a polymer's modulus in terms of its stiffness, which strongly influences how we can use it. It is obvious that structural materials, such as plastic shelving, gas distribution pipes, and pallets must exhibit high rigidity. It may not be so obvious that many fibers exhibit extremely high values of modulus; for instance Kevlar™ or ultra oriented polyethylene fibers used in personal body armor are among the stiffest materials known. On a weight for weight basis, these fibers can be stiffer than steel.

*Yield Phenomena*

When we begin to stretch a semicrystalline polymer it deforms affinely, that is, each element of the sample within the gauge region experiences identical stress and strain. As we continue to stretch the sample, we reach a point at which affine deformation ceases and the sample yields. At this point, it typically develops a local region of reduced cross-sectional area, known as a neck, as shown in Fig. 8.4 c). Yielding involves irreversible morphological rearrangement; crystallites may be destroyed and new ones form, or they can be sheared in such a way that their constituent chains slide past one another. In some cases, microvoids form within the sample. Once a sample has yielded it does not return to its original dimensions when the load is removed. Further extension results in the expansion of the necked region at the expense of the un-necked portion of the specimen. We can readily observe necking if we stretch a high density polyethylene grocery sack crossways. The necked regions undergo voiding, scattering more light and becoming more opaque than the original film. Another polymer that undergoes evident yielding is isotactic polypropylene. The neck turns white due to microvoiding and becomes flexible. We take advantage of this phenomenon to make integral hinges in polypropylene storage containers. The container and its lid are molded in a single piece; when the joint between them is flexed it, yields to form a strong and pliable hinge.

In many applications we consider the yield point of a polymer to be its point of failure. We base this definition on whether the material is still fit for its end use once it has yielded. Thus, if the teeth on a polymer gear wheel yield, they will change shape and may become useless. Alternatively, packaging film may yield, but still maintain its barrier properties and be fit for continued use.

### Tensile Strength

We normally define the tensile strength of a polymer as the engineering stress level at which it breaks. At this point two things can occur: polymer chains can slide past one another and disentangle, or the chains spanning a plane across the sample can break. The chains making up a glassy amorphous polymer are likely to break because they have limited freedom of motion. In rubbery amorphous polymers the chains have greater freedom of movement and slide past one another. When the polymer is not crosslinked, and is of moderate molecular weight, chain slippage can proceed to the point that failure occurs due to chain disentanglement. In crosslinked and highly entangled systems, chain slippage is limited and molecules break. In semicrystalline polymers that deform in a ductile fashion (i.e., those in which the amorphous matrix surrounding the crystallites is above its glass transition temperature) the situation is likely to be complex, both chain disentanglement and breakage can occur simultaneously. High tensile strength and high elastic modulus often go hand in hand.

The tensile strength of a polymer often determines its suitability for a particular application. Thus, for ropes, twines, and yarns we typically select polymers with high tensile strengths. Similarly, we demand good tensile strength from packaging films, office equipment cases, tubing and pipes. In contrast, the tensile strength of polymers in other uses is of secondary concern. Such applications include dairy product tubs, dashboard covers, and photographic film canisters.

### Elongation at Break

The elongation at break of a sample is the strain at which at which it breaks. This value varies widely depending on polymer type and processing conditions. Glassy amorphous polymers typically exhibit low elongations at break because their chains cannot slide past one another. In rubbery amorphous polymers the situation is somewhat different. High molecular weight polymers, which are highly entangled, and highly crosslinked samples cannot extend greatly prior to break. However, as the crosslink density decreases the samples can deform more readily and extension at break increases. Orientation also influences elongation at break. When we stretch a sample perpendicular to its molecular orientation direction, we generally observe a higher elongation at break than when it is stretched parallel with its orientation. This is because the molecules in oriented polymers are already stretched out and cannot extend much further before they become taut and break.

### 8.2.1.2   Compressive Testing

We perform compressive testing in a similar apparatus to the one illustrated in Fig. 8.3. Anvils are used in place of the gripping jaws, and the crosshead is lowered, rather than raised. Dynamic testing often contains a compressive testing component.

We prepare force versus compression plots in a similar fashion to force versus elongation plots. We generally perform compressive testing over a much more limited range of strain than tensile testing. Samples typically take the form of thick pads, which do not break in the same manner as tensile specimens. The limit of compressive strain can approach 100% for low density foams, but is much less for other samples. The most common property that we obtain from this mode of testing is compressive modulus.

The samples most commonly tested in compression are foams and rubbers, which experience compressive forces during use. Very often, the polymer foams that experience compression are not readily visible to us, even though they are all around. Polymer foams are widely used in carpet underlay, upholstery, shoe insoles, backpack straps, bicycle helmets, and athletic pads. Solid rubbers are much more visible, including automobile and bicycle tires, gaskets and seals, soft keys on calculators, and shoe soles.

### 8.2.1.3   Hardness Testing

We test the hardness of polymers by applying an indenter to their surface with a known force and noting the depth to which the tip penetrates the sample. These tests typically fall into one of two categories. In the first, the depth of penetration is read directly from a dial on the instrument, calibrated in arbitrary hardness units. The farther the tip penetrates the sample, the lower is its hardness. The second type of test involves impressing a pyramidal indenter tip against the sample with a known force and measuring the depth to which it penetrates. In practice we measure the dimensions of the indentation and calculate the depth of penetration and compressive modulus based on the tip geometry.

### 8.2.1.4   Flexural Testing

We perform flexural testing on polymer rods or beams in the same basic apparatus that we use for tensile or compressive testing. Figure 8.6 illustrates two of the most common flexural testing configurations. In two-point bending, shown in Fig. 8.6 a), we clamp the sample by one end and apply a flexural load to the other. In three-point bending, shown in Fig. 8.6 b), we place the sample across two parallel supports and apply a flexural load to its center.

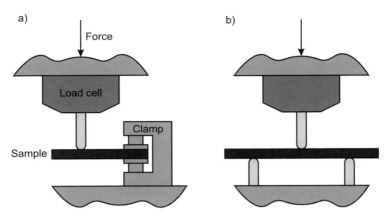

**Figure 8.6**   Types of flexural testing:
a) two-point bending (cantilever), b) three-point bending

We normally perform flexural testing over a limited range of strain, generally only sufficient to determine the sample's flexural yield point. From an end use point of view, this is a reasonable place to halt the test. If we are using a polymer in an application where it is exposed to flexural loads, such as car body panels or bulk liquid storage tanks, it is generally considered to have failed at the point where it no longer deforms reversibly. We obtain a value of flexural modulus from the force versus deformation curve, much as we do for the tensile equivalent.

*Heat Distortion Temperature*

We use a variant of flexural testing to measure a sample's heat distortion temperature. In this test, we place the sample in a three point bending fixture, as shown in Fig. 8.6 b), and apply a load sufficient to generate a standard stress within it. We then ramp the temperature of the sample at a fixed rate and note the temperature at which the beam deflects by a specified amount. This test is very useful when selecting polymers for engineering applications that are used under severe conditions, such as under the hoods of automobiles or as gears in many small appliances or inside power tools where heat tends to accumulate.

## 8.2.2   Toughness Measurement

We define the toughness of a polymer as the amount of mechanical energy required to deform it to failure under a specific set of conditions. Most of the tests that we use to measure a sample's toughness involve subjecting it to some type of high speed load and measuring or observing its response. The cost of the sophisticated equipment required to measure stresses and strains at high speeds is quite high. Most industrial testing laboratories cannot justify this expense and instead rely on simple devices that measure the total energy absorbed by a sample as it fails under high speed deformation. It is on these less sophisticated tests, which are often designed to mimic failure in end use, that we shall concentrate in this section.

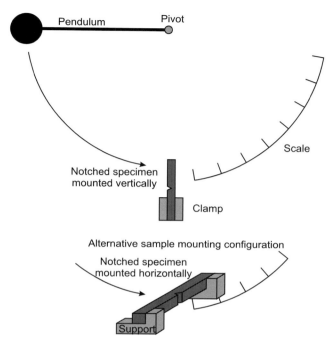

**Figure 8.7**   Schematic illustration of a pendulum type impact testing device

Most commercial toughness measurements are performed using one of two general configurations. In the first, we subject a sample to an abrupt load with the aid of a pendulum, as
shown schematically in Fig. 8.7. We record the angle through which the pendulum swings
after it breaks the sample. We compare this value with the angle through which the pendulum
swings when no sample is mounted, and thereby obtain the energy absorbed by the sample
as it breaks.

In the second type of test we drop an impacter on a sample mounted horizontally, as illustrated
schematically in Fig. 8.8. If the sample fails, we replace it with a fresh one and remove one of
the weights from the impacter rod. We define failure as a crack or break visible to the naked
eye. Conversely, if the sample survives the impact, we increase the weight and test a fresh
sample. Alternatively, we can increase or decrease the distance through which the impacter falls,
according to whether or not the sample withstands the impact. We proceed in this manner,
either increasing or decreasing the impact energy, until we have tested twenty samples. This
is known as a "staircase" or "up and down" test. We calculate the material's impact strength
from the number of trials at each specific impact energy level. When impact testing thin
films, we can measure the speed of the impacter before and after penetrating the sample. The
speed differential is proportional to the energy absorbed by the sample. We routinely perform
toughness tests at ambient and sub-ambient temperatures. A common lower temperature
is −40 °C, which represents the temperature that a sample may be exposed to when used
outdoors in extreme winter conditions.

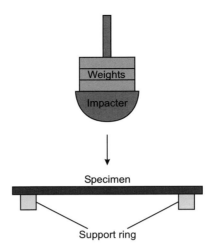

**Figure 8.8**   Schematic illustration of a falling weight impact testing device

### 8.2.2.1    Impact Beam Testing

We perform impact beam testing with the equipment outlined in Fig. 8.7. We select the weight of the pendulum based on previous experience and one or more initial trials. Our goal is to record angles that are in the mid-range of the angular scale. Some glassy amorphous polymers are highly susceptible to notches, defects, and scratches, which can drastically reduce their toughness. In such cases, it is important that we test both notched and un-notched samples in to order fully assess their toughness characteristics. We introduce a standard notch into specimens using a cutter specifically designed for the purpose.

We can mount samples either vertically or horizontally. When we mount the sample vertically, the pendulum strikes the upper half of the sample on the notch side, as shown in the upper part of Fig. 8.7. To mount a sample horizontally, we place it on a pair of "L"-shaped supports, as shown in the lower part of Fig. 8.7, with the notch pointing away from the pendulum, which strikes the sample in the middle of the opposing side.

Samples fail in a variety of ways. Glassy amorphous polymers and semicrystalline polymers below their glass transition temperatures may shatter or break cleanly into two pieces. Rubbery amorphous and semicrystalline polymers tested above their glass transition temperatures exhibit significant deformation on either side of the plane of rupture. It is this zone of deformation that is largely responsible for absorbing energy; the larger the deformation zone, the greater the impact resistance. Deformation can take the form of stretching or voiding. When voids form we observe whitening in the deformed region. We can toughen glassy amorphous polymers by blending an incompatible rubbery amorphous polymer with them. These samples absorb large amounts of energy by voiding in the vicinity of the plane of rupture. We use toughened glassy polymers in applications where we need stiffness and good impact resistance, such as automobile bumper covers, phones, and kitchen appliances.

### 8.2.2.2   Impact Plaque Testing

We test the impact resistance of polymer plaques using the configuration shown in Fig. 8.8. We subject circular injection molded plaques to the shock of a falling weight with a hemispherical impacter. This test is also known as the "Dart Drop Test". Samples can fail in a brittle or a ductile manner. Brittle samples often shatter. Ductile samples can split or a small disk may be punched out of their center. This test provides results that are analogous to those obtained from un-notched impact beam testing.

### 8.2.2.3   Film Tear Testing

We measure the tear resistance of polymer films using a pendulum type testing device the principle components of which are shown in Fig. 8.9. We mount a film sample vertically in a pair of clamps, one of which is fixed to the frame of the instrument. The other jaw is attached to a quadrant-shaped pendulum. We slit the sample in the region between the clamps with a small blade mounted on the instrument. When we release the pendulum, it swings downwards and tears the film, much as we might tear a piece of paper. We determine the film's toughness by measuring how much the upswing of the pendulum is reduced compared to when no sample is in place.

A sample absorbs energy in two ways: molecules that transect the plane of rupture must be broken and material adjacent to the path of the rupture deforms. Glassy amorphous polymers generally tear quite cleanly, with minimal deformation around the rupture. Rubbery amorphous and semicrystalline polymers above their glass transition temperatures exhibit deformation to a greater or lesser extent. As with impact testing, we observe larger deformation areas in tougher materials.

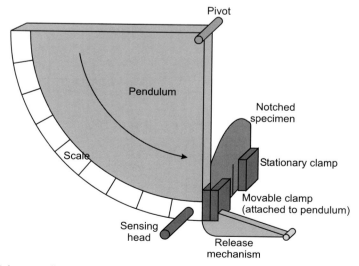

**Figure 8.9**   Schematic illustration of a pendulum type tear testing device

The tear strength of a polymer film is often highly anisotropic, that is, its value depends on the testing direction. The anisotropy of a film is a function of its orientation. When we attempt to tear a film perpendicular to the orientation of its molecules, we have to break many more molecules to create a rupture than when we tear it parallel with the chains. In extreme circumstances, a film can become "splitty", requiring many times more force to tear it in one direction than the other, as we have all seen when trying to tear into snack food packages. One way to balance the tear strength of a film is to induce bi-axial orientation during processing. This involves stretching the film in mutually perpendicular directions while it is in the process of solidifying from the molten state.

### 8.2.2.4    Film Puncture Resistance Testing

Figure 8.10 illustrates the general configuration for testing a film's puncture resistance. We can perform this test in a number of ways. The probe may be attached to the crosshead of a universal testing machine, similar to the one illustrated in Fig. 8.3. We drive the probe down through the film at a constant speed, while recording the force. Alternatively, we can perform an impact puncture test, in a similar fashion to impact plaque testing.

When we test a film's puncture resistance at a constant rate, we determine its toughness from the area under the force versus elongation curve prior to the point of failure. When tested using a falling weight configuration, we can use the staircase method to determine the film's puncture resistance. Alternatively, we can measure the speed of the falling impacter before and after puncturing the film and thereby calculate the energy absorbed by the film.

When the probe makes contact with the film, it generates a radial stress field around the point of contact. If the film is isotropic, it deforms in a uniform ring around the probe, as shown in Fig. 8.11 a). If the film is oriented, it deforms in a non-uniform manner. When the film is mildly oriented, the deformation area becomes ellipsoidal, as we see in Fig. 8.11 b), with its long axis

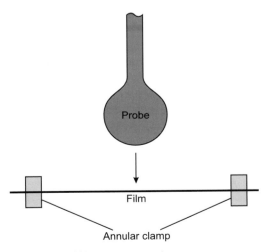

**Figure 8.10**   Schematic illustration of film puncture testing

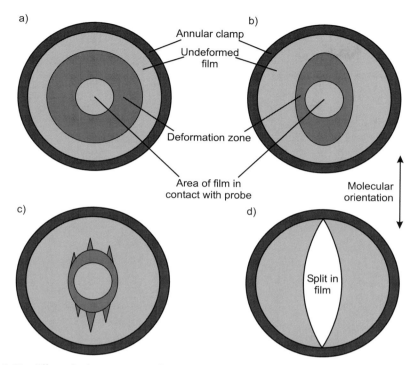

**Figure 8.11**    Effect of orientation on deformation zones observed during puncture testing, viewed from above:
a) isotropic film,
b) low film orientation,
c) moderate film orientation,
d) high film orientation

parallel with the orientation direction. At moderate orientation levels we see more complex areas of deformation that combine an ellipsoidal ring with local triangular yield zones, the apexes of which are aligned with the film's orientation, as shown in Fig. 8.11 c). In cases where the film is highly oriented, it can split parallel with the molecular orientation, as illustrated in Fig. 8.11 d). In all cases, rupture of the film initiates in the vicinity of the probe.

We commonly see the effect of puncture resistance in packaging films and bags. For instance, the film used to package meat must resist bones poking out and shoppers' fingers poking in. Likewise, ice and garbage bags must exhibit good puncture resistance.

## 8.2.3    Creep

Creep is the long-term deformation of a polymer under a sustained or intermittent load at stresses below the yield point. We see creep at work when we place furniture on carpet. If we

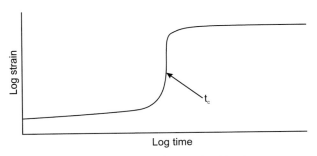

**Figure 8.12**   General form of strain versus time under sustained load

set a bookcase down on carpet and remove it immediately, the synthetic polymer fibers spring back up instantly. However, if we leave the same bookcase on the carpet for a prolonged period, the fibers take on a permanent set and we see an indentation when we remove the bookcase. Other examples of creep include the permanent expansion of pipes under pressure and the deformation of plastic storage shelves under long-term load. Creep can take place over periods ranging from a few hours to many years. Creep involves the re-arrangement of molecular segments to relieve local stress. The polymer is essentially flowing at a very slow rate. Creep is more pronounced at higher loadings and at elevated temperatures, which increase local molecular motion. We primarily observe creep in amorphous and semicrystalline polymers above their glass transition temperatures.

Figure 8.12 illustrates the general effect of creep, plotted as a function of log strain versus log time. Under applied load a sample gradually deforms until a critical time ($t_c$) after which it deforms rapidly. If creep is allowed to go unchecked, the sample may break abruptly.

We do not commonly observe the effects of creep in our daily lives for three reasons. Firstly, creep generally takes place over such a long period of time that we are unaware of its gradual effects. Secondly, the majority of polymeric items are subjected to loads for a relatively short period of time. Lastly, most polymer engineers who design durable items are well aware of the effects of creep and design their products accordingly. It is a tribute to their skill that we are largely unaware of this potentially disastrous mode of polymer failure.

## 8.3    Optical Properties

We can divide the optical properties of polymers into those relating to light transmission and those involving light reflection. Haze and transparency fall into the first category, while gloss is the principal reflective property. We observe haze in a sample when it scatters incoming light away from its original path. Transparency is the ability of a sample to transmit light without scattering it from its optical axis. Gloss is a two-part phenomenon consisting of specular (mirror-like) reflection and diffuse light scattering from the surface of a material.

### 8.3.1    Haze

The haze that we perceive within a sample is due to it scattering incoming light away from its original optical axis. Haze has two components, surface and internal scattering. Figure 8.13 shows how surface roughness causes haze. Surface roughness has many causes, including physical damage, such as scratches and abrasion, and other defects. In semicrystalline polymers, spherulites that form just below the polymer's surface generate localized pockets of higher density material, which result in "sink marks" on the polymer's surface. This is the reason that we typically observe more haze in high density polyethylene films, which have relatively high degrees of crystallinity, than we do in low density polyethylene films, which have lower degrees of crystallinity.

We can easily reduce surface haze by wiping a hazy film, such as the high density polyethylene bags found in the produce section of supermarkets, with a cloth moistened with a drop of oil. The layer of oil forms a smooth coating with a refractive index similar to that of the polymer and the haze is reduced accordingly. (One of us knew an unsavory individual who demonstrated this effect by rubbing a hazy film against his greasy forehead!)

Figure 8.14 shows how haze is generated by inhomogeneities within a sample. Light is scattered as it crosses the boundaries between phases that have different refractive indices. Inhomogeneities within a sample can arise from various sources. In semicrystalline polymers, the crystalline and amorphous phases have different densities and therefore different refractive indices. Light is scattered at the boundaries of spherulites, cylindrites, and other supermolecular features. Blends of polymers scatter light at the interfaces between the two phases. Likewise, scattering occurs at interfaces between polymers and transparent inorganic fillers.

Voids within a sample are a major cause of internal haze. We see the effect of voiding when we stretch polymers, such as high density polyethylene and isotactic polypropylene, that have distinct yield points and clearly defined necks (as discussed earlier in this chapter). The

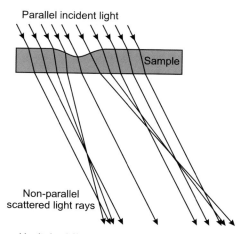

**Figure 8.13**    Haze generated by light diffraction from surface defects

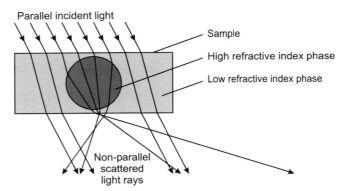

Parallel incident light

Sample

High refractive index phase

Low refractive index phase

Non-parallel scattered light rays

**Figure 8.14**   Haze generated by light scattered by an inhomogeneity within a sample

necks that form in such polymers appear white because they contain a high density of voids that scatter light.

The relative contributions of bulk and surface phenomena to the overall haze level change with a sample's thickness. In thin samples, surface roughness assumes greater importance than in thicker samples where bulk inhomogeneities are the principal cause of haze.

Haze is measured with a hazemeter, which determines the amount of light transmitted by a sample at angles that deviate from the incident light path relative to the total amount of light striking it according to Eq. 8.3.

$$\text{Haze (\%)} = \frac{T_s}{T_i} \times 100 \qquad\qquad (8.3)$$

Where:

$T_s$ = intensity of scattered light transmitted by the sample
$T_i$ = intensity of incident light

### 8.3.2    Transparency

When we say that a sample is transparent we mean that a significant portion of light that strikes it passes through without being scattered, reflected, absorbed by dyes, or blocked by pigments and fillers. When we look through a transparent sample, we can clearly see objects that are more than a few inches beyond it. In contrast, if we place a hazy polymer sample *in contact* with a newspaper, we may be able to read the text, but not if we raise the polymer a few inches above the paper. In general, amorphous polymers are more transparent than semicrystalline ones. In semicrystalline polymers, transparency increases as the degree of crystallinity decreases; thus, ultra low density polyethylene is more transparent than low density polyethylene, which in turn is more transparent than high density polyethylene.

We quantify transparency as the ratio between the intensity of the light that is directly transmitted through the sample relative to the intensity of the incident light according to Eq. 8.4

$$\text{Transparency (\%)} = \frac{I_t}{I_i} \times 100 \tag{8.4}$$

Where:

$I_t$ = intensity of unscattered transmitted light
$I_i$ = intensity of incident light

### 8.3.3  Gloss

Gloss has two principal components, specular and non-directional (diffuse) reflection of light. Specular reflection is that portion of the light that is reflected from a sample at an angle of reflection equal to the angle of incidence, as illustrated in Fig. 8.15. Specular gloss increases as the surface finish of the polymeric item becomes smoother, thus it increases as surface haze decreases and is controlled by the same factors. Gloss improves as the angle of incidence ($i$ in Fig. 8.15) increases.

We measure gloss using the basic configuration illustrated in Fig. 8.15. The observer is replaced by a photocell that measures the amount of light reflected from the surface. We calculate gloss as the ratio of the intensity of the reflected light relative to the incident light at specific angles, typically 20, 45, or 60°, according to Eq. 8.5.

$$\text{Gloss (\%)} = \frac{I_r}{I_i} \times 100 \tag{8.5}$$

Where:

$I_r$ = intensity of reflected light
$I_i$ = intensity of incident light

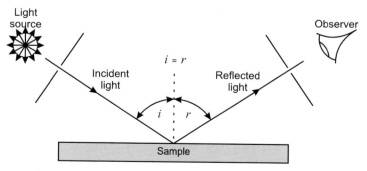

**Figure 8.15**   Specular reflection observed when incident and reflected light angles are equal

Reflective haze is the diffuse reflection of light at angles that are not equal to the angle of incident. Reflective haze is responsible for the reduction in contrast that we observe when viewing objects by reflection. Non-directional reflection of light by the surface of a polymer gives rise to the phenomenon that we know as sheen, which is most noticeable when we view a matte surface at very shallow angles.

## 8.4     Surface Contact Properties

We can divide the surface contact properties of polymers into two categories: those in which the surface remains intact when it comes into contact with another surface and the polymer surface's resistance to damage. Friction and cling fall into the first category. Surface damage can be caused by erosion, abrasion, or cavitation.

### 8.4.1     Friction

Friction is the resistive force that we experience when we try to slide one object over the surface of another. The coefficient of friction is the ratio of the lateral force required to slide the surfaces past one another relative to the force holding them in contact. Polymers exhibit two coefficients of friction: the static coefficient of friction is a measure of the force required to initiate movement, the dynamic coefficient of friction is a measure of the force required to sustain movement at a constant rate. In general, the force required to initiate sliding is somewhat greater than that required to maintain a constant rate of movement.

The coefficients of friction of a polymer depend on many variables, including the chemical composition of the material against which it is sliding, surface roughness, sliding speed, temperature, and frictional heating. The relationships controlling friction are complex and varied, so it is difficult to generalize with regard to how most of the factors affect the coefficients of friction. Friction generally increases as a polymer's temperature rises, because it becomes softer and the viscous component of its nature plays a greater role. We see this effect most clearly when we compare the coefficients of friction of a polymer measured below and above its glass transition temperatures.

#### 8.4.1.1     Coefficient of Friction Determination

Figure 8.16 shows the principal components of a coefficient of friction tester. Polymer samples in the form of thick sheets or molded plaques are attached to the base and a sled with standard dimensions, weight, and surface properties is drawn over the surface. The load beam measures the force required to initiate movement and sustain motion at a given rate of crosshead travel. Thin films can be taped to the sled and drawn across a contact surface that has known properties.

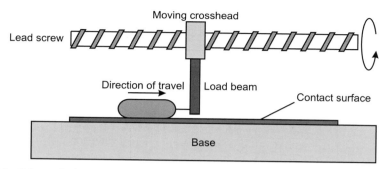

**Figure 8.16**    Schematic diagram of a coefficient of friction tester

The static and dynamic coefficients of friction are defined by Eqs. 8.6 and 8.7, respectively.

$$\mu_s = \frac{F_s}{M} \tag{8.6}$$

$$\mu_d = \frac{F_d}{M} \tag{8.7}$$

Where:

$\mu_s$ =  static coefficient of friction
$F_s$ =  lateral force required to initiate movement
$M$ =  mass of the sled

and:

$\mu_d$ =  dynamic coefficient of friction
$F_d$ =  lateral force required to sustain movement

## 8.4.2    Cling

Cling is the tendency of films to adhere to themselves and other surfaces without the benefit of adhesives. We exploit this phenomenon when we wrap left-over food in cling film or "unitize" a load of boxes on a pallet by winding stretch wrap around them. Under other circumstances cling is unwelcome. It is cling that makes it difficult for us to open kitchen trash bags. When cling occurs on a roll of film it is known as "blocking". A blocked roll may be difficult or even impossible to unwind, rendering it essentially useless. From a commercial point of view, we try very hard to avoid blocking of film on rolls.

Cling occurs when the surface of a polymer film conforms intimately to another surface or itself. We see cling most often when films are extremely thin, smooth, or soft, all of which improve the likelihood that they will conform to another surface. We can increase cling by incorporating a low molecular weight compound (such as a non-crystallizable oligomer above its glass transition temperature) that migrates to the surface of the film.

### 8.4.3    Wear Resistance

Wear is the removal of surface material by one of three mechanisms: erosion, abrasion, or cavitation. Erosion is the removal of a polymer's surface by abrasive materials carried in a fluid medium. We see this type of wear in plastic pipes used to transport waterborne slurries of minerals in mining operations and in vacuum transfer pipes used to convey powders in a stream of air. Abrasion is the result of two surfaces sliding against each other. We commonly observe abrasion of polymers in the fabrics of our clothes and upholstery. Cavitative wear is caused by voids in a liquid medium collapsing against a surface. It is essentially an impact process. Cavitation is a relatively uncommon cause of wear in polymers. Pump impellers are one of the few applications where polymers must resist this type of wear.

Abrasive wear of polymers has two components: material can be removed by the rasping action of a countersurface or it can be sheared off viscoelastically by a countersurface to which it adheres. The precise balance of mechanisms depends on the characteristics of the counterface and the conditions under which the abrasion takes place. Many polymers exhibit excellent wear resistance, which in combination with their low coefficients of friction suit them for applications where lubrication is either impossible or undesirable. We use wear resistant polymers in such diverse applications as bushings in business machines, pump seals, and replacement hip and knee joints.

#### 8.4.3.1    Determination of Wear Resistance

We employ two basic approaches to rate a polymer's wear resistance. In the first, we expose a polymer surface to a standard set of abrasive or erosive conditions and examine the surface for visual evidence of wear. We primarily use this method to qualitatively rank materials. In the second approach, we expose samples to wear inducing conditions and determine wear resistance in terms of weight loss as a function of time.

Figure 8.17 shows how we can test the mar resistance of polymers. In this test, we rain abrasive grit down on an angled polymer sample at a known rate for a given period of time. We compare samples that have undergone testing under identical conditions to prepare a comparative ranking of mar resistance.

Figure 8.18 shows the basic configuration of a "pin" type abrasion tester. The weighted pin creates a circular wear pattern on a polymer disk as it rotates. After a given time or a given number of revolutions we remove the disk and weigh it to calculate its weight loss. We may then return it to the instrument for continued testing. After several testing cycles we should have enough data to calculate the steady state wear rate.

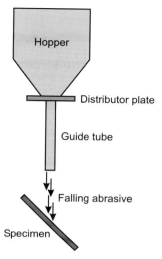

**Figure 8.17** Schematic illustration of apparatus used to evaluate the mar resistance of polymer surfaces

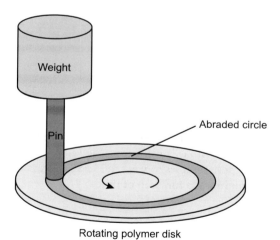

**Figure 8.18** Schematic illustration of abrasive wear resistance testing apparatus

## 8.5     Barrier Properties

We employ polymers as barriers in a wide variety of applications. In some cases, the polymer is used as a standalone material, providing its own physical support. In others it is used as a coating or a liner, supported by another material. Polymers excel in barrier applications because we can readily form them by many methods into all kind of shapes. A cursory examination of household and everyday products reveals some of the myriad barrier applications where we use polymers. In the bathroom, we find toothpaste tubes made from multilayered polymer films. Here, we also find bottles containing all sorts of products, ranging from mundane household cleansers to expensive creams and lotions. In the nursery, we find disposable diapers, which for obvious reasons must exhibit good barrier properties. Moving into the kitchen, we see such obvious polymer barrier applications as one-gallon milk jugs, freezer bags, soda bottles, and storage containers. Less noticeable are polymeric coatings inside cardboard juice containers and the plastic drainage pipes under the sink. In our cars, various fluids, including refrigerant, coolant, battery acid and gasoline are stored or transferred in polymeric tanks and hoses. Naturally, the tires on our cars must exhibit excellent resistance to gaseous diffusion. On the flip side, vacuum hoses must keep air out if our engines are to run efficiently.

Polymeric barriers play two roles: they must keep their contents from leaking out at an unacceptably high rate and they must prevent contaminants from entering. When certain molecules diffuse through a package faster than others, the composition of the contents can change noticeably. This is why our toothpaste becomes less minty if we store it too long, which allows some of the fragrant oils to escape. Alternatively, excessive oxygen diffusing into a packet of potato chips (crisps) will partially oxidize their fat and turn them rancid.

The rate of diffusion of different types of molecules through polymers varies. Polar molecules diffuse most readily through polymers that have dipole moments. Thus, water diffuses through nylon 6 more than an order of magnitude faster than it does through polyethylene. Conversely, alkanes diffuse through polyethylene faster than they do through nylon 6. When we need to prevent the diffusion of different types of molecules through the walls of a package, we commonly employ layers of different polymers. For example, tomato ketchup bottles are routinely made of five, seven, or even more layers.

On a related theme, we can take advantage of the diffusion of molecules incorporated into a polymer to its surface. Thus we can blend lubricants, bactericides, and antiblock agents into polymers, which slowly exude to the surface where they perform a useful role.

### 8.5.1     Permeation Through Polymers

Permeation of small molecules through polymers takes place in four steps. In the first stage, the permeating molecules, know as the diffusants, wet or adsorb onto the polymer's surface. Secondly, the diffusant molecules dissolve in the polymer. In the third step, the molecules diffuse down a concentration gradient towards the opposing surface. Finally, the diffusant molecules desorb or evaporate from the surface, or are absorbed into another material.

We can express the permeability of a diffusant through a polymer in terms of its diffusion rate and solubility according to Eq. 8.8.

$$P = D \times S \tag{8.8}$$

Where:

$P$ = permeability (permeation coefficient)
$D$ = diffusion coefficient
$S$ = solubility coefficient

We report permeability in terms of volume of diffusant passing through a film of unit thickness in a given time under a given pressure differential, such as:

$$\frac{cm^3 \cdot mm}{cm^2 s \cdot bar}$$

Diffusant molecules pass through a polymer in a series of jumps from one molecular sized hole to another. These holes are transient, opening and closing as chain segments move with respect to each other due to thermal agitation. In general, diffusion rates increase as the size and complexity of the migrant molecules decreases. Thus, small gas molecules, such as oxygen and nitrogen, diffuse at a faster rate than liquids. Similarly, straight chain alkanes diffuse faster than branched alkanes of equivalent molecular weight. Diffusion rates also increase as a polymer's free volume increases, its degree of crystallinity decreases, and its temperature increases. We can view diffusion as a thermally activated process with an activation barrier that obeys an Arrhenius relationship according to Eq. 8.9.

$$P = P_0 \exp\left(\frac{-E_p}{R T}\right) \tag{8.9}$$

Where:

$P$ = permeability
$P_0$ = intrinsic permeability
$E_p$ = apparent activation energy
$R$ = the gas constant
$T$ = absolute temperature

Migrant molecules can only diffuse through the amorphous regions of a semicrystalline polymer. Thus, they must follow a tortuous route between and around crystallites. We can take advantage of this effect to improve a film's barrier properties by controlling its crystallization so that lamellae form with their planes parallel to the film's surface. Thus, migrating molecules are opposed by a series of lamellae that are perpendicular to the diffusion gradient, around which they must travel in order to pass through the film.

## 8.5.2     Barrier Property Analysis

When measuring the barrier properties of polymers we are generally interested in liquid or vapor permeability depending on the specific application.

### 8.5.2.1     Liquid Permeability

We measure the permeation rate of liquids through bottles by filling them with the liquid of interest and placing them in a controlled atmosphere chamber. At intervals we remove the bottles, weigh them and return them to the chamber. We repeat this procedure over a period of days, or even weeks, until their rate of weight loss reaches a steady value. We calculate the permeability factor from Eq. 8.10.

$$P_t = \frac{R\,T}{A} \tag{8.10}$$

Where:
$P_t$  =  permeability factor
$R$  =  mass loss per day
$T$  =  average wall thickness of bottle
$A$  =  surface area of bottle

### 8.5.2.2     Vapor Permeability

We test the permeability of polymer films or sheets to various vapors and gases by mounting the polymer between chambers that contain different concentrations of the migrant molecules. We can determine the permeability from pressure changes, volumetric changes, or by microanalytical techniques that measure the concentration of the migrant molecules in a stream of gas flowing across the low concentration side of the barrier.

When determining the permeability of films to water vapor, we seal a desiccant into a small cup with the polymer covering the opening. We weigh the cup before placing it in an oven at controlled temperature and humidity. After a given period of time we remove it and weigh it a second time. We calculate the film's water vapor transmission rate based on the area of the cup's mouth and the time that it was in the oven.

## 8.6     Electrical Properties

All commodity polymers (that is those manufactured and sold in high volume) act as insulators because they have no free electrons to conduct electricity. Some low-volume polymers, such as polyacetylene, are conductive or semi-conductive, but their applications are specialized and their use limited. In this section, we shall concentrate on the properties of commodity polymers, because these materials represent the vast majority of polymers used in electrical applications.

We can divide commodity plastics into two classes: excellent and moderate insulators. Polymers that have negligible polar character, typically those containing only carbon-carbon and carbon-hydrogen bonds, fall into the first class. This group includes polyethylene, polypropylene, and polystyrene. Polymers made from polar monomers are typically modest insulators, due to the interaction of their dipoles with electrical fields. We can further divide moderate insulators into those that have dipoles that involve backbone atoms, such as polyvinyl chloride and polyamides, and those with polar bonds remote from the backbone, such as poly(methyl methacrylate) and poly(vinyl acetate). Dipoles involving backbone atoms are less susceptible to alignment with an electrical field than those remote from the backbone.

We can define the principal electrical properties of polymers in terms of four characteristics: electrical resistance, capacitive properties, dielectric strength, and arc resistance. We can change the surface characteristics of a polymer by subjecting it to a corona discharge generated by a strong electrical field. Lastly, we must also consider the influence of other physical properties on the application of polymers in electrical applications.

### 8.6.1     Electrical Resistance

The electrical resistance of a polymer has two components: volume resistivity and surface resistivity. Volume resistivity comes from the bulk chemical composition of the polymer. In general, the volume resistivity of a polymer increases as the dipole moments of its constituent bonds decrease. Thus, polyethylene and polypropylene exhibit very high volume resistivity. Volume resistivity can be reduced by the presence of contaminants, such as anti-oxidants, catalyst residues, and water. We would not choose polymers that absorb water, such as polyamides and polyesters, for applications requiring the highest resistance. Oxidation reduces volume resistivity by increasing the concentration of polarizable bonds, such as those found in carbonyl and hydroxyl groups. We define surface resistivity as the resistance between two linear electrodes placed on the surface of a material such that they form opposing sides of a square. We report surface resistivity in units of ohms/square. Surface resistivity is strongly influenced by contamination. Moisture, dirt, grease, oxidation, etc. all reduce surface resistivity. Electrical resistance decreases as temperature rises, according to Eq. 8.11.

$$\rho = \rho_0 \, \exp\left(\frac{\Delta E}{2\,k\,T}\right)$$

(8.11)

Where:

$\rho$ =  resistivity at temperature $T$

$\rho_0$ =  limiting resistivity at low temperature

$\Delta E$ =  energy gap between filled and unfilled electronic orbitals

$k$  =  the Boltzmann constant

$T$  =  absolute temperature

We measure volume resistance by placing electrodes with known dimensions on opposite sides of a specimen of known thickness. Volume resistivity is calculated from Eq. 8.12.

$$r_v = \frac{R_v\ A}{t} \qquad (8.12)$$

Where:

$r_v$  =  volume resistivity

$R_v$  =  volume resistance

$A$  =  area of measuring electrode

$t$  =  sample thickness

We obtain surface resistance in a similar fashion to volume resistance, except that the electrodes are placed on the same side of the sample. We calculate surface resistivity from Eq. 8.13

$$r_s = \frac{R_s\ P}{g} \qquad (8.13)$$

Where:

$r_s$  =  surface resistivity

$R_s$  =  surface resistance

$P$  =  perimeter of measuring electrode

$g$  =  distance between electrodes

## 8.6.2   Capacitive Properties

The dielectric constant of a polymer ($K$) (which we also refer to as relative electric permittivity or electric inductive capacity) is a measure of its interaction with an electrical field in which it is placed. It is inversely related to volume resistivity. The dielectric constant depends strongly on the polarizability of molecules within the polymer. In polymers with negligible dipole moments, the dielectric constant is low and it is essentially independent of temperature and the frequency of an alternating electric field. Polymers with polar constituents have higher dielectric constants. When we place such polymers in an electrical field, their dipoles attempt to align themselves with the field. At temperatures below the polymer's glass transition temperature, dipoles are hindered from aligning with the electrical field, especially when they involve atoms that are part of the polymer's backbone. When the temperature is raised above the glass transition, the dipoles are free to align with the electrical field and the dielectric constant increases.

In a slowly alternating electric field a polymer's dipoles have time to align themselves with the electric field as its polarity changes. Thus, the dipoles are susceptible to the electric field and the polymer displays a high dielectric constant. As the electric field's frequency increases, the dipoles' alignment begins to lag behind the changes in polarity, i.e., it becomes less susceptible to the electric field, and the dielectric constant falls. Dipoles involving backbone atoms lag more than those on side groups, which have greater freedom of movement. At extremely high frequencies the electric field alternates so fast that the dipole alignments lose all registration with the field and the dielectric constant reaches its minimum value.

The dissipation factor of a polymer (which we also refer to as tan δ) is the ratio of energy lost to the energy stored when it is placed in an alternating field. The dissipation factor is analogous to a mechanical tan δ describing rheological behavior. The dissipation factor at a specific frequency is defined according to Eq. 8.14.

$$D = \frac{I_r}{I_c} \tag{8.14}$$

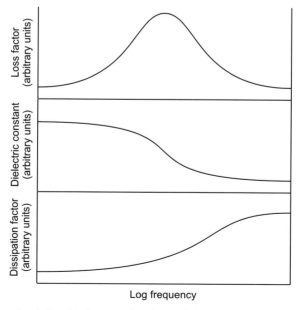

**Figure 8.19** Schematic relationship between loss factor, dielectric constant, and dissipation factor

Where:

$D$ = dissipation factor (tan $\delta$)

$I_r$ = loss factor (dielectric absorption)

$I_c$ = dielectric constant

The loss factor is highest at frequencies corresponding to the greatest phase difference between the alternating field and the dipole moments within the sample. Figure 8.19 illustrates the general relationships of dissipation factor, loss factor, and dielectric constant with the frequency of an alternating electrical field. High dissipation factors result in power losses in capacitors, which causes the capacitors to heat up. This phenomenon is especially important at higher frequencies.

We measure the capacitive properties of a polymer in a capacitor that is constructed so that we can compare the properties of the test material relative to a vacuum.

### 8.6.3    Dielectric Strength

The dielectric strength of an insulator is the electric field strength at which it physically breaks down and begins to conduct electricity. High values are required when the material will experience high electrical stresses, such as those found in the insulation surrounding power transmission cables. Dielectric strength decreases as temperature and humidity increase. It also decreases with time, due to the creation of conductive paths on the surface of the material.

We measure dielectric strength by placing electrodes on opposite sides of a sample of known thickness and ramping up the potential difference. When the sample ruptures and begins to conduct electricity we note the breakdown voltage. The dielectric strength is reported as the voltage gradient at which failure occurs.

### 8.6.4    Arc Resistance

The arc resistance of a polymer is the length of time for which we can apply an electrical discharge to its surface before it breaks down and begins to conduct electricity. Breakdown typically takes the form of conductive carbonaceous tracks that are caused by oxidation due to locally high temperatures. Polymers with low polarity typically fare better in this test than polar polymers. Arc resistance is reduced by dirt, moisture, and other surface contaminants.

We can measure arc resistance on dry surfaces or when they are wetted by an electrolyte dripping on the surface between pairs of sharp electrodes. We report arc resistance as the time it takes to form a conductive track into which the arc disappears.

### 8.6.5    Corona Treatment

Due to the low polarity of polyolefin molecules, it is difficult to get anything to adhere to their surfaces. We can see this when we try (in vain) to glue two pieces of polyethylene together. Another result of the low surface energy of polyolefins is that it is difficult to print on their surfaces. We can readily demonstrate this by trying to write on the surface of a plastic grocery sack or a food storage container with a felt tip pen. Most inks bead up and will not wet the surface. In order to obtain a printable polyolefin surface, manufacturers may use corona treatment prior to printing.

We corona treat films by passing them over an electrically grounded chilled metal roller, above which we mount a bar electrode in close proximity to the film. We apply a high voltage, typically 5 to 50 kV, between the electrode and the roller. The air or other gas molecules between the electrodes ionize and attack the surface of the film. The ions abstract hydrogen atoms from carbon-hydrogen bonds at the polymer's surface, leaving radicals attached to the backbone. These radicals in turn react with oxygen molecules in various ways to produce ketones, esters, aldehydes, etc., which increase the polarity of the surface and hence improve printability.

### 8.6.6    Factors Influencing Polymers in Electrical Applications

When selecting polymers for use as insulators, in capacitors, and other electrical applications, we must consider factors other than their interaction with electric fields. The following examples illustrate some of the other factors that we must consider.

Our principal concern is often the polymer's mechanical properties. For instance, the requirements of the handle of an electrician's screwdriver are very different from those of wire insulation. In the former application, we are free to choose stiff polymers of many types, including glassy amorphous polymers. In contrast, wire insulation must be flexible, which limits our choice to ductile polymers.

We must also consider service temperature, not just under standard operating conditions, but also at extremes, where breakdown could have catastrophic consequences. For instance, the insulation surrounding electrical transmission cables must withstand temperatures associated with temporary current overloads. If we were to use regular polyethylene to insulate such cables, the insulation could melt, exposing the conductor. Accordingly, we routinely use crosslinked polyethylene to insulate high-voltage cables. Even if its crystallites were to melt, the crosslinks would prevent the polymer from flowing and exposing the conductor.

In storage capacitors it is important that we keep the dielectric constant to a minimum. This is not just a question of selecting the most appropriate polymer; we must also ensure that contaminants that could raise the dielectric constant are kept to a minimum. Polymer manufacturers sell special grades of polypropylene (which is invariably the polymer of choice) that they describe as "ultraclean". These resins are made with catalysts that are extremely active and thus leave very little residue in the polymer. Such resins typically contain little or no anti-oxidant and are extruded under conditions and using equipment that are designed to minimize oxidation.

## 8.7        Weather Resistance

We are interested in the effect of weathering on polymers for two distinctly different reasons. We may wish to retard it, so that our products survive longer in outdoor applications, or we may wish to accelerate it, so that products degrade rapidly when exposed to the elements. In either case, we need a way of predicting the response of polymers to the factors that produce measurable changes in their chemical and physical characteristics. Ideally, we would like to be able to obtain these results in as short a period of time as possible.

Some of the weathering factors that we must consider are time, temperature, radiation levels, and the chemistry of the environment, including moisture, salt and pollutants. Given the wide variety of polymers and the range of environmental factors to which they are exposed, it is impossible to generalize their response. A general protocol for determining weathering resistance is to measure the properties of interest in samples that have been exposed to weathering for various lengths of time and compare them to the properties of unexposed samples. We may also analyze chemical changes, including crosslinking, loss of anti-oxidants, stabilizers and other additives, oxidation, and changes to molecular weight distribution. We have three general classes of environment to which we can expose samples: natural outdoor weathering, accelerated outdoor weathering, and artificial weathering.

### 8.7.1        Natural Outdoor Weathering

To test the outdoor weathering resistance of polymers, we expose them on racks to the elements. Florida and Arizona are popular locations for this type of testing, due to their routinely high temperatures and abundant sunshine. Florida is chosen when humid conditions are desired, while Arizona is preferred when a dry atmosphere is required. Samples may be exposed at any angle ranging from horizontal to vertical, with their faces of interest pointing towards the equator. If we wish to simulate the effect of sunlight on roadside trash, we may place the samples at an angle 5° to the horizontal, which represents the average angle of litter lying on the ground.

Samples exposed outdoors are not subject to a controlled environment; temperature, light level, humidity, rainfall, and even wind fluctuate from day to day and seasonally throughout the year. It is important to monitor and record weather conditions during the test.

### 8.7.2        Accelerated Outdoor Weathering

If we wish to accelerate the outdoor weathering of polymers we can use a series of large reflectors that concentrate sunlight onto the racks supporting the samples. Motors adjust the angle of the reflectors to ensure that the radiation remains focused on the samples. This type of testing inevitably raises the temperature of the samples significantly. To reduce the role of increased temperature it is standard practice to cool the samples with fans blowing air onto the back of them.

### 8.7.3    Artificial Weathering

We perform artificial weathering in environmental chambers containing racks of samples that face and rotate around a light source. A variety of light sources is available, including ultraviolet fluorescent lamps, carbon arcs, and xenon arcs. None of these light sources perfectly mimic sunlight, so filters may be used to modify the frequencies of radiation falling on the sample. Temperature and humidity within the chamber are controlled; they may be held at constant values or they may be cycled to simulate the diurnal cycle. Samples may be intermittently sprayed with water to simulate the effect of rain and dew. Artificial weathering has the advantage that it can be run twenty four hours per day.

## 8.8    Conclusions

In this chapter we have discussed the five principal categories of polymer testing relating to mechanical, optical, surface contact, barrier, and electrical properties. As we have seen, for each category there are numerous tests from which we can select to determine the properties of interest. The tests that we select are determined by the requirements of the final application. It is by judicious selection of tests and interpretation of their results that we develop products meeting increasingly demanding requirements. Once we have developed a commercial product we apply these tests to assure the quality of ongoing commercial production.

## Review Questions for Chapter 8

1. What is meant by the term "mechanical property"? Give examples of three mechanical properties of polymers.

2. Why do the temperature and speed of testing matter in describing the mechanical properties of a polymer?

3. Why do the tensile behaviors of rubbery amorphous, glassy amorphous and semicrystalline polymers differ as they do?

4. How do we determine the tensile modulus, tensile yield point, elongation at break and tensile strength of a polymer? What characteristics of the polymer define these properties?

5. Define compressive modulus. Why is this property commonly performed on materials such as foams and rubbers?

6.   Compare flexural and tensile modulus testing methods. Why is it reasonable to say that both methods describe the "rigidity" of a material?

7.   Describe how we perform toughness testing on polymers with both the impact beam and the falling weight methods. Explain the different manners by which an amorphous polymer will fail above and below its glass transition temperature.

8.   Why do we perform tear testing on polymeric films? What creates anisotropy in a film's mechanical properties at a molecular level? Why is this anisotropy so much more important in films than in thicker parts?

9.   Define haze and gloss. What attributes of a polymer generate these observed properties?

10.  How do we measure the wear resistance of polymers? In what applications would you predict this property to be a critical measurement?

11.  The barrier properties of polymers depend on the type of permeant in contact with the polymer. Why is this so? How does a permeant cross a polymer barrier?

12.  What do resistivity, capacitance and dielectric strength properties depend on? How do we measure each of these electrical properties?

13.  What is meant by "weather resistance"? What environmental factors can alter the properties of polymeric materials?

# 9     Polymer Degradation and Stability

## 9.1     Polymer Lifetime Experiences

Polymeric products experience a wide range of environmental conditions over their lifetime. During manufacture and processing they endure high temperatures, high pressures, and variable shear regimes. Once they attain their final shape, energy in the form of heat, electricity, and electromagnetic radiation may bombard them. Chemically reactive species such as oxygen, residual catalysts, unreacted monomer, solvents, acids, and bases can attack the polymer. Finally, polymers can experience physical manipulation such as pulling, bending, or impact from foreign objects. Each of these influences has the potential to alter the chemical structure of the polymer. Any change to the polymer's chemistry is likely to affect its macroscopic properties. The processes that lead to the chemical change of the polymer are all examples of degradative processes. The study of degradative processes allows us to understand both how they occur and how to prevent them so we can improve the performance of polymer products.

For most applications, engineers combat the degradative effects of processing and in-use conditions by adding chemical species that stabilize polymers. Some additives interact directly with prodegradants (species that act to speed up degradation); chemical stabilizers (specifically antioxidants), heat stabilizers, and UV stabilizers all work in this way. Another way to address the problem is to add a chemical species that changes the receptivity of the polymer to the effect of shear and heat, by altering the viscosity of the material during processing. Process aids, plasticizers, and waxes act in this way. Alternatively, there are cases where the degradation of the properties is actually desirable, such as in agriculture films that are tilled back into a field at the end of a growing season, the plastic rings that hold together six pack aluminum beverage cans and ablation shields on rockets.

In this chapter, we will explore the effect of various environmental conditions and the polymers' chemical composition and structure on its stability. We will examine the mechanisms associated with degradation and will explore several well studied polymeric systems. Finally, we will discuss the role of additives and their effects on properties.

## 9.2     General Descriptions of Degradation

When discussing degradation, it is most instructive to break the process down into three different categories: chain scission, which results in reduced molecular weight of the final polymer, crosslinking, which connects neighboring polymer chains, thereby increasing the polymer's average molecular weight, and charring.

Chain scission occurs when energetic sites, such as backbone carbons near an electron-withdrawing group or an imperfection (such as a double bond or incorporated contaminant) are attacked by reactive species present in the polymer. Examples of reactive species include radicals, acids, bases, strong oxidizers, molecular oxygen, and charged or neutral metal species. These reactive species chemically attack the polymer backbone at the reactive site and break the bond, while incorporating themselves or a part of themselves into the polymer chain. The result is a reduction in the polymer's average molecular weight, reduced tensile strength, and embrittlement. Often, these processes are autocatalytic, meaning that once degradation begins, it will continue at an ever-increasing rate. This autocatalytic effect can be devastating to a polymer because, if left unchecked, it can create a charred mess or a soupy liquid!

Crosslinking occurs when a reactive site on one polymer chain reacts with another chain to create a link between the two. This process is identical to that described in Chapter 2, where we saw an overall increase in the average molecular weight of the polymer. Typically, crosslinking occurs in localized areas of a melt during processing and are observed as "gels" or "fisheyes" in the final product. The names gels and fisheyes refer to the fact that these spots typically are pigment-free and look clear and raised relative to the rest of the part. These defects are especially problematic in thin films or bottle walls. When the effect is widespread, the polymer's tensile strength and toughness increase while flexibility decreases.

The term charring refers to the complete degradation of a polymer after which there is no longer any polymeric character to observe. Charring results from chain scission reactions that are left unchecked and is the typical process by which thermosets degrade. The resulting material is typically black and brittle.

## 9.2.1     Mechanisms of Thermal Degradation: Chain Growth Polymers

When any material is heated, the internal energy of that material is increased. In a polymer, this increase in internal energy results in an increased rate of rotation of any freely moving groups (such as the pendant methyl groups in polypropylene), an increase in the vibrational energy of the bonds, and an increase in the translational movement of any species that are mobile. This extra energy creates a situation where intermolecular forces are weakened, bonds are more easily broken, and the mobility of absorbed species is increased, allowing them to migrate throughout the polymer to react with energetic sites. The general mechanisms of the degradation of chain growth polymers follow the same steps that we saw in polymerization. There is an initiation step followed by propagation, branching, and termination steps.

During initiation, degradation begins at sites on the polymer that are susceptible to nucleophilic attack. These reactive sites are typically associated with nearby electronegative groups that enhance the reactivity of carbon atoms, as shown in Fig. 9.1. In general, we represent the initiation step as shown in Fig. 9.2.

Initiation creates free radicals that can react with additional loci on the polymer chains in the propagation step, resulting in the transfer of the radicals from one chain to another. Since there are often many different types of reactive loci on a polymer chain, we typically have a variety of mechanistic steps to consider when describing propagation. Some of the propagation

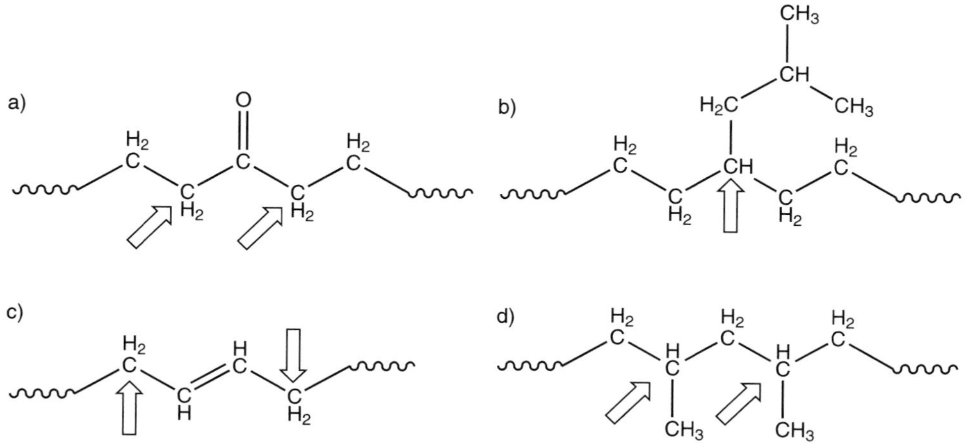

**Figure 9.1**   Examples of carbons activated by neighboring electrophilic groups:
a) methylene groups adjacent to a carbonyl group,
b) tertiary carbon atom at a branch point,
c) carbon atoms adjacent to a double bond and
d) tertiary carbon atoms adjacent to methyl groups in polypropylene

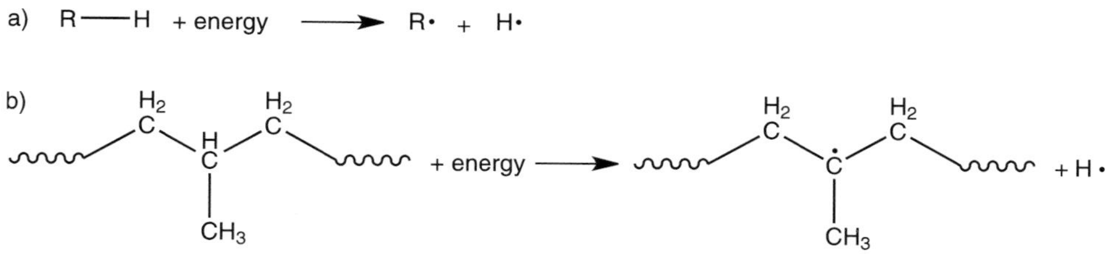

**Figure 9.2**   Examples of the initiation of degradation:
a) general case and b) scission of a tertiary carbon hydrogen bond in polypropylene

reactions involve the incorporation of the radical species into the polymer chain. Another mechanism allows the chain to break as seen in Fig. 9.3.

During propagation, the number of radical species present in the polymer remains constant. Branching creates additional radicals by the decomposition of the products of the initiation reaction. Such reactions are typically associated with oxygen-containing sites introduced during initiation. An example is the decomposition of a hydroperoxyl group to create a hydroxy radical and a carbonyl radical, as illustrated in Fig. 9.4. Termination results when two radical species meet one another and react to form a bond, as shown in Fig. 9.5. This process reduces the number of radical species in the material, thereby reducing the rate of the degradation. In the early stages of degradation, termination is rather rare as it requires two radicals, which are initially at low concentrations, to meet one another. As more radicals are produced, the termination step occurs more frequently slowing the rate of degradation.

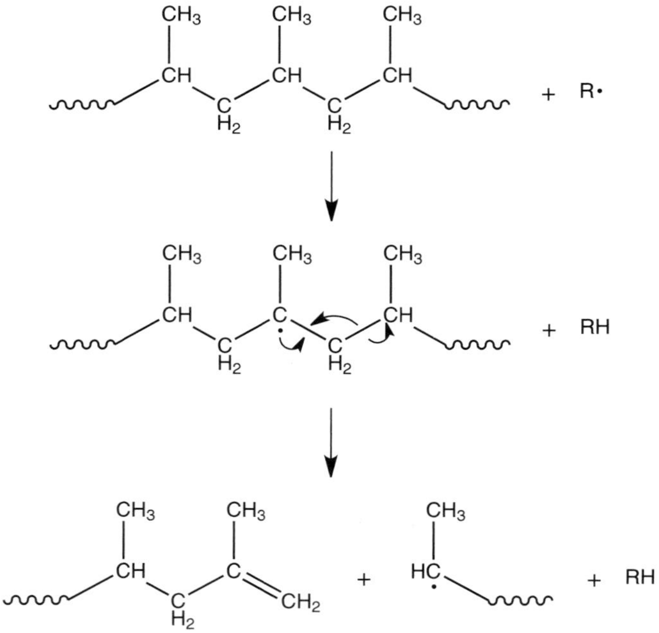

**Figure 9.3**   Example of chain scission in polypropylene

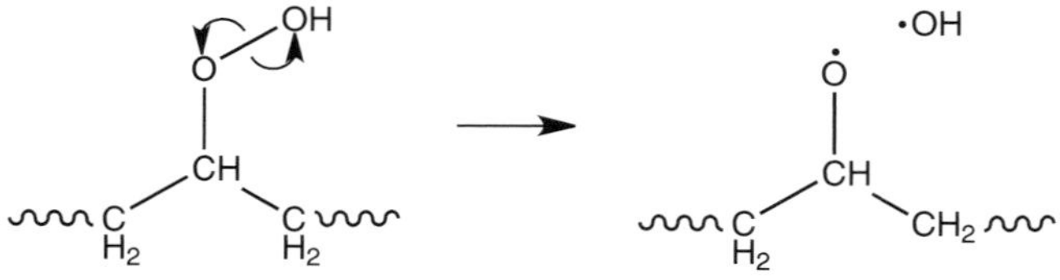

**Figure 9.4**   Decomposition of a hydroperoxyl group to form carbonyl and hydroxy radicals

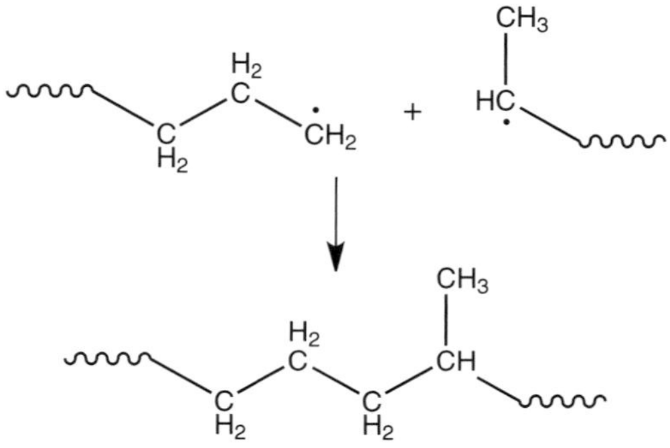

**Figure 9.5**   Termination reaction between two radicals

### 9.2.1.1    Chemical Degradation of Chain-Growth Polymers

The most common chain-growth polymers, i.e., polyethylene, polypropylene, and polystyrene, are chemically inert towards most species, because they have very low dipole moments. When chemical degradation occurs with these polymers, it typically results from impurities, such as unsaturated sites in the polymer chain. Oxygen in the form of ozone or dioxygen can attack the double bond. The process by which these species react is a complex series of mechanistic steps, involving radical formation and the subsequent reaction of radicals with the polymer chain. Any reaction involving the incorporation of oxygen in a polymer is referred to as an oxidation process. These chemical degradative processes are particularly important on our planet as oxygen is ubiquitous.

Another reaction mechanism that occurs in some chain-growth polymers is solvolysis. In this type of reaction, a species reacts with a C-X bond, where X represents a halogen, and breaks it. Specifically, this becomes important when describing the degradation of polyvinyl chloride. Acidic species act to remove the chlorine atom, forming hydrochloric acid.

### 9.2.1.2    Kinetics of Degradation: Chain Growth Polymers

Not only do we need to know how susceptible a polymer is to degradation, we also need to understand how quickly the polymer will succumb to its effects. From a processing standpoint, we need to see how long the polymer can be held at elevated temperatures before excessive degradation occurs. From an end-use perspective, we need to predict the lifetime of the final product as a function of the environment it will encounter. For these reasons degradation kinetics are extensively studied.

Since the depolymerization process is the opposite of the polymerization process, the kinetic treatment of the degradation process is, in general, the opposite of that for polymerization. Additional considerations result from the way in which radicals interact with a polymer chain. In addition to the previously described initiation, propagation, branching and termination steps, and their associated rate constants, the kinetic treatment requires that chain transfer processes be included. To do this, a term is added to the mathematical rate function. This term describes the probability of a transfer event as a function of how likely initiation is. Also, since a polymer's chain length will affect the kinetics of its degradation, a kinetic chain length is also included in the model.

## 9.2.2    Mechanisms of Degradation: Step Growth Polymers

Just as the polymerization mechanisms differ between step growth polymers and chain growth polymers, the degradation mechanisms also differ. Step growth polymers form via a chemical reaction between monomers, creating a linkage through reaction between two groups. These reactions are exothermic and develop an equilibrium. Since the process is defined by the equilibrium, it is possible for the reaction to progress in the forward or reverse direction, i.e., creating more high polymer or a reaction back to the monomers. The response of this system

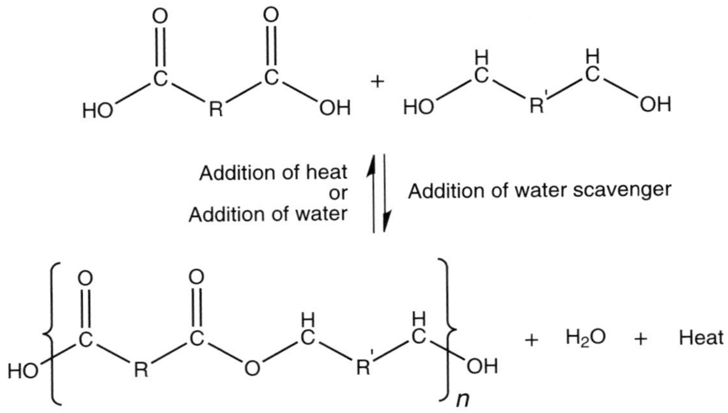

**Figure 9.6**   Principal effect of Le Chatelier's Principle

to stress will depend on Le Chatelier's principle, introduced in Chapter 3, which states that any system will respond to an imposed stress in a way to relieve that stress. Thermal and chemical degradation can both be explained using this principle. An example of this can be seen in Fig. 9.6. The general condensation reaction of a polyester is a reversible equilibrium reaction. The effect of heat and water is to drive the reaction back to monomer so the processing of this type of polymer is very sensitive to its conditions!

### 9.2.2.1    Thermal Degradation of Step Growth Polymers

The addition of heat shifts the equilibrium concentrations away from the products and back towards the reactants, the monomers. This is one reason why processing these types of polymers is often more difficult than processing products of chain growth mechanisms. The thermal degradation process can be dramatically accelerated by the presence of the low molecular weight condensation products such as water. Polyester, as an example, can depolymerize rapidly if processed in the presence of absorbed or entrained water.

### 9.2.2.2    Chemical Degradation of Step Growth Polymers

The chemical nature of the main-chain linkages of step-growth polymers makes this class of polymers particularly reactive to a wide variety of chemical species. Solvolysis reactions break the C-X bond at the polymer linkage bonds. These types of reactions are often pH-dependent, so the stability of the polymer is highly dependent on the acidity or basicity of the prodegradant.

### 9.2.3    Photodegradation

Light energy absorbed by polymers can initiate degradation. Chromophores in the polymeric system, either those that are part of the polymer itself (such as the styrene group in polystyrene) or those associated with incorporated contaminants, such as alkene or carbonyl groups, absorb energy from the incoming light. Chromophores commonly associated with polymers, such as conjugated double bonds, aromatic rings, and nitrogen-containing groups are the primary species responsible for photodegradation of polymers. The type of chromophore defines the wavelength of light to which the polymer will be susceptible. The energy absorbed with the photon creates an electronically excited state in the chromophore, which then can induce radical formation in the polymer, a primary photochemical process, or transfer its energy to another receptor species. When the energy is transferred to a receptor species via a secondary photochemical process the receptor can utilize this energy in one of two ways: it can transfer that energy again or it can form a radical species. Degradation results from the interaction of the radicals with the polymer. Degradation occurs in two distinct ways: either through crosslinking or chain scission. The process occurring in any polymeric system depends on the type of polymer and the availability of oxygen in the surrounding atmosphere or in the polymer.

The mechanisms of photochemical degradation of a polymer are rather difficult to study as the observed degradation is typically a result of the effect of light on the contaminants in the polymer rather than the pure polymer. Studies typically look for the degradation products, either low molecular weight volatiles or radicals, or work to identify the wavelength of light absorbed by the system via spectroscopic methods.

### 9.2.4    Solvent Effects

Degradation processes in polymers occur in regions where there is high mobility of the prodegradants and degradation products. For this reason, degradation is observed in amorphous regions of semi-crystalline polymers rather than in the crystalline regions. For the same reason, the effect of solvents and mobile phases on a polymer's stability can be rather dramatic.

Polymer surfaces often come into contact with gases and liquids during their use. These fluid phases can be intimately involved in the degradation of the polymer. This is a result of the fluid phase's ability to transport degradation products away from the initial site of degradation and move the species to another site, promoting degradation at that site. For example, the radical products of the degradation of a polyolefin can migrate in the air to another site on the surface of the polyolefin, initiating degradation at that site. We see a similar effect when corrosive species are in contact with a polymer in a fluid phase. The fluid phase allows the corrosive species to migrate to the polymer's surface where it can react. Sometimes the fluid phase acts as a solvent penetrating the wall of the plastic part. This increases the mobility of the polymer chains, allowing reorganization and increasing the accessibility of the polymer to reactive species. This is especially important in food packaging when greasy, spicy, or acidic foods are wrapped in plastic film. The grease can penetrate the polymer, while the corrosive flavoring ingredients can chemically degrade the polymer.

## 9.2.5 Mechanical Degradation

Although polymer degradation is typically associated with heat or chemical attack, degradation can also occur as a result of mechanical strain, which causes chain rupture. The scission creates radicals, thereby further promoting degradation throughout the part. Bond rupture is not the predominant mechanism by which non-crosslinked polymers relieve imposed strains. This is because the breaking of a backbone bond requires an enormous amount of energy relative to that needed to reorganize the chains so that the individual chains can slide past one another. It happens enough, though, that the degradative products can be found at higher concentrations in areas of continued strain, relative to the bulk polymer.

## 9.2.6 Preventing Degradation

Now that we have an idea of how polymer degradation occurs, we need to spend some thought examining how to prevent degradation from destroying the polymer with which we are working. The science (and art) of successfully defining the recipe of a polymer material, known as formulation, requires an understanding of the types of stabilizers available and the best ways to use them under a specific set of processing and end use conditions. Stabilizers come in three varieties based on the type of degradation they prevent: chemical stabilizers (typically antioxidants), thermal stabilizers, and ultraviolet radiation-stabilizers. The first two act by removing the reactive species created during the initiation step. They are radical scavengers that react with radicals to either create a less reactive radical species or to create a termination reaction, thereby eliminating the radical species. Ultraviolet radiation stabilizers interact with the photons which may cause degradation of the polymer. They either block the photons from reaching the polymer or they absorb the high-energy radiation and emit a lower frequency of light, which does not have the energy necessary to initiate bond breakage. In addition to describing the type of process they are mediating, stabilizers are often referred to as either primary or secondary stabilizers. Primary stabilizers react at the site of degradation, immediately halting the reaction by chemically sequestering the degradant. The slower migrating secondary stabilizers, when they collide with a reacted primary stabilizer, accept the degradant and bind it irreversibly. The regenerated primary stabilizer can then react with another product of the degradation process. It is not always necessary to use both a primary and secondary stabilizer, so it is not unusual to see formulations containing only a primary stabilizer.

### 9.2.6.1 Chemical and Thermal Stabilizers

Chemical and thermal stabilizers both inactivate the byproducts of degradation processes, preventing them from causing further damage to the polymer. Their chemical structure and mobility in the part define their effectiveness in any given polymeric system. The most common type of chemical stabilizers are antioxidants.

*Antioxidants*

Antioxidants are species that accept the reactive byproducts of oxidation reactions. They are typically hindered amines or phenols that accept radicals, inactivating them and preventing further effects of oxidation. The level of antioxidant used in a polymeric item depends on the expected lifetime of the final part, the environment in which the part will be used, and the susceptibility of the polymer to oxidation. Figure 9.7 shows two common antioxidants used in polyolefins.

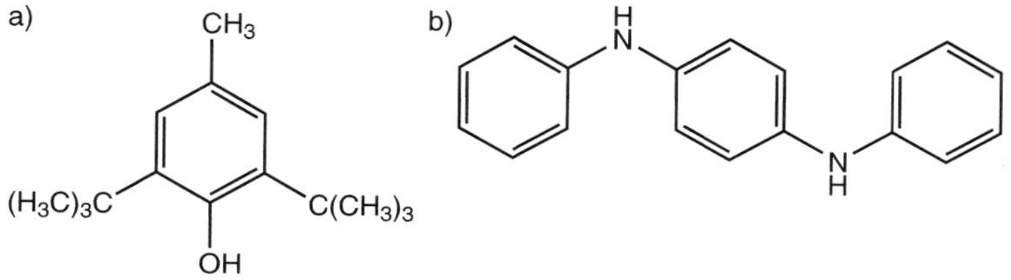

**Figure 9.7** Examples of common antioxidants:
a) 2,6-t-butyl 4-methyl phenol (butylated hydroxy toluene) and
b) diphenyl-*p*-phenylene diamine

*Thermal Stabilizers*

Certain polymers are particularly unstable during processing and use. The most notorious of these is polyvinyl chloride. The highly reactive chlorine atom easily forms radicals under high shear and heat conditions and can then migrate through the polymer, abstracting a proton and forming hydrogen chloride. The process is autocatalytic, so it is essential that we eliminate the radicals before the evolution of hydrogen chloride damages the processing equipment and degrades the resin further. Several different types of stabilizers are use to accomplish this, but all act to remove the radicals. Examples of the thermal stabilizers used in polyvinyl chloride include members of the organo-tin mercaptide family which have a general structure of $(RSn(SR')_3)$ where R and R' represent organic functional groups. Lead oxides $(PbO_2)$, and poly-basic lead sulfates $(nPbO \cdot PbSO_4 \cdot H_2O)$ are also used as thermal stabilizers. In addition, some bases have been incorporated in polyvinyl chloride formulations to react with the hydrogen chloride.

### 9.2.6.2    Ultraviolet Radiation (UV)-Stabilizers

Ultraviolet radiation stabilizers can either block the light from reaching most of the polymer or can absorb the light and then release it as heat. An example of an ultraviolet radiation blocker is carbon black. This material blocks the ultraviolet radiation light from penetrating far into the polymer by physically blocking the radiation. An example of an ultraviolet radiation absorber is benzotriazole. This molecule absorbs ultraviolet radiation light in the range of 250 to 400 nm and then re-emits the energy at a longer frequency and lower energy, where it cannot cause damage to the polymer. Blocking stabilizers are typically pigments that can be incorporated into the polymer. The presence of the pigments prevents the light from penetrating far into the polymeric part so that degradation will only occur near the surface. Ultraviolet radiation stabilizers that absorb the radiation contain strong chromophores, which dissipate the energy they absorb to the surroundings as heat.

### 9.2.6.3    Processing Aids

Several processing aids are used in polymers to reduce the viscosity of a polymer melt, thereby reducing the frictional heat generated during viscous flow. They also prevent chain rupture by allowing the chains to move more freely over one another during processing. Finally, they coat the metal equipment over which the polymer flows during processing, reducing the likelihood of localized "dead spots" in the equipment. The residence time of any "caught" polymer is dramatically longer than desired, leading to localized degradation processes, which then catalyze further degradation throughout the melt. Typical processing aids are long chain fatty acids or low molecular weight (compared to the high polymer) waxes. Processing aids have several disadvantages associated with them. Since they increase the mobility of the polymer chains, they make the polymer more susceptible to degradation in the long run. Also, they tend to migrate to the surface of the polymer (a phenomenon called blooming), and can deposit on processing equipment which creates non-uniform surface properties. Their use must be carefully monitored, as a little bit is often good, but it is very easy to fall into the "more is better" mentality, leading to absolutely abysmal final properties.

## 9.3    Conclusions

The loss in properties and the financial cost associated with the degradation of polymers leads to a need to understand the processes by which these chemical reactions occur and how to prevent them. As we have seen, degradation can be affected by numerous factors at all stages in a product's lifetime. Therefore, careful engineering and thought must to go into developing polymer products. This is especially important with regards to the processing conditions, the types of stabilizers we use and the expected lifetime of the final product.

# Review Questions for Chapter 9

1.  In many polymers, degradation begins at defects in the polymer chain. Why is this true?

2.  How do chain scission and crosslinking occur? What effect do they have on a polymer's properties?

3.  How do free radicals interact with a polymer to cause degradation? Why are many heat stabilizers chemicals that react to quench the free radicals?

4.  What environmental conditions can promote degradation of polymers?

5.  Why are nylons and polyesters susceptible to hydrolysis?

6.  What is the role of a chromophore in polymer photodegradation?

7.  Why does polymer degradation typically occur in the amorphous region of semi-crystalline polymers rather than in the crystalline regions?

8.  What is mechanical degradation? Why is it less commonly encountered relative to thermal and photo-oxidative degradation?

9.  How does the primary mechanism of stabilization differ between ultraviolet radiation stabilizers and thermal stabilizers?

10. How do processing aids reduce the degradative effects of processing on polymers? What issues can arise with improper use of processing aids?

# 10    Polymeric Mixtures

## 10.1    Introduction to Polymeric Blends and Mixtures

So far, our discussions have focused on the properties and behavior of pure polymeric systems. This chapter explores the world of polymeric systems that contain more than one component. There are several reasons that we would wish to incorporate additional materials into a pure polymer. Typically, we need to alter the properties of the original polymer for specific end-use requirements. We have already seen this in Chapter 9 on degradation; stabilizers are incorporated into the resin to prevent the loss of properties due to processing at high temperatures and shears. In addition to stabilizers, we can incorporate many other types of materials. Minerals or glass fibers act as reinforcing agents. Low molecular weight organic waxes and liquids reduce the viscosity of the melt or the rigidity of final parts. Finally, incorporating other polymers can alter a polymer's behavior to create new materials with enhanced properties. When we mix two or more polymers we call the product a blend. If the secondary component is anything other than a polymer we create a mixture. In both types of system we must consider the thermodynamics of mixing and the rheological behavior of the polymer(s) during processing as we create the mixtures or blends. In this chapter, we will introduce the concepts behind the creation of mixtures and blends, the issues that often arise in their manufacture and some examples of why we use them in the first place.

## 10.2    Polymer Blends

Every polymer has its own strengths and weaknesses. For example, polycarbonate is really tough at low temperatures and resists deformation under loads at high temperatures but it is extremely sensitive to crazing and cracking under stresses. On the other hand, acrylo-nitrile-butadiene-styrene copolymer (which is better known as ABS) exhibits excellent impact strength and resistance to weathering though it distorts under loads at high temperatures. A blend of 20% acrylo-nitrile-butadiene-styrene copolymer and 80% polycarbonate creates a material that balances the best of both materials; it exhibits excellent resistance to deformation, high impact strength, and improved stress crack resistance relative to the unblended polycarbonate. This synergistic result comes about by a careful choice of the blended components and processing conditions under which the blend is formed. To describe a blend, we must understand the way in which the materials interact on a molecular level, the conditions under which a blend will form, and the rheological considerations associated with working with the final material.

## 10.2.1  Miscibility

When we mix two polymers, the resulting material may be categorized as either a miscible or immiscible blend. Miscible blends result when the mixing conditions and the polymer chemistry of the components permit the two polymers to completely mix at a molecular level. For example, a blend of two different linear low density polyethylenes, such as ethylene-co-octene and ethylene-co-butene, would be miscible. The polyethylene molecules are so similar that the chains completely interpenetrate to create a homogenous mixture of the two. On the other hand, attempts to create a blend of a linear low density polyethylene and polypropylene will not result in a molecularly intermixed material. Instead, the two components will remain as separate phases with an inhomogeneous structure. This type of blend, an immiscible one, is the most common type encountered in polymer blending. The fact that two polymers do not molecularly interpenetrate can be related to two effects: the free energy of mixing and rheological considerations.

### 10.2.1.1  Free Energy of Mixing

Our everyday experiences of creating liquid solutions leaves us poorly equipped to understand the behavior of polymers as we attempt to blend them together. For example, in our kitchens it is quite easy to mix up a pitcher of fruit punch from a powdered mix. The powder dissolves in the water, we stir a bit and we have ourselves a drink. The reasons why this works so well are numerous. First of all, water is very fluid, the molecules themselves have a great deal of freedom to move around the container. This freedom means that an individual molecule may experience millions of interactions with a dissolved (or dissolving) solute per second. Typically, the strength of the interactions between the powder in the drink mix and the water are about the same as the net effect of the interactions of the powder with itself and the water with itself. This means that, from a purely enthalpic perspective, there is little desire to create a solution. Knowing that the criteria for spontaneity is Gibbs free energy, all is not lost. Recall that the change in Gibbs free energy ($\Delta G$) includes both an enthalpic ($\Delta H$) and an entropic ($\Delta S$) term moderated by temperature ($T$) as shown in Eq. 10.1.

$$\Delta G = \Delta H - T\,\Delta S \tag{10.1}$$

This equation allows us to explain the solution formation based on entropic effects. In the juice example, the entropic effect is overwhelming. The degree of disorder after mixing the powder into the water is huge compared to that of the separate materials before mixing. This means that the value of the Gibbs free energy change is negative because of the high entropic contribution, making the process spontaneous.

Now, let's look at a polymeric system. To begin with, the motion in polymer chains is hindered. The massive size of the polymer itself and the intermolecular forces within the chains create an inflexible system, especially when compared to the aqueous systems with which we are most familiar. Secondly, the entropy of mixing is not actually as great as that seen in typical solution formation. Polymers are inherently highly entropic, so the benefit of mixing them together is modest. Therefore, any two polymers that form a miscible blend depend primarily

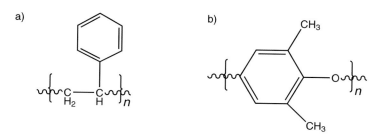

**Figure 10.1** Chemical structures of polystyrene and poly(phenylene oxide)

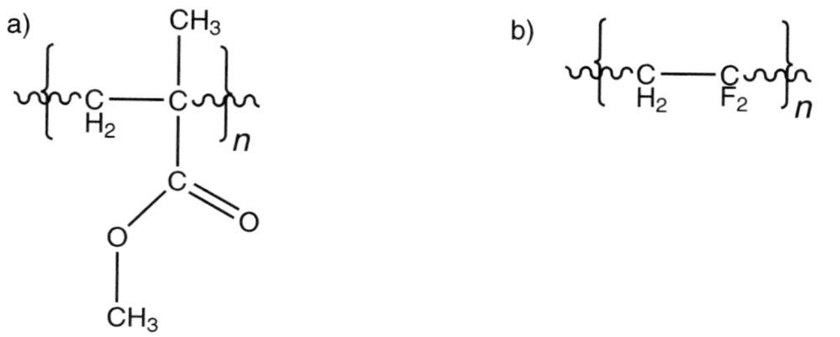

**Figure 10.2** Chemical structures of: poly(methyl methacrylate) and poly(vinylidene fluoride)

on the enthalpic effects to do so. The enthalpic effect relies on intermolecular forces between the components of the polymers and will create a favorable situation when the strength of interactions between the two polymer chains is very high. For example, miscible blends of polystyrene and poly(phenylene oxide), the chemical structures of which are shown in Fig. 10.1, can be created because of the strong interactions between the phenyl groups in both polymers.

Likewise, poly(methyl methacrylate) and poly(vinylidene fluoride), the chemical structures of which are shown in Fig. 10.2, make a miscible blend because of the strong specific interactions between the oxygen atoms on the methacrylate and the fluoride group in the vinylidene fluoride group.

These examples show that the criterion for a miscible blend is strong enthalpic interactions. Sometimes it is possible to create a miscible blend from polymers that do not form strong associations with one another by using a copolymer rather than a homopolymer. The copolymer is chosen to create strong interactions between the components, thereby promoting miscibility. An example of this is the copolymerization of styrene with p-(hexafluoro-2-hydroxyisopropyl)styrene to create a blend with polycarbonate. The highly electronegative fluorine atoms create a strong interaction with the polycarbonate, as illustrated in Fig. 10.3, allowing the blend to form.

One of the difficulties in working with miscible blends is that they are often very temperature-sensitive. The temperature sensitivity depends on the components of the blend, but once you pass the region of stability, the two components will separate and the material becomes an

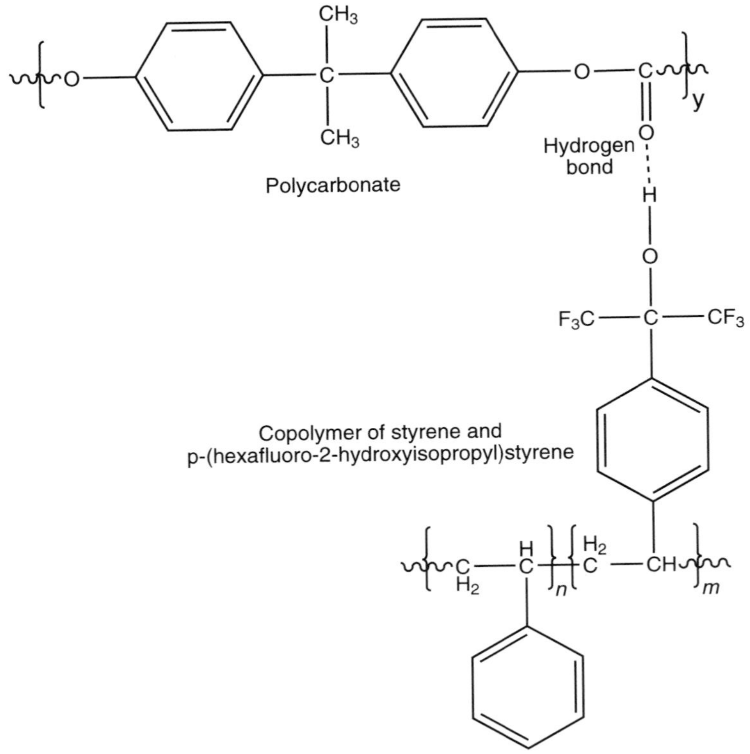

**Figure 10.3**   Example of a strong hydrogen bond linking polycarbonate with a copolymer of styrene and *p*-(hexafluoro-2-hydroxyisopropyl)styrene

immiscible blend with two phases. Figure 10.4 shows a phase diagram for a pair of polymers that can form a miscible blend at an intermediate temperature but will phase separate at both high and low temperatures, for example polystyrene and poly(vinylmethylether). Typically, phase separation, which is a slow process, occurs as the temperature of the blend is raised due to inhomogeneous concentration gradients that affect the free energy of mixing. The gradient change can be enough to create immiscibility allowing the phases to separate, though it can also occur through a "freezing out process", leading to a rather tentative balance between the two limits.

As stated earlier, miscible blends are the exception rather than the rule. Immiscible blends are much more common. The thermodynamics of mixing create a positive free energy of mixing, leaving the two components as separate phases. Typically, one is discrete (the one that is in lower concentration), while the other is continuous. Figure 10.5 shows a schematic diagram of several possible morphologies of discrete phases present in a continuous phase of an immiscible blend. The discrete phases create specific morphologies based on their crystallinity. For example, polyvinyl alcohol will form crystalline lamellae on cooling and,

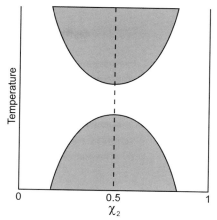

**Figure 10.4**   Idealized phase diagram for pairs of polymers that form miscible blends at intermediate
temperatures

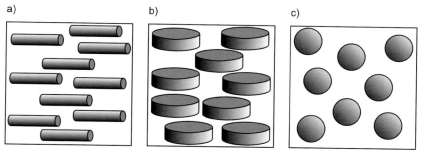

**Figure 10.5**   Schematic diagram of various discrete phase morphologies:
a) rods, b) platelets and c) spheres

when mixed with poly(ethylene terephthalate), will form plate-like structures in the cooled part. These plate-like structures are responsible for the low gas permeability of soda bottles. The two components of the blend interact through their boundaries and the strength of the blend will be dramatically affected by the surface energy of the two polymers at their boundaries. The behavior of the blend will normally be dominated by the major phase which is typically continuous. An immiscible blend is characterized by two glass transition points characteristic of each of the individual components.

### 10.2.1.2  Polymer-Polymer Interaction Parameters

One way to describe the interaction between two polymers is through their interaction parameter, $\chi'_{1,2}$. The interaction parameter is a thermodynamic term which encompasses the entropic and enthalpic effects associated with creating a blend. The interaction parameter concept is based on small molecule solution chemistry, which has been modified to describe the solution formation between two polymers. The starting point of the model is to say that the polymer consists of many small groups connected together. The small groups are like beads on a string and the segments, like beads, can touch one another when they are near enough. The number of segments interacting with any one segment provides us with a coordination number, z. By knowing the coordination number and the strength of interaction between any two segments $a$ and $b$ ($\varepsilon_{a,b}$), we can determine the interaction parameter. Mathematically we describe it according to Eq. 10.2.

$$\chi'_{1,2} = \frac{z\,(\varepsilon_{1,1} + \varepsilon_{2,2} - 2\,\varepsilon_{1,2})}{2\,k_b\,T} \tag{10.2}$$

This parameter then contains information about the strength of the interaction between the segments but does not take into account entropic effects. Since we are interested in determining if two polymers will blend together spontaneously, we need to determine the free energy of mixing, $\Delta G_m$, as described before. The way to do this is through the Flory-Huggins equation, as shown in Eq. 10.3.

$$\frac{\Delta G_m}{R\,T} = n_1\,\ln\phi_1 + n_2\,\ln\phi_2 + n_1\,\phi_2\,\chi_{1,2} \tag{10.3}$$

Where $n_1$ and $n_2$ represent the number of moles of the major component and minor components respectively, $\phi_1$ and $\phi_2$ represent the volume fraction of the major and minor phase, and R is the ideal gas constant. The first two terms represent the entropy of mixing, while the last term refers to the enthalpy of mixing due to the strength of the interaction as defined by the interaction parameter. Since the entropy of mixing is typically negligible, mixing will only occur spontaneously if the interaction parameter is negative. A negative interaction parameter arises when there are strong intermolecular forces between the components of the blend.

## 10.3   Properties of Blends

Though both miscible and immiscible blends are composite materials, their properties are very different. A miscible blend will exhibit a single glass transition temperature that is intermediate between those of the individual polymers. In addition, the physical properties of the blends will also exhibit this intermediate behavior. Immiscible blends, on the other hand, still contain discrete phases of both polymers. This means that they have two glass transition temperatures and that each represents one of the two components of the blend. (A caveat must be added here in that two materials that are immiscible with very small domain sizes will also show a single, intermediate value for $T_g$.) In addition, the physical properties are often a mixed bag based on the relative composition of the two phases, the form of the polymer in the phase, and the connectivity between the two materials.

### 10.3.1   Properties of Miscible Blends

The two polymers comprising a miscible blend are intimately mixed. This molecular contact results in strong interactions between the two polymers, resulting in a material with intermediate properties between those of the components. Since miscible blends form under quite tenuous conditions, they are often unstable. The polymers participating in the blend typically have a limit to their mutual solubility, so they can separate rather easily creating an immiscible blend. Interestingly, it is possible to create a thermodynamically unstable miscible blend, which we call a metastable blend. As you can imagine, these materials can easily be made immiscible through the addition of heat or stress or other types of triggers that unbalance the system.

### 10.3.2   Properties of Immiscible Blends

Immiscible blends have phase boundaries between the two components, therefore the surface energetics of the boundary, as well as the size and continuity of the domains, contribute to the properties that they exhibit. Sometimes it is necessary to add a compatibilizing agent to an immiscible blend. These are surface active materials that specifically reside between the two phases, creating interactions to help improve the mixing of the two materials. The compatibilizing agents are necessary when the surface energies of the two materials are very different. Surface energy describes the interactive strength of a surface. If the surface is covered with many highly polar groups, the interactive strength of this surface is high creating a high surface energy. If, on the other hand, there are solely nonpolar groups, such as methyl groups, the surface energy is low. Any attempt to mix two materials with very different surface energies will fail because one surface has a high interaction strength and would prefer to interact with something having a similar high strength. A low energy surface does not provide the stabilization that would be achieved if there had been polar groups. Therefore the two phases will be difficult to mix, as they would rather self associate than form a high area of surface that is incompatible. The effect can be compared to mixing oil and water. Water has a very high surface energy, while the surface energy of oil is very low, therefore the two are incompatible. An example of an incompatible polymeric pairing is polyvinyl chloride and low density polyethylene. Although their surface energies are not as different as those of water and oil, the degree of mismatch is high enough to prevent good mixing. Therefore, the addition of something that has both polar and nonpolar components would stabilize the interface. Two examples of successful compatibilizers for this blend are chlorinated polyethylene or a styrene-butadiene-styrene block copolymer.

Once a blend has been successfully manufactured, it is possible to describe the properties. A good illustration of these properties can be found with a blend of polypropylene and rubber. Polypropylene is brittle at temperatures below its glass transition temperature of approximately 0 °C. One way to compensate for this is to add a rubbery copolymer, such as ethylene-propylene rubber (EPR), as a minor phase. The presence of the rubber greatly increases the impact strength of the polypropylene. A widely accepted explanation for the improved physical properties of this type of blend interprets the interface between the two components as a stress concentrator. In the blend, the concentrated stresses can dissipate in the rubbery matrix of the rubber and through localized crazing around the surface of the rubber. The disadvantage of this system is that the presence of the ethylene-propylene rubber reduces the tensile strength and modulus of the system, relative to pure polypropylene. It has been shown that there is an ideal size for the rubbery phase component and that domains that exist above this size reduce the strength of the polymer blend to unacceptable levels, while smaller domain sizes lead to no strengthening effects.

### 10.3.3     Rheological Effects in Manufacturing and Use of Blends

So far, we have seen that the chemistry of polymers used in a blend determines whether or not a blend will form. However, there is another factor that we must consider when making blends. Specifically, we need to understand the rheological behavior of both polymers to determine whether we can make them mix together well. There are cases, such as blends between a high molecular weight polymer and one that has a low molecular weight, that are known to be miscible, but resist melt blending. Blending can be accomplished by dissolving both polymers in a solution and then evaporating off the solvent. If we attempt to form the same blend by mixing them in the molten state, we may find that it is impossible. The reason for this difficulty lies in the rheological behavior of the two components. Imagine a system comprised of two polymers that individually have very different viscosities. If we add the fluid polymer to the viscous one, the fluid will just sit on top of the melt and slip away as the mixing paddle tries to force incorporation. This is the same issue that we face when we try to incorporate true fluids into polymer melts. Therefore, in cases where the rheological behaviors do not match, blending becomes extremely difficult.

If we are trying to produce an immiscible blend, we still encounter difficulties. Even though we are not attempting to achieve intimate molecular interactions, we still desire good distribution of the components. This may be very difficult if we cannot force any incorporation of the two components. In addition, since there are still two distinct phases in an immiscible blend, we have the potential for further segregation, even after mixing occurs. This is a result of the response of a system under shear to reduce the associated stresses. Specifically, the areas with the highest stresses are adjacent to the walls of equipment through which a material flows. The best way to dissipate the stresses is for the low viscosity material to remain at the walls. Therefore, these materials travel to the outside and segregate themselves to reduce the overall stresses in the system. This can be disadvantageous as it leads to weak areas in the final product. The best way to get two polymers to mix is to choose materials with similar rheological behaviors under the conditions at which the blends will be created. Another option is to always use a solvent deposition process, but this is a commercially unfeasible.

## 10.4    Polymer-Additive Mixtures

Many polymer resins incorporate additives to either improve properties, or reduce the cost to manufacture the final product. The type of additive, the means by which it is introduced, and the processing it will experience all determine the effectiveness of the additive. Both solid and liquid additives pose specific problems in their incorporation. In general, liquids are so fluid relative to the molten polymer that they tend to resist incorporation into the melt; instead, they slip over the highly viscous melt, even during intense mixing. Solid powdered additives tend to clump together, much as flour does when you add it to eggs and milk when making a cake. Fibrous materials want to bundle, while individual fibers may be easily broken during mixing. It is for these reasons that the equipment used to incorporate additives into a polymer is often highly specialized in order to achieve good mixing without damaging either the polymer or the additive. The following sections describe the specific types of additive used, the reasons for their use, and the means by which we characterize the final mixture.

Two important terms used to describe additives in polymeric mixtures are "distribution" and "dispersion". When mixing polymers with additives, we want to create a system in which the additive is both well distributed and dispersed. Distribution refers to the even placement of the additive throughout the polymer. For example, a well made batch of chocolate chip cookies has good distribution of the chips if every bite has a chocolate chip in it. A poorly distributed cookie would have all its chips on one side. Dispersion, on the other hand, refers to the separation of the individual components of a solid additive into its smallest parts. Figure 10.6 illustrates both good and bad distribution and dispersion.

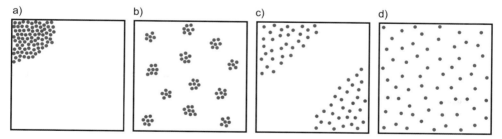

**Figure 10.6**    Examples of various types of mixing:
a)  poor distribution and poor dispersion,
b)  good distribution and poor dispersion,
c)  poor distribution and good dispersion and
d)  good distribution and dispersion

### 10.4.1    Liquid and Low Molecular Weight Solid Additives

We routinely add liquid and low molecular weight solids to polymers to provide color, stabilization, or processing stability. Specific examples include dyes, plasticizers, and waxes. Often the biggest difficulty in mixing a liquid additive into a polymer is getting the two phases to interact. The liquid resists incorporation into the high viscosity polymer, even when mixing is thermodynamically favorable. One way of promoting mixing is to place the liquid additive in an intermediate molecular weight carrier, which will accept the liquid, and then adding the mixture to the polymer melt. The closer the rheological properties of the components, the better the resulting mixing will be. Another option is to introduce the material into the polymer under extremely high shear conditions. Dyes are incorporated into the polymers used to make carpet fibers in batch mixers. These are large heated blenders that run for extended periods of time under high shear. It is also possible to inject a high pressure liquid directly into a melt at a point where it is also under high pressure. This prevents the separation and forces the liquid to mix with the polymer since the liquid has no room to leave the system. Finally, some polymers can be premixed with the liquid additives allowing the resin to absorb the additives prior to entering the melting equipment. Polyvinyl chloride manufacturers use this strategy to incorporate liquid plasticizers into powdered resin in a preblending mixer. They prepare the preblend immediately before the material is needed in the processing equipment by heating the polymer and then adding the liquids to the hot swirling powder. The amount of time that the materials are in contact determines how much of the liquid absorbs into the polymer. Once the liquids are absorbed and distributed throughout the powdered resin, they can be fed into the processing equipment. Plasticized polyvinyl chloride exhibits very different properties from the unplasticized material. It becomes very flexible and is used in a wide range of application, such as upholstery, toys, shower curtains and wire insulation. The unplasticized resin is rigid and would not be suitable for these applications.

### 10.4.2    Solid Additives

Solid additives, such as minerals, glass fibers, wood fibers, wood flour and pigments, create specific issues when we incorporate them into polymers. Such additives have a high surface area, creating a problem with forcing the polymer to envelope each particle, since an interface is formed. The difficulty experienced in forming an interface depends on the surface energy of the polymer, the surface energy of the additive and the size of the additive. Since many polymers are nonpolar or only slightly polar with low surface energies, the additives that incorporate most easily are those that also have low surface energies. Talc, a nonpolar mineral, is relatively easy to incorporate in polypropylene, which is also nonpolar, and is a common additive in injection molded polypropylene automotive interior parts. Most additives, though, are polar, exhibiting a high surface energy. Glass fibers, titanium dioxide, calcium carbonate and wood fibers are all rather difficult to incorporate in polypropylene. To counteract this problem, additive manufacturers apply surface coatings to the polar material. These coatings are amphiphilic surfactants that bond well to the polar mineral with their polar heads, creating a nonpolar surface with their hydrocarbon tails. The polymer then interacts with the nonpolar tails of the surface treatment, rather than the high surface energy additive. Without these surface treatments, dispersion of these materials becomes very difficult, leading to non-uniform physical properties and appearance. Polyolefin-based mixtures containing calcium carbonate rely on the surface treatment of the mineral to create a homogenous mixture. The most common treatment, stearic acid, contains a hydrophilic carboxylic acid head group, which interacts with the calcium carbonate, and a twelve carbon atom aliphatic chain. The long chain lowers the surface energy to match that of the polyolefins, allowing the mineral to mix into the polymer.

## 10.5    Conclusion

In this chapter we have discussed the thermodynamic formation of blends and their behavior. Both miscible and immiscible blends can be created to provide a balance of physical properties based on the individual polymers. The appropriate choice of the blend components can create polymeric materials with excellent properties. On the down side, their manufacture can be rather tricky due to rheological and thermodynamic considerations. In addition, they can experience issues with stability after manufacture due to phase segregation and phase growth. Despite these complications, they offer polymer engineers and material scientists a broad array of materials to meet many demanding application needs.

We have also examined how mixtures are made. Mixtures, like blends, present the manufacturer some challenges. Incorporating anything into a viscous liquid is difficult because of the need for both an even distribution of the materials and good dispersion. Despite the manufacturing drawbacks, mixtures can provide novel properties relative to those of a pure resin.

## Review Questions for Chapter 10

1.  How do entropic and enthalpic effects determine the miscibility of polymers in blends? Which term dominates the formation of a miscible blend?

2.  In Fig. 10.4 there is a temperature range in which two polymers form a miscible blend. Why do the polymers separate into distinct phases at both high and low temperatures?

3.  Why does the minor component in an immiscible blend prefer to form spherical morphologies in the absence of applied stresses? How do the less stable shapes of rods or platelets form?

4.  How do the polyvinyl alcohol platelets reduce gas permeability in polyethylene terephthalate soda bottles?

5.  For the interaction parameter defined in Eq. 10.2, why are the signs on $\varepsilon_{1,1}$ and $\varepsilon_{2,2}$ positive while that of $\varepsilon_{1,2}$ is negative?

6.  What analytical methods could you use to determine whether a blend was either miscible or immiscible?

7.  How do compatabilizers work to stabilize immiscible blends?

8.  Why do polymers with different viscosities present specific challenges when manufacturing blends?

9.  What are dispersion and distribution? Why is it important to achieve both good dispersion and distribution when forming a mixture?

10. What methods are used to introduce liquid additives into polymers?

11. How does surface treatment of polar solid additives lead to better dispersion and distribution of the additive in the polymer matrix?

# 11 Extrusion Processes

## 11.1 Introduction

In order to make useful products from polymers we typically follow a three-step process in which we sequentially melt, shape, and cool the polymer. Naturally, given the wide variety of polymers available and their myriad applications, variations on this general process abound. In this chapter, we will concentrate on extrusion, some variant of which is used in the majority of commercial fabrication processes.

Extruders play a key role in many conversions processes; we use them to melt solid polymers and pump the resulting molten material to a die or a mold. After the molten polymer has been molded to the desired shape, it is cooled to form a solid product. We can feed the output of an extruder to continuous forming processes to create films, pipes, and fibers. Alternatively, we can employ discontinuous molding processes to create discrete items, such as soda bottles, lenses for single use cameras, or bathtubs.

## 11.2 Principles of Extrusion

The role of an extruder is to convert incoming solid polymer pellets or powder to a homogeneous melt that is pumped to a forming device such as a die. Figure 11.1 shows a longitudinal cross-section through a single screw extruder, illustrating its principal components. A drive unit turns a long screw within a heated barrel. Polymer pellets are fed into the gap between the screw and the barrel from a feed hopper mounted above the machine.

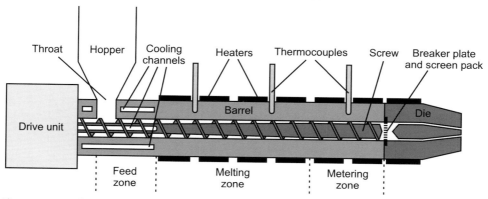

**Figure 11.1** Schematic diagram showing the principal components of a single screw polymer extruder

As the screw turns, it conveys the pellets from the feed zone towards the melting zone. A combination of external heating and mechanical work melts the polymer as it is transported towards the metering zone. The metering zone pumps a uniform stream of molten polymer to a forming device, such as a profile die. Other types of extruders that employ two or more screws are commonly used for compounding polymer blends. The principles of twin screw extrusion will be discussed in Chapter 12.

Polymer extruders are based on the Archimedean Screw, which has been used for more than 2,000 years to convey liquids, powders, and slurries. In order to understand the principle of an extruder, we can think in terms of a long bolt onto which we thread a nut. If we turn the bolt and prevent the nut from rotating, the nut will move along the length of the bolt. In a similar fashion, we can think of the extruder screw as the bolt, the polymer as the nut, and friction between the polymer and the barrel wall as the force that prevents the nut from rotating. Thus, as we turn the extruder screw we will force the polymer along the length of the barrel. This principle is illustrated in Fig. 11.2, in which we see a cut away of part of an extruder. The screw consists of a helical flight wound around a shaft. A small pile of material

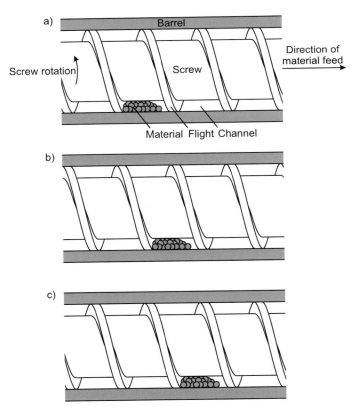

**Figure 11.2**   Principle of material transport in an extruder:
a) starting position, b) one half revolution, and c) one full revolution

is trapped in the channel between two turns of the flight. As the screw turns in the direction shown, the flights appear to move from left to right, transporting the pile of material in the same direction.

In practice, the entire channel of a single screw extruder is filled with polymer. The polymer enters the feed section in its granular form, which is conveyed towards the melting zone. Most polymers are fed into extruders in the form of pellets, which resemble short sections of a rod or roughly spherical beads with dimensions of approximately 2 to 5 mm. In the melting zone the pellets gradually melt and the material is densified by the exclusion of air. The melting zone also homogenizes the molten polymer. The final section is the metering zone, which pumps the molten polymer. Unlike conventional pumps, which work on the principal of positive displacement, an extruder is a viscosity pump. In this case, material conveyance is achieved by the action of the flight rotating against the viscous polymer melt, which adheres to the barrel wall.

Extruders vary widely in their dimensions. Laboratory scale extruders can have screws with diameters as small 6.35 mm. At the other extreme, the screws of extruders used to create continuous profiles with large cross-sectional areas, such as large diameter pipe or the insulation surrounding power transmission cables, can be up to 300 mm in diameter. Screw lengths range from approximately 20× to 30× the screw diameter. The depths of the channel decreases from the feed zone to the metering zone. Channel depths can range from 0.02× to 0.07× the screw diameter. Longer screws typically have shallower channels than shorter ones. Compression ratios, that is the ratio between the channel depth in the feed zone to that in the metering zone, vary from approximately 2 to 4. Screw speeds range from approximately 60 to 300 rpm, depending upon extruder size and polymer characteristics. Smaller extruders generally run at higher revolution rates than larger ones.

The relative lengths of the feed, melting, and metering zones depend on the rheological characteristics of the molten polymer. A general purpose screw, which could be used to process polyethylene and polypropylene, may have feed, metering, and melting zones that are approximately equal in length. Low viscosity melts, such as those of polyamides, need a relatively longer metering zone to build enough pressure to ensure a consistent extrudate. Highly viscous molten polymers have relatively short melting and metering zones to prevent excessive pressure building up at the end of the barrel. The correct choice of screw design is crucial for optimum product quality and output rate.

## 11.2.1    Feeding

Polymer granules are fed by gravity from the hopper into the feed throat of the extruder. Many commercial products are made of blends of two or more polymers or a polymer and a "masterbatch", which contains pigments, dyes, or other additives. To accommodate this, we employ multiple subsidiary feed hoppers equipped with gravimetric feed mechanisms to ensure that the correct ratio of components is let down into the main hopper. To prevent "bridging" of polymer pellets across the feed throat (imagine The Three Stooges trying to go through a doorway simultaneously), which would prevent polymer from entering the barrel, we sometimes mount a feed auger in the throat of the hopper or vibrate the hopper. Prior to extruding hygroscopic polymers, such as polyesters and polyamides, it is standard practice to dry them. We do this in a continuous dryer, which pumps warmed dehumidified air through a bed of polymer pellets in a small silo. If we were to neglect the drying step, the water absorbed into the polymer would vaporize within the extruder, which would degrade the polymer and result in generally poor extrudate quality.

After the pellets enter the barrel, they are conveyed towards the feed zone by the rotation of the screw. To ensure consistent feeding it is essential that the polymer does not begin to melt within the feed zone. If this happens, the pellets can clump and block the feed throat or adhere to the screw. In order to prevent premature melting, we commonly cool the barrel and the screw in the feed zone. The conveyance of material within an extruder requires that friction between the granules and the barrel wall exceed that between the screw and the granules. We can achieve this friction differential by machining longitudinal grooves in the barrel walls of the feed zone or cooling the screw to a lower temperature than the barrel. The channel depth within the feed zone is constant.

## 11.2.2  Melting

As the polymer granules enter the melting zone (which we also refer to as the transition zone), their temperature rises and they begin to melt. The heat required to melt the polymer comes from two sources, external heaters and mechanical work from the action of the screw. When we extrude viscous polymers at high rates we may need to cool the barrel to remove some of the heat induced by working the molten polymer.

Melting of the polymer does not occur instantaneously; it starts at the barrel wall where frictional and external heating are greatest, forming a molten film, as shown in Fig. 11.3. A pool of molten polymer forms ahead of the trailing flight as it wipes the molten film from the barrel wall. The wiping action results in circulation within the melt pool. As the polymer is conveyed along the barrel, the width of the molten pool of polymer grows until the solid granules are completely melted. The depth of the channel decreases in this zone, which helps compact the polymer and exclude air, which is forced back along the extruder and out through the feed throat. Degassing may be aided by the addition of vacuum ports that draw off excess gases.

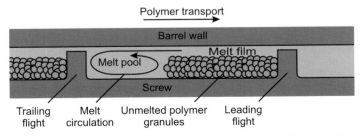

**Figure 11.3**   Cross-section through a melt channel in the melting zone of an extruder

In addition to melting the polymer, mixing also occurs in the transition zone, which produces a homogeneous output. Mixing primarily occurs due to circulation within the melt pool. We may improve homogenization by various mixing strategies. These include sections of the screw having reverse flights, interrupted flights, and "pins" set into the wall of the barrel. We will describe some of these mixing elements in more detail in Chapter 12.

### 11.2.3  Metering

The final zone of an extruder is the metering section (also known as the pumping zone). The principal function of this zone is to ensure a steady output of molten polymer at a constant pressure. The longer the metering zone, the greater the pressure built up within it. Channel depth is constant to ensure a uniform transport rate, which helps reduce pressure fluctuations arising in the mixing zone.

The build up of pressure is aided by the use of a breaker plate, which is a thick disk of perforated metal positioned just downstream of the tip of the screw. A screen pack, comprising layers of woven metal mesh, is placed on the upstream side of the breaker plate. The screen pack increases back pressure within the metering zone and acts as a filter to catch any extraneous material that may have entered the extruder.

## 11.3  Profile Extrusion

We use profile extrusion to make continuous products that have fixed cross-sectional dimensions, such as pipes, house siding, refrigerator door gaskets, and windshield wiper blades. During profile extrusion the molten output from an extruder is pumped to a die where it is formed to approximately the desired cross-sectional profile. As the molten polymer leaves the die, we apply the final forming step and simultaneously cool it to yield the product in its solid state.

Figure 11.4 schematically illustrates one of several configurations used to extrude pipe. Polymer is pumped from the metering section of an extruder into an extrusion die where it flows around a mandrel. The mandrel is held in the center of the die by a series of radial fins, which we collectively refer to as a "spider". The fins of the spider are streamlined so that the molten polymer flows around them smoothly and merges on the downstream side with the minimum of flow disruption. When the polymer tube emerges from the die, it is still molten and would collapse under its own weight if we did not support it. A common way of preventing the tube from collapsing is to inflate it with air that we inject through the mandrel. Air is prevented from escaping by a plug inserted in the end of the pipe or a "floating" plug that is attached to the die with a chain running through the pipe. The final sizing of the polymer tube is achieved as it passes through an external calibration tube, which also serves to cool it. Alternatively, we can draw the pipe over an extended internal mandrel, which cools, sizes and polishes its interior.

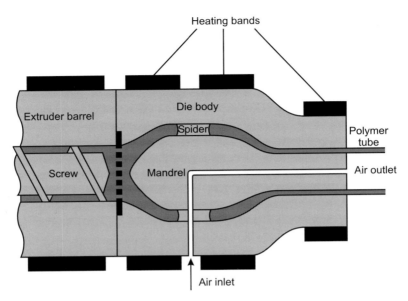

**Figure 11.4**    Pipe extrusion

We use variants of profile extrusion to produce tubing with diameters of less then 1 mm and pipes with diameters exceeding 1 m. Wall thicknesses can vary from a few tens of micrometers up to several centimeters. Extruded window and door frames are more complex than pipes. Such profiles are largely hollow with internal ribs and fins that reinforce and divide the interior into two or more channels. We use solid rubber profiles in applications such as door seals and windshield wipers. We can produce foamed extrudates by incorporating a blowing agent, such as butane or carbon dioxide, into the polymer in the molten state. As the polymer exits the die, its internal pressure drops and the dissolved gas expands to form bubbles within the product. Examples of foamed extrudates include pipe insulation and automobile door gaskets.

## 11.4    Film and Sheet Casting

We use the casting process to make polymer films (less than 0.3 mm thick) and sheets (more than 0.3 mm thick) for such diverse end uses as cling wrap, merchandise bags, roofing membranes, landfill liners, and the interior walls of refrigerators. During chill roll casting, molten polymer is extruded as a curtain from a slot die onto a chilled metal roller where it solidifies. The product is transported over a series of rollers to a winder where it is wound up.

Figure 11.5 schematically outlines the process of film casting. The molten output from an extruder is pumped through a heated pipe to the top of a slot die, whose exit is pointed

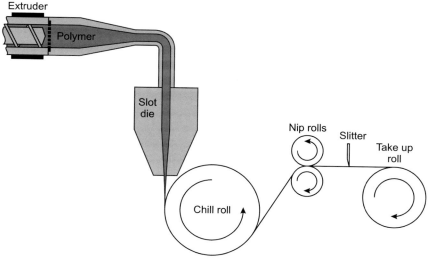

**Figure 11.5**  Film casting

downwards. In the die, the molten polymer stream is flattened within a manifold until its width is hundreds of times its thickness. The precise dimensions of the flow channel within the die are critical. If the design of the channel does not match the rheological characteristics of the molten polymer, we can end up with too little or too much polymer reaching the ends of the die, which results in an uneven product. The molten polymer emerges from the die between a pair of die lips. Typically, one of these is rigid and the other is designed to be somewhat flexible. We control the die gap and hence the thickness of the film by applying local pressure to the flexible lip at numerous points along its length. When we are using simple or small dies, we can adjust the die gap manually. On more sophisticated dies a thickness gauge constantly scans the product and provides feedback to a computer that automatically controls the adjusters on the flexible lip. Other variables that we can use to control film thickness are extruder output rate and the draw ratio, which is the ratio of material speed at the chill roll to its speed as it emerges from the die.

As the molten web emerges from the die, its speed increases because the surface of the polished metal chill roll is pulling the film many times faster than the melt curtain is moving as it exits the die. As a result, the thickness of the molten web is reduced greatly as it is drawn down. The acceleration within the melt curtain aligns the polymer molecules in the direction of material transport. The orientation of the polymer's molecules is frozen into the product as it rapidly solidifies on striking the chill roll. We can increase the orientation within the sample by closing the die gap or increasing the draw down rate. The polished surface of the chill roll imparts a smooth surface to one side of the film. The solidified film makes a partial wrap around the chill roll before being drawn off by a pair of nip rolls and transported to the winding station. Slitter blades cut off the outer edges of the film, which are inevitably thicker than the rest of the film. When desired, we can slit the film into a series of narrower cuts that are wound up individually.

We use the casting process to make films with various levels of complexity. Simple products include cling wrap and many packaging films. Such products generally comprise a single layer of polymer, which may be a blend of two or more different polymers based on similar monomers. We can also make much more complex products that have multiple layers of different polymers. To accomplish this, we feed the die from two or more extruders. The cavity of such a die is quite complex, often with separate channels for each of the melt streams, which are brought together just before exiting the die. Using this technology, we can combine the properties of several polymers into a single product. Thus, a potato chip bag may be made from a film containing five or more different layers, including polyethylene terephthalate, for strength and gloss, and polyethylene vinyl alcohol, which acts as an oxygen barrier. The principal polymer layers may not be compatible; to prevent them from delaminating, we employ intermediate tie layers comprising thin layers of what is essentially an adhesive. As we commonly wish to print packaging films, it may be necessary to treat their surface to accept inks. This may be done with a corona discharge that generates active gas species above the film that chemically attack its surface, thereby raising its surface energy. We can cast polymer films as thin as 10 μm and as wide as 5 m. High speed packaging film lines can run at upwards of 300 m/min. The majority of polymer films are used a single time and then discarded.

Polymer sheets are generally used in more durable applications than polymer film. Due to their thickness we commonly make sheets at much slower line speeds than films. This gives the molecules more time to relax after emerging from the die. Sheeting is typically less oriented, and therefore has more balanced tensile properties, than films. Polymer sheeting may be used directly or may undergo a secondary molding process such as thermoforming. Direct usage of polymer sheeting includes thermoplastic elastomer membranes, used to waterproof flat roofs, and geomembranes made of linear low density polyethylene, which are used to line garbage dumps and toxic waste pits. Flexible sheeting is wound onto rolls. Stiffer materials are cut into individual sheets as they reach the end of the production line. We can use these stiff sheets directly in their flat form or can thermoform them into three dimensional structures such as refrigerator liners, car bumpers, and bath tubs, as we shall describe in Chapter 16. More often than not, such complex products comprise multiple layers that together provide the necessary balance of stiffness, impact resistance, barrier properties, and surface characteristics.

## 11.5     Film Blowing

We use film blowing to make thin films from polymers that typically have good melt strength, such as low density polyethylene. We use blown films extensively to make trash and merchandise bags and many types of packaging. During film blowing, molten polymer is pumped upwards out of an annular die to form a molten tube that is inflated with air to expand its diameter prior to its cooling.

Figure 11.6 illustrates the general configuration of a film blowing operation. Molten polymer from the extruder is pumped into an annular die, where it is distributed around a tubular melt channel before emerging vertically as a relatively thick-walled molten tube. The top of this tube is closed off by nipping it between a pair of rolls mounted several meters above the die. We inflate the tube between the die and the nip rolls by injecting air through the center of the die. Once the bubble has reached the desired size, we shut off the air flow to trap a fixed volume of air within the bubble. As the molten polymer tube emerges from the die, it is expanded circumferentially by the trapped air and stretched vertically by the nip rolls that have a surface speed several times that of the molten polymer as it emerges from the die. An annular air ring blows air onto the outside of the bubble just above the die to promote cooling. As the polymer ascends from the die, it cools and begins to crystallize. We observe the onset of crystallization as a distinct "frost line", where the polymer becomes hazy due to the formation of supermolecular structures and surface roughness that scatter light. The bubble continues to travel upwards for several meters until it reaches the collapsing frame, which flattens the bubble prior to the nip rolls. The product that emerges from the nip rolls is a flattened tube. The flattened tube can be wound up directly or it can have its edges slit off and the resulting pair of films can be wound up separately.

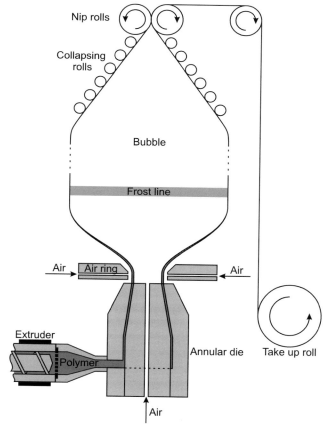

**Figure 11.6**   Film blowing

We have several ways of controlling blown film thickness. By lowering or raising the tapered core of the die we can increase or decrease the die gap and hence the film's gauge. As we increase the blow up ratio – that is the ratio of the bubble diameter to the die outlet diameter – by inflating the bubble more, we decrease the film thickness. Likewise, if we increase the take up ratio – that is the film speed at the nip roll relative to its speed as it emerges from the die – we also reduce the film thickness. Changing the blow up and take up ratios alters the balance of orientation within the final product. We can change the direction of orientation from being primarily in the machine direction to being primarily in the transverse direction. As a general rule, the properties of blown film are more balanced than those of cast film.

The majority of blown films are made from ethylene-based resins. We can make blown films with thicknesses as low as ten micrometers and diameters of up to five meters, which can yield flat films up to fifteen meters in width. We use blown films in many of the same applications as cast films, especially packaging and bag making. Blown films are generally not as uniform in thickness as cast films and therefore are often used in less demanding situations. We make extensive use of blown polyethylene film in agricultural applications, such as green house skins, ground cover to control weed growth, and covers for bales of hay. It is also widely used in the construction industry as a moisture barrier and in decorating as a drop cloth.

## 11.6    Fiber Spinning

We use the spinning process to make polymer fibers and filaments that can be converted into fabrics and cordage. During fiber spinning, molten polymer is pumped through holes in a plate to form a multiplicity of strands that are rapidly stretched and cooled. The finished product comprising oriented fibers is either wound up on spools or converted directly into a non-woven fabric.

Figure 11.7 illustrates the general configuration of the fiber spinning process. Molten polymer with a relatively low molecular weight and viscosity is pumped from the extruder to a die called a spinneret. The polymer is forced through numerous holes in the horizontal output plate of the spinneret to form molten strands that are rapidly drawn down towards a set of rolls. The vertical transport and stretching of the molten strands may be assisted by air flow down the tube in which the strands are drawn. The air also serves to cool the strands, which solidify before they reach the first of the rolls. Further drawing of the fibers in the solid state is achieved by a set of godet (pronounced *goday*) rolls. The polymer strands are wound in a spiral around each godet roll, each of which rotates faster than the preceding one. The net result of drawing in the molten and solid states is a bundle of highly oriented fibers that is wound up on a spool.

We control fiber properties by changing the relative speeds of different stages of the process. Orientation is increased and fiber thickness decreased by increasing the final take-up speed relative to the rate at which the molten polymer strands leave the spinneret. To produce high modulus fibers we generally adopt conditions that maximize orientation. Fiber diameters

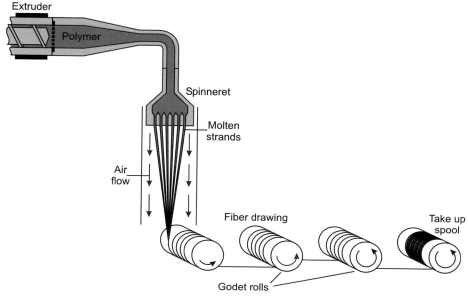

**Figure 11.7**   Fiber spinning

range from less than one micrometer up to a millimeter or more. Not all fibers have circular cross-sections; we routinely manufacture elliptical, ribbed, lobed, and hollow fibers.

We can manufacture fibers from a wide range of polymers. Polyamides, polyesters, and polypropylene can be woven or knitted into fabrics, ranging from those as coarse and strong as those used in back packs, luggage, and sails, to soft and highly flexible fabrics used in sweaters, shirts, and other apparel. Polymer filaments and yarns can be twisted or woven to make string, twine, cords, and ropes.

## 11.6.1   Non-woven Fabric Production

We can make non-woven fabrics directly from the spinning process without recourse to weaving. To do this, we use the equipment shown schematically in Fig. 11.8. Molten polymer is pumped through a very wide spinneret that produces a curtain consisting of thousands of fibers, which are blown down onto a porous conveyor belt. As the fibers land on the conveyor belt, they form an entangled web that is conveyed to a pair of heated rollers. One of the rolls, known as the anvil roll, is smooth, while the other is covered with a pattern of raised embossing points. As the entangled web passes between the rollers, its fibers are fused between the raised points of the embossing roll and the anvil roll. The resulting product has many of the characteristics of a woven fabric, but is generally not as strong.

Non-woven fabrics can be produced very economically, which suits them for many single-use health and hygiene applications, such as the outer cover and body contacting layers of

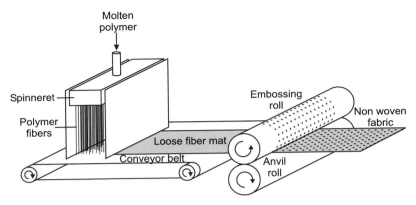

**Figure 11.8**    Manufacture of non-woven fabric

disposable diapers, components of feminine hygiene products, surgical drapes, and pre-moistened wipes. Other outlets include filters and furniture upholstery.

## 11.7    **Extrusion Coating**

We use extrusion coating to apply thin layers of polymer to the surface of non-polymeric substrates such as cardboard or aluminum foil. Extrusion coated materials are extensively used in food packaging. Products include the coated cardboard used to make milk cartons and the coated aluminum foil used to seal dairy product tubs. The process of extrusion coating has much in common with film casting.

Figure 11.9 illustrates the general configuration of the extrusion coating process. A molten curtain of polymer is pumped from a slot die to fall onto a substrate as it makes a partial wrap around a rubber coated lay-on pressure roll. The substrate conveys the polymer into contact with a chilled metal roller, where it is quenched under pressure to the solid state. In the case of relatively rough substrates, such as thin cardboard, the polymer layer is pressed against the substrate's surface to form a mechanical bond. When coating a metal foil, we commonly use a polymer, such as polyethylene vinyl acetate, that incorporates polar substituents that are attracted to the high energy metal surface. The polymers used for extrusion coating typically have relatively low molecular weights and thus have low melt viscosities. We can readily draw down such polymers to form very thin coatings that may be less than ten micrometers in thickness.

The coating of polymer that we put on paperboard serves two purposes when we convert it into juice and milk cartons. It prevents the contents from leaking out and it acts as an adhesive so that we can heat-seal the coated paperboard to itself to form a container. Low density polyethylene is the polymer of choice for this application; it has sufficient melt strength to

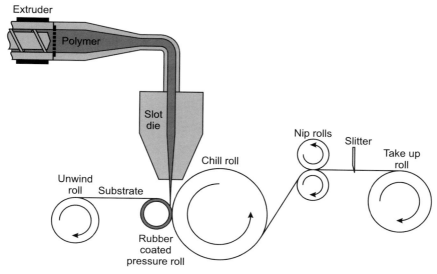

**Figure 11.9**   Extrusion coating

be drawn down to a thin layer, it begins to melt at a relatively low temperature during heat sealing and it is an effective barrier against aqueous liquids. In the case of coated aluminum foil, the polymer layer primarily acts as a hot melt adhesive, facilitating the production of an impermeable package.

## 11.8    Wire and Cable Coating

We can coat wire and cable of any gauge with insulating layers of polymer. In a coating die, we create a tube of molten polymer that adheres tightly to a core of wire or cable that is fed through the center of the die. Insulated wires and cables are used to carry a wide range of voltages and currents in applications ranging from microelectronics up to the nationwide power transmission grid.

Figure 11.10 shows a cross-section through a wire coating die that illustrates the general configuration of the process. We pump molten polymer from an extruder into a crosshead tubular die that distributes the flow into an annular cavity. Wire is fed into the rear of the wire guide, comprising the core of the die. The molten polymer encapsulates the wire within the die ring prior to its emergence from the die. In the case of thin insulation layers, the product is cooled by exposure to air. When the insulation is thicker, we cool it in a water bath, which contains external calibration plates to ensure that the finished product has the desired cross-section.

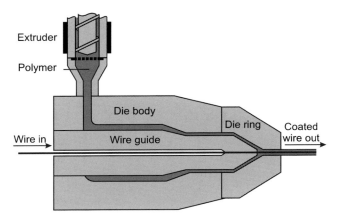

**Figure 11.10**   Wire coating

We control the thickness of the polymer coating by a combination of extruder speed, haul-off rate, and the diameters of the wire and die ring opening. The wire guide and die ring are interchangeable to accommodate a range of wire gauges and insulation thicknesses.

We use a wide variety of polymers to insulate electrical components. Low and linear low density polyethylene, and plasticized polyvinyl chloride are widely used for their flexibility in smaller wire gauges. When cables have to withstand higher service temperatures, very high voltages, or are used in safety critical applications, we often use crosslinked polyethylene insulation. In such cases we coat the wire with polyethylene into which is blended a crosslinking agent, such as an organic peroxide. As the coated cable emerges from the die, it passes into a curing tunnel where the temperature is raised to the point where the crosslinking agent decomposes to create radicals that crosslink the polyethylene. Other polymers and additives are used in specialized applications. Wiring used on airplanes and in naval vessels incorporate fire retardants into their insulation. Cables that have to withstand abrasion may be coated with a fluorinated polymer. Foamed polymers are used to surround co-axial cables to minimize their weight and improve their flexibility.

## 11.9    Conclusion

In this chapter we have introduced the basic principles underlying polymer extrusion. As we have seen, extruders have two primary purposes; they convert polymer granules into a homogeneous molten stream that is pumped to a forming device. The forming devices can take many forms to make a wide variety of continuous products. Among the continuous products made by extrusion are films, pipes, and fibers, which are used in all manner of products that make modern life more convenient and comfortable. Each of the processes that we have described has multiple variants, both major and minor. In each of the sections we have outlined a single configuration that illustrates what we feel to be the most important features. A brief examination of everyday polymer products will reveal that variations on these basic processes abound.

## Review Questions for Chapter 11

1.  How does an extruder convert polymer granules or pellets to a homogenous melt? What forms of energy are used to perform this process?

2.  Why does the channel depth of the screw generally decrease as you move from the feed zone to the metering zone of an extruder?

3.  In general, the feeding, melting and metering zones of an extruder are maintained at different temperatures. A standard temperature program will maintain the feeding zone at a temperature that is below the melting zone, which is below the metering zone. Provide an explanation of why this type of program is used.

4.  Why is it sometimes necessary to cool an extruder barrel, even though you need to maintain high processing temperatures?

5.  Sketch the opening of a profile die that would produce the following products:
    a)  pipe
    b)  a gutter
    c)  the gasket on a refrigerator

6.  What types of products require the use of a mandrel die? Why?

7.  One of the cardinal rules in die design is to avoid all sharp corners. Why?

8.  What is the role of a tie layer in processes which create multilayer films?

9.  What process variables control the degree of orientation in a film manufactured via cast film extrusion and blown film extrusion?

10. What sort of problems would you expect a processor to encounter while coating polyethylene onto a cardboard substrate?

# 12 Compounding

## 12.1 Introduction

We often alter virgin polymers by adding stabilizers, colorants, fillers, and other functional additives in order to improve their properties for a specific application. This process, known as compounding, requires us to mix additives uniformly throughout the polymer. This deceptively simple sounding task often poses quite a challenge to the compounder. The success of the process relies on both the even distribution and dispersion of the additives into the resin. Since polymers are high molecular weight, viscous melts, they resist the incorporation of low molecular weight, low viscosity additives. This means that compounding often requires that we impose a great deal of shear to achieve the necessary mixing. Unfortunately, high shear tends to degrade polymers, so compounders must perform a balancing act between distributing and dispersing the additives and maintaining the polymer's integrity.

Before the polymer and additive can be mixed together, the compounder must first introduce the components into the mixing equipment. Since additives come in many different forms, either solid or liquid, and different shapes and sizes, the means of their introduction must be matched to the material being added.

After compounding, the mixture has to be formed into usable-sized pellets, suitable for further processing. The pellets must be uniform in size and smooth so that they are easy to pour and handle. Compounded pellets are typically approximately the size of a grain of wheat with shapes ranging from spherical to cylindrical.

## 12.2 The Goal of Compounding

When we mix an additive into a polymer, our aim is to obtain a homogeneous mixture. Two things must happen to realize this: dispersive and distributive mixing as shown in Fig. 12.1. Dispersive mixing separates the additive from itself, allowing it to interact with the polymer matrix. This is accomplished by applying a shear gradient across the undispersed material. Distributive mixing then incorporates the dispersed additive evenly throughout the melt. To do this, the regions of high additive concentration need to be folded into those regions that are additive deficient. This process is analogous to kneading dough when making bread. The trick in compounding is to achieve both dispersion and distribution without damaging the additive or the polymer to the point that the final product does not exhibit the properties necessary for its intended use.

The ease with which dispersion and distribution occur depends on both the form of the additive (whether it is a liquid or a solid) and the viscosity of the resin. In general, for additives

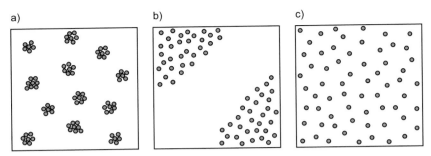

**Figure 12.1**   Pictorial representation of various types of mixing:
a) distributive, b) dispersive and c) combined distributive and dispersive

that are liquid at compounding temperatures, the more closely the viscosity of the additive matches that of the molten polymer, the more readily the materials will mix. In the case of solid additives, their cohesive strength, particle size, moisture content, and surface energy all help determine the ease with which incorporation occurs. Highly cohesive powders are very difficult to separate and therefore require high shear. Very fine powders also cause difficulties because of their high surface area. To achieve good mixing, the individual particles must be fully wetted by the polymer to achieve dispersion. Finally, the general statement that "like dissolves like" is applicable in polymer compounding. We need to expand it to "like dissolves or disperses like". Additives whose surface energy match that of the polymer will blend much more easily than those that do not. For this reason, many polar minerals such as calcium carbonate and titanium dioxide are coated with amphiphilic materials to make their surface chemistry more closely match that of the resin to which they will be added.

## 12.2.1   Types of Compounds

Compounders produce one of two types of mixtures: fully compounded resins or concentrates. A fully compounded resin is the complete formula, with all additives at the required concentrations, for the final application. For example, a compounder would take the recipe from the customer who makes black polyethylene trash bags. From this recipe, they would manufacture a compound by adding the appropriate amounts of carbon black, slip agents, and antistatic agents to polyethylene in their mixing equipment. Their customer would then feed this fully compounded resin directly to their film blowing equipment, without adding any other ingredients. For the film manufacturer this means that only one ingredient is needed, reducing the complexity of their process but also limiting their flexibility. A concentrate is just what its name says: it is a concentrated formulation of the necessary additives, which is diluted in resin at the converter's manufacturing site. Taking the same trash bag example as above, stabilized polyethylene from the resin supplier would be mixed with a carbon black color concentrate, which might also contain the slip agents and antistatic agents. The film producer would feed the concentrate and as-received resin into their film blowing equipment. Though this increases the complexity of the process overall for the film producer, it usually reduces their manufacturing costs because compound will always cost more than the virgin resin. It also allows the film processor to alter their recipe during production to address any processing related problems.

### 12.2.2  Types of Additives

As discussed in Chapter 10, a wide variety of additives is used in the polymer industry. Stabilizers, waxes, and processing aids reduce degradation of the polymer during processing and use. Dyes and pigments provide the many hues that we observe in synthetic fabrics and molded articles, such as household containers and toys. Functional additives, such as glass fibers, carbon black, and metakaolins can improve dimensional stability, modulus, conductivity, or electrical resistivity of the polymer. Fillers can reduce the cost of the final part by replacing expensive resins with inexpensive materials such as wood flour and calcium carbonate. The additives chosen will depend on the properties desired.

## 12.3    Compounding Equipment

In order to compound material we first must introduce the materials to be compounded into the mixing equipment. This is accomplished through feed systems. The type of feeder, the timing of addition, and its location in the system will depend on the resin and the additives. Once the materials are in the mixing equipment, they are homogenized to produce the final compound. Finally, the material has to be shaped into a useable form. This typically occurs by passing the material through a die which pelletizes the homogenized melt into pellets about the size of a grain of wheat.

### 12.3.1  Additive Feeders

A feeder introduces a material to a mixing system at the appropriate time and location to achieve good mixing. It consists of a hopper, which is a large container that holds the material until it is used, and the conveying equipment, which actually introduces the material to the mixing equipment, as shown in Fig. 12.2. Usually, the additive conveying equipment is coupled, via computer systems, with the resin feeders' conveyors so that the ratio of additive to resin is constant. There are several types of feeders, which are chosen to maximize the control with which the additive is introduced. For liquids, we typically use a volumetric metering pump. The pump feeds the liquid into the polymer melt at a preset volume per pound of resin. Solids are typically added gravimetrically. There are feeders that allow the solids to pour into the mixing equipment under gravity. Others introduce the additives into the mixing equipment by forcing the material into the melt in a region of the system under high pressure. This last method is typically used for cohesive solids or liquids that are difficult to disperse.

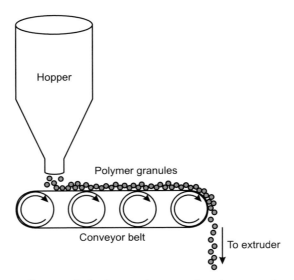

**Figure 12.2**   Schematic diagram of a feed system incorporating a hopper and a conveyor belt

When selecting a feeder, there are several considerations to keep in mind. For liquid additives, pumps must be able to consistently deliver the liquid against its own resistance to flow. Highly viscous liquids can present serious challenges when introducing them into a melt under high pressures. Additionally, some liquids can react with moisture in the air or degrade at high temperature, requiring hoppers that are cooled and fully enclosed or purged with an inert gas. For solid additives, it is important to keep the feeder at a temperature below the melting point of the solid. This is particularly important when they are low melting point waxes. Since the temperature above an extruder may exceed 50 °C, hoppers often are equipped with cooling systems to moderate their temperature. Finely divided powders are the bane of the compounding industry. Handling them creates copious amounts of dust, which can permeate the factory, contaminating other processes if proper care is not taken. They must be stored in specially designed hoppers with very steep slopes on the sides to prevent the material from "bridging" or "rat holing". Bridging occurs when the overlying material sticks together and hangs up in the neck of the hopper after the underlying material has been used, as shown in Fig. 12.3. Rat holing occurs when a slug of material falls out leaving a hole. Both processes create surging of the additive since there is a deficiency when the material is bridged and then a surge of the additive when (and if) the bridge collapses. To address this, some powder hoppers incorporate vibratory shakers to prevent the bridges from forming.

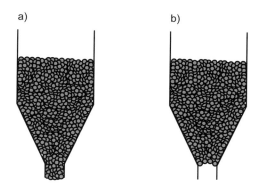

**Figure 12.3**    Schematic diagram showing a cross-section through a feed hopper:
a) full hopper and b) bridge preventing material feed

## 12.3.2    Types of Mixing Systems

There are three main types of mixing systems: batch mixers, high shear continuous mixers, and extruders. Regardless of the system, the goal is always the same; to achieve both distributive and dispersive mixing. Our choice of a mixing system depends on the resin and the additives. Batch mixers have been used successfully for at least 50 years for manufacturing small volumes of specialty materials such as pigment concentrates. The advantages of a batch mixer include highly controllable mixing for small samples and considerable flexibility. The main disadvantage, though, is that the amount of material that can be processed in each run is small. This makes this method uneconomic for large-scale production. Continuous mixers, on the other hand, have the potential for enormous output. For some systems, a single large mixer can process 10,000 to 15,000 kilograms of compound per hour. This makes them a poor choice for low volume usage specialty materials, but a perfect match for material-intensive processes. A compromise between the two methods is the single or twin screw extruder. A typical twin screw extruder can produce 1,000 to 2,000 kilograms of compound per hour while still being able to run small batches for specialty materials. Much larger extruders can process material at the same rates as continuous mixers, but these are less commonly used by specialty compounders. Such large extruders may be found in resin manufacturing plants where they receive dry polymer powder directly from the reactor into which they incorporate stabilizers prior to pelletization.

### 12.3.2.1  Batch Mixers

Batch mixers typically consist of a large chamber heated by oil or electricity, containing intermeshing mixing blades that homogenize the resin with the additives. A much simplified schematic of a batch mixer is shown in Fig. 12.4. A weighed amount of polymer pellets drops into the chamber and is mixed and heated until molten. Once the polymer melts, the required quantities of additives are introduced. The material then mixes until the additives are fully

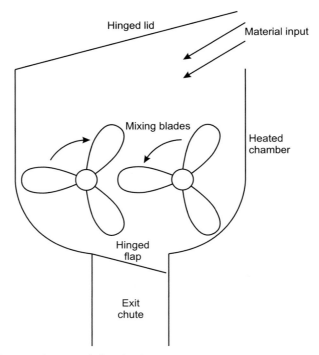

**Figure 12.4**    Schematic diagram of a batch mixer

dispersed and distributed into the resin. The mixture is dropped out of the bottom of the mixer and transferred into a pelletizing system. A common pelletizing system comprises a single or twin screw extruder coupled to a pelletizing die. The extrusion step not only pelletizes the mixture but also removes air and water trapped during the mixing process.

### 12.3.2.2  Continuous Mixers

A typical continuous mixer has two parts. The first is a high shear mixer into which resin and additives are fed. The second is a single screw extruder into which the compound from the mixer drops while still molten. A diagram of a typical design is shown in Fig. 12.5. The mixer consists of a long barrel with an hour glass profile, inside which sit two mixing screws, such as those shown in Fig. 12.6. These screws do not intermesh, creating large volume mixing regions where the polymer can be beaten by paddles and blades. Initially, the resin is plasticated by both shearing forces and heat at the feed throat. After plastication, the additives are introduced. The blades of the mixer push the material from the feed throat to the outlet of the mixer but the actual path that the resin takes is quite tortuous. As the melt interacts with the blades and mixing elements, it is pushed both backwards and forwards by mixing elements designed to redirect the flow. This action creates excellent mixing of the additives into the resin. At the outlet, the melt is dropped into a single or twin screw extruder where it

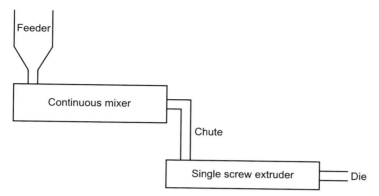

**Figure 12.5**   Schematic diagram showing a continuous mixer feeding a single screw extruder

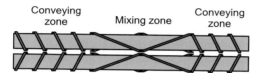

**Figure 12.6**   Example of mixing screws in a continuous mixer

is further homogenized and devolatilized. From there it passes into a pelletizing die to create small compound pellets.

### 12.3.2.3  Single Screw Extruders

Single screw extruders can be used for compounding, though they are usually reserved for compounds that are relatively easy to mix. To achieve the necessary mixing, several specialized mixing elements are added to a standard extrusion screw. A representative compounding single screw design incorporating mixing elements is shown in Fig. 12.7. In general, the polymer enters the screw at the feed throat and is heated until molten. The melt passes to the compression zone where the additives are introduced. The additives are incorporated by passing through mixing elements. The mixing elements of the screw are designed to impose high shear forces while splitting and recombining the melt stream repeatedly. Examples of mixing elements for a single screw extruder can be seen in Fig. 12.8. The combination of the shearing and the splitting and recombining of the melt creates both distributive and dispersive mixing. After mixing, the material enters a metering zone which creates an even flow of material to the die. It is common to remove excess gas from the molten resin via a vent port in the barrel wall in the metering zone. The vent port, either at atmospheric pressure or under a vacuum, allows volatiles to escape the melt, preventing the formation of voids in the compound. Often, after venting, the melt can pass through two or more sets of compression, mixing, and metering zones to further homogenize it. We refer to extrusion processes with multiple mixing regions as multistage extrusion.

**Figure 12.7**    Schematic diagram of a single screw incorporating two mixing elements

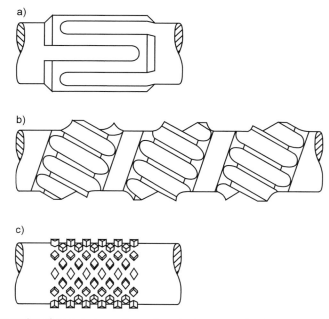

**Figure 12.8**    Examples of mixing elements used on single screw extruders:
a) Maddock mixer, b) Saxton mixing section and c) pineapple mixing section

### 12.3.2.4  Twin Screw Extruders

A twin screw extruder consists of a barrel with a figure of eight profile encasing two intermeshing screws that rotate in either the same direction (co-rotating) or in opposite directions (counter rotating) to achieve material conveyance and mixing. Figure 12.9 shows schematics of both configurations. Co-rotating extruders have screws with the same pitch and thread, while counter rotating screws have opposite threads. Twin screw extruders mix resins more effectively than single screw extruders. Better mixing arises from the high shear introduced by the movement of the screw against the barrel and by the movement of the screws relative to one another. Twin screw extruders are most often equipped with co-rotating screws. Counter rotating extruders, though less common, can be effective mixers if designed correctly.

During extrusion, the blend components are fed into the feed throat of the extruder. Compounding extruders are starve fed. This means that material is conveyed away from the feed throat more rapidly than the raw materials are fed into the barrel. Therefore the volume of raw material introduced is always less than the total available volume of the flights

a)

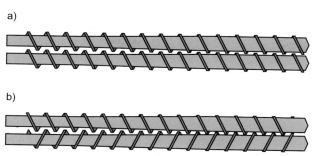

b)

**Figure 12.9**   Twin screw extruder configurations:
a) co-rotating screws and b) counter-rotating screws

of the screws. In the feed section, the motion of the screw against the barrel walls heats the polymer while conveying it towards the higher shear sections of the screw. The first of these is a compression zone. In the compression zone, the polymer is compacted by entering narrow clearance screw elements, which force the entire resin to melt. Once the resin is molten, the remaining additives are forced into the melt. Typically, the additives added at this point are those that would interfere with the melting process. For example, liquids would lubricate the surface of polymer pellets, reducing their interaction with the barrel wall and lessening shear heating. Additionally, solid powders that are very cohesive are typically injected directly into the melt under pressure, to promote the dispersion process. From the metering zone the melt passes to a mixing zone. In the mixing zone, intermeshing screw elements with very tight clearances and tortuous pathways shear the polymer melt and fold it into itself as it passes through, as seen in Fig. 12.10. Often, there are several different mixing regions. Downstream

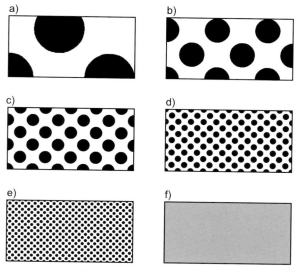

**Figure 12.10**   Representation of melt mixing in the mixing zone channels of an extrusion screw

of the mixing regions, you typically find either atmospheric or vacuum assisted vents. These allow the removal of entrapped moisture, air, and degradation products from the melt. After mixing is accomplished, the melt is passed to the die via deeper grooved channels that pump a steady melt stream to the die.

### 12.3.3    Melt Filtration

Before the melt leaves a continuous mixing system or extruder it is often filtered to remove contaminants. Filtration can be achieved by placing stainless steel wire mesh screens immediately before the die, through which the melt must pass. Typically, several screens with different mesh counts are stacked to form a screen pack. Screens with large openings and thicker wires support screens with smaller openings. Screen packs capture oversized materials, gels of degraded polymer, unmelted polymer (if there is any), and foreign contaminants, including flakes of metal that have worn off the processing equipment and paper fibers from the bags in which the additives were transported. The finer the mesh of the screens, the more resistance they offer to the flow of the compound. Since the screens impede the flow of the melt, pressure builds up behind the screen pack, which increases mixing efficiency. This effect is exacerbated when the screens become plugged with the contaminants. To prevent undue pressure build-up, screen packs are replaced when pressure within the extruder exceeds a given threshold.

### 12.3.4    Pelletization

The final step in compounding involves taking the molten mixture of resin and additives, now called compound, and forming it into pellets. These pellets are typically conveyed to the converting equipment by pneumatic systems. To ensure that the material feeds into the converting equipment consistently, compounded pellets must be uniform in size, flow easily, and preferably have a smooth, slick surface.

To achieve a compound that meets these feeding requirements, compounders form their product into small pellets. To obtain these small pellets, they use specialized dies consisting of a metal plate with many small holes through which the molten compound is forced. This part of the process is very similar to how spaghetti is formed from pasta dough. To generate small pellets, the extruded strands are chopped into small pieces by rotating blades.

There are two means by which we can make pellets. The first, referred to as strand pelletization, produces pellets by chopping cooled strands of the extrudate with rotating cutting blades as shown in Fig. 12.11. The extrudate is cooled by immersing it in a water-filled trough. The solidified strands are dried with forced air and then chopped into pellets. Another option, often used for high output continuous systems, utilizes pelletizing dies as shown in Fig. 12.12. In these systems, the extrudate is forced out of the die and immediately cut by pelletizing knives which rotate flush against the die face. There are two major variants of these systems. The first cuts the extrudate in air and then flings the pellets into water to cool them quickly.

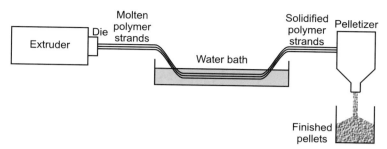

**Figure 12.11**   Strand pelletization system

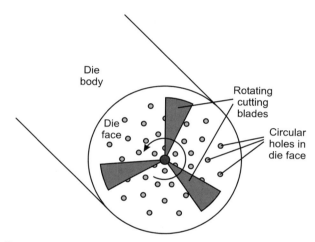

**Figure 12.12**   Pelletizing die

In the second system, the die face is immersed in a chilled water bath. A strong current of water carries the pellets away from the die as soon as they are cut by the blades.

Pellets produced by each system have characteristics shapes. Pellets created with strand pelletization are typically short cylinders with well defined edges since they were cut when in the solid state. Underwater pelletization yields pellets that are more rounded.

## 12.4    Additional Considerations

In this chapter we have summarized a very complex subject in a few pages. There are several additional considerations that must be taken into account when making compounds. In addition to considering the resin's melt behavior, the feeding requirements of the additives, the required heat profiles, processing speeds and the pelletizing process, compounders need to account for the impact of the resin and the additives on their equipment. For example, certain resins such as polyvinyl chloride and fluoropolymers generate acidic gases that are extremely corrosive. When processing these materials it is necessary to use hardened or chromium plated metals that are less susceptible to corrosive attack. Other polymers are extremely moisture-sensitive. Polyesters are notoriously sensitive to water and will depolymerize if the moisture in the melt exceeds 0.005%. For these polymers, compounders use intensive drying equipment that warms and desiccates the polymer prior to processing. Finally, most mineral additives are extremely taxing on processing equipment. Calcium carbonate and titanium dioxide are both very hard minerals making them exceptional abrasives. They can scour a barrel thereby reducing the efficiency of the compounding system. Using hardened steel or chromium or nickel plated surfaces in contact with the melt will minimize these effects.

## 12.5    Conclusions

The process of compounding additives into resins requires that the equipment be able to both disperse and distribute the additives into the resin system. These processes can be achieved by shearing and mixing the resin with the additives through mixing processing equipment. Typically, compounding of large quantities of masterbatch or concentrate is accomplished through twin-screw extrusion or continuous mixers. Smaller quantities may be made with batch mixers. The final compound is typically supplied in the form of small pellets to converters who process the compound into useful items via various molding operations, which we shall explore in the next three chapters.

## Review Questions for Chapter 12

1.  Explain the differences between dispersive and distributive mixing. Why does dispersive mixing rely on shearing processes, while distributive mixing relies on iteratively folding the polymer melt back on itself?

2.  Calcium carbonate used as an additive in polymeric systems, is a hydrophilic, finely divided powder. When used as a component in polyolefins, the calcium carbonate's surface is modified by coating it with a thin layer of stearic acid. Why would compounders prefer to use the stearic acid coated material over the uncoated mineral?

3.  Give three reasons that a manufacturer would choose to use a concentrate rather than a fully compounded material?

4.  What type of feeding system is required to incorporate a low viscosity fluid into a high molecular weight polymer? What issues might arise during the compounding process due to the form of the additive?

5.  Create a table listing two or more advantages and disadvantages of each of the following compounding systems: batch mixers, continuous mixers, single screw extruders and twin screw extruders.

6.  What is the primary mechanism by which a mixing element in a single screw extruder creates mixing?

7.  Why is most compounding of powder additives with polyolefins achieved via either continuous mixers or twin screw extruders?

8.  Why is melt filtration necessary and how is it accomplished? What would an operator observe if a melt filter were to rupture?

9.  Why does the shape of compounded pellets differ between those created by underwater versus strand pelletization systems?

10. A manufacturer observed that the compound produced by one twin-screw extruder exhibited much better dispersion than that produced in a similar extruder making the same compound. The only difference between the two extruders was the age of the screws and barrel. What could be causing the observed difference?

# 13    Injection Molding

## 13.1    Introduction

Injection molding is a common process that we use to convert polymer granules to solid objects. Unlike the products made by continuous extrusion processes, discussed in Chapter 11, products made by injection molding are discrete objects, produced in individual mold cavities. We encounter injection molded products of all sorts in our daily lives, ranging from combs, bottle caps, and ballpoint pens to car steering wheels, camera bodies, and the keys on our computers.

The basic concept of injection molding is quite simple; we inject molten polymer into a cavity mold, which defines the shape of the object that we wish to make. After allowing sufficient time for the polymer to cool to its solid state, we open the mold and remove the product. In practice, the process is much more complex. There are many variants on the basic process, such as injecting more than one polymer into the same mold, creating products with hollow cores, and encapsulating non-polymeric items. Products made by these techniques include tooth brush handles, automotive air intake ducts, and screwdriver handles. Injection molded items range in size from tiny gears used in miniature electronic items, weighing less than 0.01 g, to railroad ties made from recycled polymer that can weigh over 100 kg.

## 13.2    Principles of Injection Molding

Figures 13.1 illustrates the basic principles of injection molding. The hardware consists of two main components, an extruder and a mold. The extruder screw executes two types of motion; it rotates around its long axis and reciprocates along its length. The extruder screw rotates to melt and homogenize the polymer, as it does during profile extrusion. As the screw rotates, it moves backwards away from the mold, building up a reservoir of molten polymer ahead of it. Melt in the reservoir is injected into the mold when the screw is thrust forward. During injection, the mold is held closed by a sturdy clamping mechanism. The mold opens in order to remove the product then closes again, ready for the cycle to start again.

a)

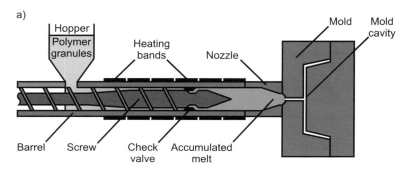

b)

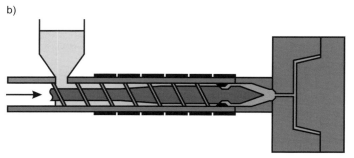

c)

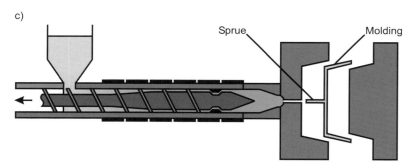

**Figure 13.1**   Schematic diagram of injection molding:
a) mold closed and melt accumulated, b) injection and packing, c) ejection

## 13.2.1   The Molding Cycle

We can explain the injection molding process with reference to the molding cycle illustrated in Fig. 13.2. The cycle consists of four stages: injection, packing, cooling, and ejection.

The injection stage starts at the twelve o'clock position in Fig. 13.2; the mold is closed at this point and a reservoir of molten polymer is accumulated ahead of the screw, as shown in Fig. 13.1 a). Molten polymer is injected into the mold when the screw is thrust forward. A check valve prevents polymer from being forced back along the flights of the screw. As the molten polymer flows into the mold, the portion that touches the wall is rapidly quenched to form a solid layer, as shown in Fig. 13.3. Molten polymer continues to flow between the walls of solidified polymer. At the advancing flow front, the molten polymer adopts a "fountain flow" pattern. If we inject too little polymer into the mold, we get what is known as a "short shot", where the mold is not completely filled. If excess polymer is injected into the mold, it can force its way into the joints between the various parts of the mold, resulting in thin fins of polymer attached to the product, which are known as "flash". At the end of the injection stage, the equipment is in the configuration shown in Fig. 13.1 b).

Once the polymer has filled the mold, the rapid forward motion of the screw ceases, but the thrust force is sustained to apply packing pressure to the polymer. During the packing stage, a small volume of polymer is forced into the mold to compensate for the increase in density of the polymer as it solidifies. If we did not maintain pressure, the polymer would shrink from the walls of the mold as it cooled. This effect would be particularly noticeable in thicker parts of the molding, where depressions known as "sinkmarks" would form. Plastic laundry baskets and crates routinely exhibit sinkmarks. By maintaining pressure, we insure optimum surface characteristics and reduce warpage in the final product that could be caused by differential shrinkage.

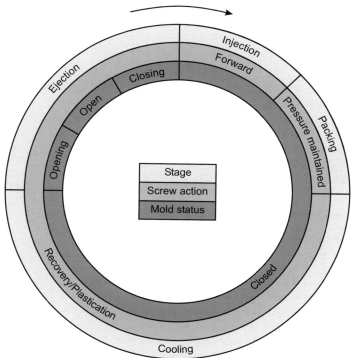

**Figure 13.2**  Simplified injection molding cycle

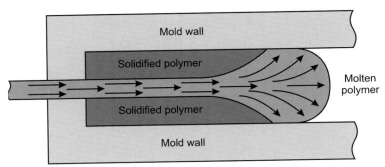

**Figure 13.3**    Cross-section through mold illustrating fountain flow during polymer injection

Cooling is the longest stage of the molding cycle. During this stage, the polymer continues to solidify as its heat is absorbed by the mold. Moldings with thick cross-sections take disproportionately longer times to cool than thinner ones, due to the insulating effect of the polymer. While cooling is taking place, screw rotation restarts, plasticating the next injection charge. As the screw rotates, it gradually moves backwards to accommodate the charge of molten polymer accumulating ahead of its tip. The screw's back travel distance controls the injection shot size.

Once the polymer has cooled to its solid state, the molding is ejected. This is accomplished with the aid of "ejector pins" that protrude from the mold walls as it opens. Small items typically drop directly into a catch pan or onto a conveyor belt below the mold. Larger items are removed manually.

The duration of the molding cycle is determined primarily by the dimensions of the molding. Smaller molds generally fill faster than large ones and the polymer cools faster, due to the molding's larger surface to volume ratio. Cycle times vary from less than a minute for small moldings to more than half an hour for extremely large products. Molding conditions and mold designs are carefully developed to minimize the cycle time in order to maximize productivity. All modern injection molding machines are computer controlled.

### 13.2.2    Mold Considerations

An injection mold performs several functions during the molding cycle. In addition to giving the polymer the desired shape, it distributes the molten polymer, cools it, and ejects the product.

Between the injection nozzle on the end of the extruder and the gate that leads into the mold cavity the polymer flows through a channel known as a "runner". It is vital that the polymer does not solidify in the runner before the mold is completely filled. We can prevent premature solidification in the runner in one of two ways: we can use a large diameter unheated (cold) runner in which the polymer solidifies after the polymer in the mold cavity, or we can use a heated (hot) runner in which the polymer does not solidify. Moldings that are produced with a cold runner system are ejected with an attached sprue, as shown in Fig. 13.1 c), which must be removed in a separate step. We typically grind sprues into granules and feed them back into the hopper. Hot runner systems are significantly more complex and costly than cold runner systems. However, they are generally preferred, because they eliminate the cost of removing and recycling sprues.

When we injection mold small plastic items, it is standard practice to mold several identical parts simultaneously in a mold containing multiple cavities. In order to insure product uniformity, we must design runners that distribute the molten polymer evenly to each cavity.

The location and dimensions of the gate through which the molten polymer enters the mold cavity are crucial to insure rapid and uniform filling. In circular items, such as margarine tub lids, the gate is located at the center of the cavity. For more complex shapes the placement of the gate requires greater consideration. Larger moldings often require two or more gates. When the flows from these gates meet, the flow fronts must be fully molten or they will not weld together properly. The weld line between flows from different gates is often a weak point and thus the relative flows must be adjusted so that this weakness does not lie in an area where maximum mechanical strength is required. When molding large flat objects, we commonly use an elongated gate that spans the width of the mold cavity. By adopting this configuration, weld lines can be completely eliminated.

The flow properties of the polymer must be matched to the mold (or *vice versa*) in order to achieve optimum product quality. In particular, we must manage the pressure drop between the gate and the furthest extent of the flow path. Articles with thin walls, such as margarine tubs, require low viscosity resins. In other cases, the situation is more complex and the cavity may need to be designed so that the product is thicker near the gate than it is at its extremities. The development of complex products is aided by computer modeling of melt flow within the mold cavity.

When an injection mold is completely filled with polymer, it is subjected to an hydraulic opening force proportional to the injection pressure multiplied by the area of the cavity. Injection molding machines are rated in terms of their clamping capacity, which resists the opening force exerted by the molten polymer. Small machines have clamping capacities of a few tons. At the other end of the scale, the largest injection molding machines can exert clamping forces of several thousand tons. Clamping forces can be generated hydraulically, electrically, or by a combination of the two.

The simplest molds consist of two parts: one fixed and one moveable. We use such molds to fabricate items, such as disposable stadium cups, compact disks, and coat hangers, which have simple designs with no undercuts. As the complexity of the product increases, so must the complexity of the mold. The molds used to produce large items, such as dishwasher tubs and automobile bumpers, may consist of multiple moveable components and weigh several tons. Even apparently simple items, such as screw tops for bottles, require a complex ejection system that unscrews the product from part of the mold. In order to facilitate cooling, virtually all molds are designed with liquid cooling channels.

## 13.3    **Products**

We will illustrate the wide range of injection molding products and process variations by discussing a few key examples.

We use co-injection molding to produce items consisting of a core of one polymer covered by a skin of another. This process requires two extruders that inject sequentially through a common gate. The first extruder partially fills the mold with the polymer that will become the skin. The second polymer is then injected through the same gate, displacing the molten core of the first polymer and pushing it ahead, to complete the formation of the skin by the process of fountain flow. We can use co-injection for various reasons; the polymer required for the skin may be very expensive or contain expensive pigments or other additives that we would not wish to waste in the core. Alternatively, we may require a high gloss surface from a glassy amorphous polymer with a semicrystalline polymer core to give the product toughness.

Foam core injection molding is a variant of co-injection molding. In this case, the second polymer foams as it is injected into the mold, either due to a drop in pressure that permits a volatile blowing agent to vaporize or by a chemical reaction that generates a gas. The net result is a foamed polymer core with a solid polymer skin. We use this process primarily to reduce product weight without compromising structural integrity. It is important to allow foam cored products to cool thoroughly before removing them from the mold. If we were to remove the molding prematurely, it might collapse under its own weight, shrink unevenly, or warp. Examples of foamed core products include surf boards and television set housings.

We use overmolding to apply a layer of one polymer to the surface of another. In this process, we mold the core of the item in one mold and then transfer it to a second mold into which a second polymer is injected to form a cover over the first. The covering can be complete, as in the case of golf balls, or partial, as in the case of comfort grips on ballpoint pens, tooth brushes, and power tools. In the case of golf balls, the core is supported within the center of the secondary mold by a series of pins that retract just prior to the end of the injection stage. Overmolding is becoming increasingly popular as a way to provide a soft and comfortable covering over a structural interior.

When a molding is required to have greater strength than can be achieved with a single polymer, we can reinforce it with fibers of glass or other more exotic materials. This can be achieved in two ways:

- We can blend short lengths of fiber into the polymer in the extruder prior to injection or
- We can feed continuous fibers into the injection nozzle along with the molten polymer.

Fiber reinforcement of polymers allows us to use them in engineering applications that were previously off limits. We use fiber reinforced injection moldings in many under-the-hood automotive applications, such as rocker box covers and front end modules, which support the radiator, cooling fan, and hood latch, that are subjected to a wide range of temperatures.

Traditionally, it was not possible to produce hollow parts by injection molding. This has changed with the invention of gas- and water-assisted injection molding. During these processes, we inject molten polymer into a mold, which solidifies against the walls. Before the core has time to solidify, we inject water or gas, which displaces the molten core through a valve at the end of the mold. In the case of water-assisted injection molding, the water polishes the interior of the molding, giving it a smooth surface finish suitable for fluid transport. The majority of new cars use air intake manifolds made by these processes.

## 13.4    Conclusion

Injection molding is one of the most widespread methods used to make polymer products. A cursory look around our homes, schools, work places or cars suffices to find an abundance of injection molded products. One of the reasons that this conversion method is so popular is that it enables us to create complex shapes in a single fabrication process.

Virtually all thermoplastic polymers can be processed by injection molding. The net result is that we can mold products of almost any shape with properties that range from flexible to rigid, for applications ranging from O-rings to motorcycle helmets, and from gear wheels in electronic equipment to outdoor furniture. As technology advances, we will see injection molded products used in an even greater range of applications.

## Review Questions for Chapter 13

1. Explain how each of the following machine control parameters can affect flash formation in injection molding:
   a) melt temperature
   b) packing force
   c) clamp pressure
   d) mold fill rate

2. Injection mold designers avoid long, narrow runners in their mold designs. Why?

3. Polyethylene is used in extrusion and injection molding processes. When used to manufacture blown films a standard polyethylene will have a melt index of approximately 1 g/10 min, while that used for injection molding typically has a melt index of 5 to 20 g/10 min. Why are the polymers used for these two processes so different?

4. What advantages are there to using a hot runner mold rather than one that has cold runners?

5.    What product defects would an injection molding operator observe if the following
      parameters were set incorrectly?
      a) cooling time in the mold
      b) mold temperature

6.    What is overmolding? How is it done?

# 14    Blow Molding

## 14.1    Introduction

We use blow molding to make plastic bottles, jars, and other containers. The resulting products are lightweight and tough with a large volume-to-weight ratio, which makes them suitable for numerous packaging and storage applications. A quick look around our kitchens, bathrooms, and garages reveals blow molded products of many types. Shampoo, liquid soap, cooking oil, peanut butter, and automotive fluids are packaged almost exclusively in blow molded containers. Other applications include automotive coolant overflow bottles, fuel tanks, and agricultural chemical storage tanks. Products range from simple radially symmetric bottles to complex items incorporating handles or with sculpted surfaces. Small bottles can weigh as little as a few grams, while industrial storage tanks can weigh over 250 kg and hold several thousand liters of liquid.

Blow molding is basically a two-step process, with two principal variants: extrusion blow molding and injection blow molding. During extrusion blow molding, we extrude a molten tube of polymer that we inflate within a hollow mold. Products made by this process include high density polyethylene milk jugs and linear low density polyethylene squeeze bottles. In the first step of injection blow molding, we injection mold a "preform", which looks like a thick walled test tube attached to the threaded neck of a bottle. The body of the preform is heated above its softening point then placed in a blow mold where it is inflated. Common products of the injection blow molding process include polyethylene terephthalate carbonated drink bottles and other clear bottles and jars.

## 14.2    Principles of Blow Molding

During blow molding we take advantage of the inherent melt strength and high viscosity of certain types of polymers – primarily polyethylene and polyethylene terephthalate. Given the appropriate average molecular weight and molecular weight distribution, we can extrude molten tubes or injection mold preform shapes that can be inflated to create thin-walled objects, much as we might blow bubbles from bubble gum.

### 14.2.1    Extrusion Blow Molding

Figure 14.1 illustrates the general principles of extrusion blow molding. A molten tube of polymer, known as a parison, is extruded vertically downwards from an extrusion die, as shown in Fig. 14.1 a). The two halves of the blow mold surround and then close on the

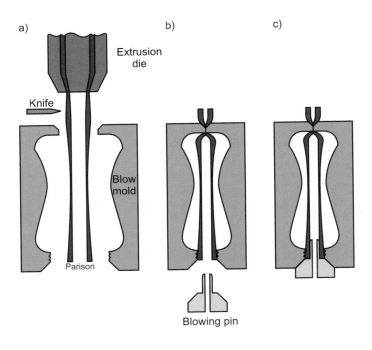

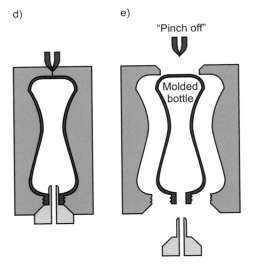

**Figure 14.1**  Schematic diagram of extrusion blow molding:
a)  mold open, surrounding parison,
b)  mold closes around parison,
c)  insertion of blowing pin,
d)  inflation and cooling,
e)  ejection

parison, pinching it closed near the top and leaving the bottom open within the neck of the mold, as shown in Fig. 14.1 b). A knife mounted just below the die severs the parison after it is captured by the mold. A blowing pin is inserted into the open end of the parison, as shown in Fig. 14.1 c), sealing off the opening. When gas is injected through the blowing pin, it inflates the still molten parison, forcing it against the interior, as shown in Fig. 14.1 d). After allowing sufficient time for the polymer to solidify, the mold opens and the bottle is ejected, as shown in Fig. 14.1 e). The "pinch off" is removed from the bottle and either scrapped or – preferably – ground up and recycled to the extruder. The pinch-off line can clearly be seen on the bottom of products such as opaque shampoo or automotive oil bottles.

The extrusion blow molding cycle is illustrated in Fig. 14.2. The extrusion component of the cycle is normally continuous. As soon as one length of parison has been captured by the mold, another length starts to form. To allow room for a new length of parison to emerge from the die, the mold moves aside as soon it has captured a parison and the knife has severed it. The mold is rapidly translated to a remote blowing station where inflation takes place. After the product is ejected, the open mold moves back under the die where it surrounds and captures another length of parison.

We can generally extrude a parison much faster than we can inflate, cool, and eject the product. When this is the case, we employ more than one mold. If we are using two molds, they shuttle back and forth alternately between their individual blowing stations and the parison capture

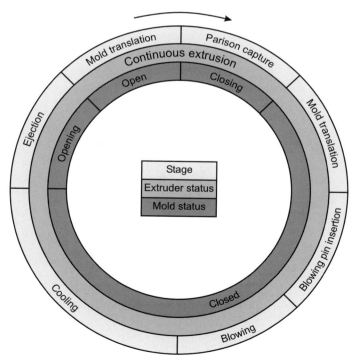

**Figure 14.2**   Simplified extrusion blow molding cycle

location below the die. To further increase the output rate, we can simultaneously extrude two or more parisons, each of which is captured at the same time within a multicavity mold. Another strategy for increasing output is to mount a series of molds on the perimeter of a wheel. As the wheel rotates, each mold in turn captures a length of parison and passes sequentially through each of the stages. When we wish to manufacture bottles with multilayered walls, we feed a concentric co-extrusion die from two or more extruders.

In order to produce bottles with uniform wall thicknesses, we commonly manipulate the wall thickness of the parison, making it thicker in some places than others, as shown in Fig. 14.1 a). The thicker regions correspond to the larger product diameters. When the parison is inflated, the thick regions stretch further, ultimately attaining the same thickness as the other parts of the product. We achieve a programmed parison wall thickness by raising or lowering the mandrel in the center of the extrusion die, to increase or decrease the die gap as the parison is being extruded. When programming a parison we must take into account die swell and the parison sagging under its own weight.

When making large products, such as 55 gallon drums, the parison tends to sag significantly under its own weight. In such cases, if we were to extrude the parison continuously it would take so long that it would sag excessively and ultimately tear off. To avoid this complication when making large products, we typically use an accumulator arrangement that stores molten polymer until there is a sufficient quantity for it to be extruded fast enough that sag is manageable. One way that we approach this situation is to use a reciprocating extruder screw, similar to the arrangement that we discussed in the chapter on injection molding. Alternatively, we can extrude molten polymer into a closed cylinder from which it is forced rapidly by a plunger.

## 14.2.2  Injection Blow Molding

We use injection blow molding to make transparent containers with precisely molded threaded necks. Figure 14.3 illustrates the principles of the process. The first step is to injection mold a preform, as shown in Fig. 14.3 a). The preform consists of a thick-walled tube with a closed bottom that is attached to a threaded neck. The blowing stage starts by heating the tubular part of the preform with radiant energy until it is soft. We transfer the softened preform to a blow mold in which it is suspended by its neck as shown in Fig. 14.3 b). Blowing and ejection take place in a similar manner to the extrusion blow molding process.

Injection blow molding produces bottles with bi-axial orientation, in contrast to the extrusion blow molding process, which yields products with primarily uni-axial orientation running around the circumference of the product. Bi-axial orientation increases product crystallinity, which imparts greater stiffness, improved barrier properties, and greater clarity; all of which are desirable in food packaging. The injection blow molding process has the added advantage over extrusion blow molding that there is no pinch-off. This virtually eliminates scrap from the system. By eliminating the pinch-off line, which can be a point of weakness, the thickness of the product can be reduced. Against this must be weighed the added cost and complexity of the process versus extrusion blow molding, which requires a single mold.

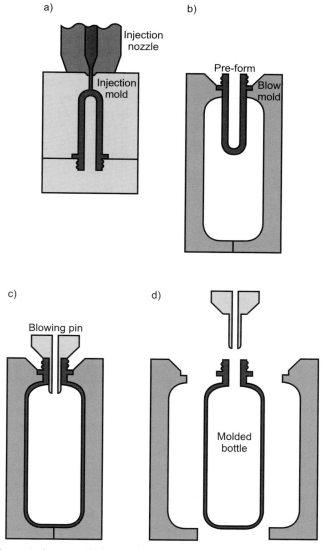

**Figure 14.3**   Schematic diagram of injection blow molding:
a)  injection molding of preform ,
b)  preform placed in blow mold,
c)  blowing and cooling,
d)  ejection

An important variant of the injection blow molding process is stretch blow molding, which is shown schematically in Fig. 14.4. In stretch blow molding, the softened preform is mechanically elongated by an extended blowing pin that stretches it to almost the full length of the blow molding cavity prior to inflation. Stretch blow molding increases longitudinal orientation and improves wall thickness uniformity. An added advantage is that we can use higher molecular weight resins with higher melt viscosities, which improves strength and barrier properties.

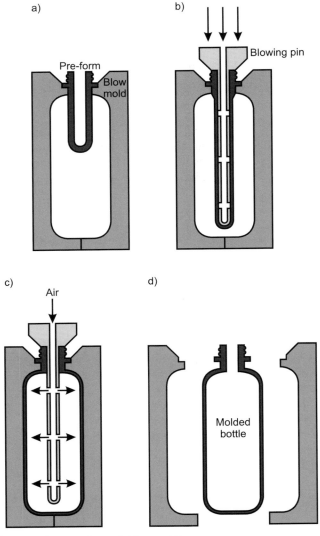

**Figure 14.4**   Schematic diagram of stretch blow molding:
a) preform placed in blow mold, b) stretching, c) blowing and cooling, d) ejection

### 14.2.3    Mold Considerations

Great care is taken that we design blow molds to avoid thin or weak regions that could result in premature failure. Molds are designed to avoid excessive draw into corners that would result in locally thin areas. For this reason, blow molded products invariably have rounded corners. Another potential source of weakness is the pinch-off line. To compensate for this fact, it is common to program the parison to produce a thickened base.

When a parison or preform is inflated, it displaces the air around it within the mold. If no provision is made to vent the mold, compression of the air around the parison or preform can raise its temperature to such an extent that it can scorch the surface of the product. To avoid this problem, we equip blow molds with vents. These can consist of slit vents at the parting line between mold halves, porous plugs of sintered metal, or small holes drilled into the cavity walls.

The blowing pressure during parison or preform inflation is typically less than 10 bar, which is more than an order of magnitude lower than the melt pressure within an injection mold. For this reason, blow molds and the clamping devices needed to close them need not be as sturdy or expensive as those used for injection molding. For short production runs or prototyping purposes we can use molds made from lightweight and easily machined materials, such as zinc or aluminum alloys. Regular production molds are commonly made of stainless steel, which does not require surface hardening.

## 14.3    Products

Polyethylene is extrusion blow molded into many common items. Larger products, or those with thinner walls, such as one gallon milk, fruit juice, or vinegar bottles, are generally molded from high molecular weight, high density polyethylene. Such commodity products are typically used for only a short time before being discarded. There is great incentive to make them from a cheap material and little incentive to make them aesthetically attractive, so they are generally unpigmented. Many of the largest blow molded products, such as agricultural storage tanks, are also made of high density polyethylene, whose strength and chemical inertness are valuable attributes. We use low density and linear low density polyethylene to make smaller bottles, which generally need not be as stiff as their higher density cousins. Products include squeeze bottles, laundry detergent bottles, and automotive fluid bottles. Such bottles are normally made opaque with titanium dioxide and are often pigmented to make them more appealing or distinctive. Children's toys are often blow molded from low density polyethylene, which is soft, lightweight, and tough.

Polyethylene terephthalate is injection blow molded to make water and carbonated drink bottles, and other liquid food packages. High stiffness, excellent clarity, and good resistance to carbon dioxide permeation are the principal attributes of these bottles. We carefully control molding conditions to promote transparency and surface gloss, which are prized for food packaging. When we examine the bottom of these bottles, we can easily see the preform's injection point.

We use isotactic polypropylene to package food products that do not require the highest clarity, such as individual servings of fruit juices or those that are hot filled with liquids, such as maple syrup, which would soften polyethylene containers. We employ stretch blow molding to improve molecular orientation and hence the strength of such containers. Isotactic polypropylene is often nucleated to increase its crystallinity and hence its strength and softening temperature. To improve its clarity we can add clarifying agents, which modify its semicrystalline morphology to eliminate large spherulites that would scatter light.

Other polymers that we blow mold include polyvinyl chloride and polycarbonate. We use the former to make cooking oil and household cleaning product bottles. The latter is used to make the large water bottles that we see in water coolers.

## 14.4    Conclusion

Blow molding is widely used to make polymeric containers of all sorts. In many markets blow molded plastic products have largely displaced such traditional materials as glass, ceramics, or metal. The inherent advantages of polymeric containers include their light weight, clarity, toughness, and versatility.

A limited range of thermoplastics accounts for the vast majority of blow molded products. Polyethylene resins are used to package many staple and low profit margin liquid foodstuffs. Liquid foods, which are packaged while still hot are typically supplied in isotactic polypropylene bottles. Where profit margins are higher and packaging must be more eye-catching, it is more common to use polyethylene terephthalate. As resin technology improves, particularly in the area of barrier properties, we can expect to see blow molded polymer packaging making further inroads into markets where glass still holds sway.

## Review Questions for Chapter 14

1.    Describe both the extrusion blow molding and injection blow molding processes. Why would one method be chosen over the other?

2.    The polymers used in blow molding must have a high melt strength. Why is this?

3.    How is orientation introduced into an injection blow molded part? Compare this to orientation in a stretch blow part?

4.    Why do blow molded products have rounded corners?

5.   Why are vents incorporated into the mold?

6.   Why are molds for blow molding generally less expensive to make than the molds used in injection molding?

# 15 Rotational Molding

## 15.1 Introduction

Rotational molding of polymers developed out of metal fabrication processes for producing hollow artillery shells. In this method, molten polymer coats the inside of a rotating mold, which is then cooled and the product removed by opening the mold. Rotational molding has a distinct advantage over injection and blow molding in that it can produce very large parts with thin walls and complex designs with no weld lines, scars from ejector pins or blemishes, and minimal molecular orientation. In this chapter, we will explore the process itself, as well as the polymers used by it and examples of manufactured products.

## 15.2 The Process

Rotational molding takes place in four distinct steps. A schematic of this process is shown in Fig. 15.1. In the first step, as shown in Fig. 15.1 a), a weighed amount of polymer is placed in a hollow mold. Secondly, the mold is transferred to a heated oven in which it rotates about one or two axes, illustrated in Fig. 15.1 b). During the rotation step, the polymer resin flows around the interior of the mold in response to gravity and at the correct temperature begins to coat the inner surface of the mold. The third step, cooling, shown in Fig. 15.1 c), takes place once the interior is fully coated. The mold is removed from the oven and is cooled with air or water. The fourth and final step, is demolding, illustrated in Fig. 15.1 d) during which the mold is opened and the final part removed.

### 15.2.1 Charging the Mold

Gravimetric hoppers feed a mold with a prescribed weight of polymer. The feed stock is either a finely divided powder or a liquid plastisol. A plastisol is a suspension of a resin powder, typically polyvinyl chloride, in a plasticizer, used to manufacture. Rotational molders use liquids and powders in their process since both flow freely. This property permits the easy addition of the materials to the mold. More importantly, they flow smoothly around the interior of the mold as it rotates. In doing so, they coat the entire surface.

Many polymers can be purchased as finely divided powders. These include polyethylene, polycarbonate, nylon, and rigid or lightly plasticized polyvinyl chloride. Highly plasticized polyvinyl chloride comes in a plastisol form. After the material is charged into the mold it is clamped shut prior to heating and rotation.

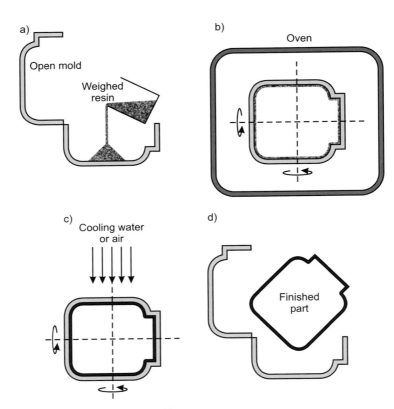

**Figure 15.1**  Four stages of rotational molding:
a)  loading mold with resin,
b)  heating and rotation,
c)  cooling and
d)  demolding

## 15.2.2    **Rotation and Heating**

The mold containing the polymer is transferred to an oven where it rotates while being heated. The rotation can occur in one of two ways. In standard rotational molding, the entire mold rotates in two directions through a 360° motion, as illustrated in Fig. 15.1 b). This process is known as multi-axial rotation; it is predominantly used to manufacture parts that do not have one dominant axis, such as storage tanks. Another method, rocking-rotational molding, rotates the mold completely around its long axis, while rocking it back and forth along a perpendicular axis. This rock-and-roll process is shown in Fig. 15.2. This method is typically used to make products with a high aspect ratio. The heat from the oven causes the polymer to sinter to the walls of the mold. Sintering is the process by which the polymer sticks to the walls of the mold, building up a layer of powder, the granules of which adhere

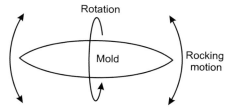

**Figure 15.2**  Rock and roll mold motion

to each other due to the melting of their surface. As heating and rotation continues the polymer granules completely melt and fuse to form a continuous layer coating the interior of the mold. The time required for this step depends on the size of the part being molded, the thermal conductivity of the polymer and the mold, the thickness of the polymer layer and the oven temperature. Once fusion is complete the mold is removed from the oven and is cooled while continuing to rotate.

### 15.2.2.1 Multilayer Molding

Sometimes, part design requires multiple polymer layers. For example, an automotive gasoline tank can consist of both nylon and crosslinked polyethylene to provide combined chemical resistance and strength. To produce multilayered parts, the first polymer, which will form the outer layer, is placed in the mold and sintered to its inner surface. The second polymer is then added to produce an interior layer. After the addition of the second polymer, the mold rotates during heating to create the next layer of the part as shown in Fig. 15.3. This can be repeated as many times as required. When choosing multiple resins, we must consider the thermal and chemical properties of each of the components. Ideally, the resins should be chemically compatible, permitting excellent adhesion between the layers. Thermal properties, such as melting temperature and thermal stability should also be similar. Finally, the degree of shrinkage of the two polymers should match closely. In situations where the shrinkage is very different, the molded part will distort considerably upon cooling.

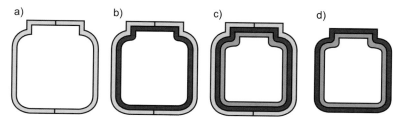

**Figure 15.3**  Multilayer molding:
    a)  mold,
    b)  mold with first layer,
    c)  mold with first and second layers and
    d)  demolded part

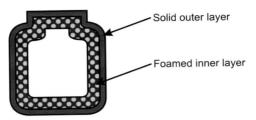

**Figure 15.4**     Rotationally molded item with foamed inner layer

## 15.2.2.2  Foamed Parts

Sometimes we wish to create products with a lower density than is available with solid polymers. To do this we incorporate additives that thermally degrade to yield inert gases that expand within the molten polymer to create a foam. Typically, the polymer used to create a foamed structure consists of two differently sized particles of the same resin. The smaller particles sinter to the inside of the mold first. The larger particles sinter to that inside layer as the mold temperature increases. These larger particles contain the blowing agent, which decomposes upon sintering, forming the pore structure. A schematic cross section of a part manufactured this way can be seen in Fig. 15.4. The outer wall has no pores, forming an impermeable skin while the inner layer comprises low density foam.

## 15.2.2.3  Crosslinked Structures

Sometimes polyethylene is crosslinked to create a final product that has enhanced thermal stability, chemical resistance, strength and toughness. The crosslinking process occurs within the mold by reaction of polyethylene with a low concentration of a radical generator. The radical generator, typically dicumyl peroxide, decomposes at elevated temperatures to create reactive radicals. Initially, the polymer coats and sinters onto the surface of the mold. The mold temperature is then increased to decompose the crosslinking agent, thereby creating radicals that crosslink the polymer. Once crosslinked, the polymer cannot be reprocessed or recycled. We will discuss the chemical process of crosslinking polyethylene in more detail in Chapter 19.

## 15.2.2.4  Introducing Fasteners and Integrated Parts

For many applications it is necessary to combine rotationally molded products with other structures. An example of this is in manufacturing a catamaran. The floats on either side of the catamaran are molded individually, as is the central section. To lock these parts together, we must connect them in some way. One way would be to drill holes into the parts and connect them with bolts and nuts. However, this can introduce weakness in the part, as well as requiring additional manufacturing steps. We can avoid this problem by fitting molds with small cavities into which a small insert, such as a nut, can be held in place during molding. The mold fills with the polymer and the polymer, ideally, sinters to the insert just as it does

**Figure 15.5**   Rotationally molded part with screw insert

to the mold cavity. When removed from the mold, the insert is fully integrated with the part. A diagram of a molded-in insert can be found in Fig. 15.5.

### 15.2.3   Mold Cooling

Forced air, water mist, or water spray cool the mold after it leaves the oven. Once the mold cools sufficiently for the polymer inside to solidify, it is opened. The cooling time must take into account the resin, the wall thicknesses, and the metal from which the mold is fabricated. Very rapid cooling of the mold can introduce stresses that can weaken or distort the final part. This is especially apparent when working with semicrystalline resins or parts with multiple resin layers.

### 15.2.4   Mold Release

After cooling, the final part is removed and the mold is recharged with resin. Since the mold surfaces are often complex, processors commonly apply mold release agents to allow the part to demold. Mold release agents are lubricants that coat the metal surface with a waxy material preventing the polymer from adhering to the mold's surface. Examples of mold release agents include silicone sprays or olefin based waxes.

### 15.2.5   Post Processing

After the part is removed from the mold, it is often necessary to perform additional post-processing functions on the product. For example, holes may need to be drilled or protrusions cut off to create an opening into a storage tank. Some parts have very thin regions from which the polymer has to be removed, for example the eye socket of a plasticized polyvinyl chloride doll head. Sometimes multiple parts need to be affixed to one another using adhesives or solvent adherents.

## 15.3    Materials Used for Rotational Molding

Rotational molding processes rely on a few resins and many different additives to manufacture a wide variety of different products. The choice of both the resin and the additives depends on the end use requirements and the processing conditions.

### 15.3.1    Resins Used for Rotational Molding

Polyethylene is, by far, the most commonly used resin in rotational molding processes. Low density, high density, and linear low density polyethylene are all used extensively. Additionally, crosslinked polyethylene is used for parts that require high chemical or heat resistance or enhanced impact resistance.

Polyvinyl chloride is also widely used. Rigid polyvinyl chloride is introduced to the mold in powder form. The material is chosen for durable constructions because of its chemical resistance and ease of processing. It incorporates functional additives and demolds easily. Plasticized polyvinyl chloride can be used to produce flexible parts such as balls and soft toy parts. The polyvinyl chloride is introduced to the mold as either a plastisol or powder. A plastisol is a suspension of granules in a plasticizing agent. When heated, the polymer granules absorb the plasticizer and fuse to form a cohesive, flexible material.

Nylon is used in specialty applications that require chemical and heat resistance and high rigidity, such as gasoline tanks. Polycarbonate is found in applications that experience high service temperatures and require high toughness, such as the transparent globes on exterior lamps. Polyethylene-co-vinyl acetate is used to produce soft, flexible parts that exhibit excellent low temperature impact strength. An example is the orange highway traffic cones used to divert traffic around construction sites.

### 15.3.2    Additives Used in Rotational Molding

The additives used in rotational molding fall into several different categories. These include processing aids, stabilizers, and those that provide specific end use attributes. Processing aids include flow enhancers, antistatic agents, and mold release agents. Flow enhancers create freely flowing powders necessary for this process. Antistatic agents prevent the build-up of static electrical charge in powdered resins. Charge build-up occurs due to inductive charging of the particles as they collide with each other during rotation. Mold release agents make it possible to cleanly remove intricate parts from the mold.

Stabilizers counter the effect of the high temperatures and the oxygen rich atmosphere experienced by the resin during the rotational molding process. Since some rotational molded parts require up to one hour of residence time in the oven, such stabilizers are essential. Without them, the polymer would lose its inherent properties, becoming unfit for the final application.

The last class of materials modifies the resin to achieve required end use properties. Impact modifiers improve the toughness of the final part, especially at low temperatures. Fire retardants, necessary for many building products, inhibit combustion by replacing the combustion supporting atmosphere with an inert (to combustion) gas. Colorants create pigmented parts. Crosslinking agents chemically react with the polymer creating a network polymer thereby improving chemical and environmental resistance.

## 15.4    Rotational Molding Equipment

Rotational molding equipment consists of the mold, the oven and the moveable arms that transfer the mold through each of the processing stages. The design of the oven is quite simple. It needs to be large enough to accommodate the mold, it must have temperature programming capabilities, and it must be able to open and close to allow the mold to enter and leave. In the next two sections, we will focus more closely on the molds and the machinery.

### 15.4.1    Rotational Molding Molds

Rotational molding differs from injection and blow molding processes in that the molds experience very low pressures during use. This means that the molds are generally much less expensive to manufacture. The mold walls can be thinner and machining tolerances are normally looser than for injection or blow molds. Additionally, the issues associated with preventing orientation in constricting regions of molds such as gates and runners are irrelevant to rotational molds. This means that there is less time and money invested in mold design itself. For these reasons, rotational molding is often used for large or complex parts that will be manufactured in limited quantities.

### 15.4.2    Rotational Molding Machines

Although the general process by which we accomplish rotational molding follows four steps, the actual equipment used to achieve this process can take many different forms. The simplest machines have one mold which is transferred in and out of the oven on a single arm. Variations on this theme include the shuttle machine, in which the arm slides in and out of the oven, analogous to how we roast marshmallows in a campfire. The swing-arm machine rotates the mold in and out of the oven. Single-arm machines are relatively inexpensive to build and operate. They provide easy control of the processing conditions for a single mold and can, therefore, produce parts of very high quality. They are limited, though, in that they only produce one part at a time. This type of machine is typically used to make small numbers of large parts.

Rock-and-roll machines are a modification of the single-arm machine. They roll a complete 360° on one axis while rocking back and forth along another axis making approximately a 45° angle change per direction of rocking. This design requires a smaller oven than an equivalent multi-axial rotation set up.

Multiple arm machines rotate molds into the different stations of the process. A three-arm machine would, for example, have one mold filling while another is heating in the oven and the third cooling in preparation of demolding. This is very efficient but requires that each stage takes the same time to accomplish. When there are three different parts, the largest part or the one with the thickest walls defines the speed for the entire process.

## 15.5    Rotational Molded Products

Rotational molding can produce a wide variety of shapes and sizes of products. Durable goods such as truck camper tops, light boat hulls, gasoline tanks, septic tanks, medical supply carts, mannequins, utility sheds, pick up truck storage boxes, agricultural and industrial storage tanks, and navigational floats can all be manufactured via rotational molding. Children's toys or components of toys such as balls, dolls' heads, toy furniture and ride-on toys also can be produced with this method. Rotational molding can be used to create products with a wide range of sizes. Small parts, fitting into the palm of your hand, include balls and other small toys. Large parts such as water tanks and pesticide drums can hold 200 to 400 liters. Then, at the far end of the spectrum, there are the truly breathtaking 80,000 liter water tanks.

## 15.6    Conclusions

Rotational molding creates a wide variety of plastic products that cannot be made effectively, efficiently, or economically by other means. What sets this method apart from others is that it can create thin-walled, hollow parts that exhibit no weld lines or scarring from ejector pins and from the process itself. It also has the advantage of having little scrap and minimal molded-in stresses, due to the low pressure and low shear rate characteristics of the process. Finally, it can be used to make parts that are very large which would be impossible to manufacture by other methods.

# Review Questions for Chapter 15

1.  What types of products are produced by rotational molding? Why? Give three examples.

2.  What are the advantages of rotational molding over injection molding? What are the disadvantages?

3.  What issues arise when using multi-armed rotational molding machines when each arm produces a different sized part?

4.  Why is the feedstock for rotational molding processes either a finely divided powder or plastisol?

5.  Why do the resins used in rotational molding processes typically contain a higher level of antioxidants and thermal stabilizers than the resins used to manufacture parts by injection molding?

6.  How can multilayer rotational molding be achieved?

7.  What is a mold release agent? Why are they used? Why are antistatic agents used?

8.  For the following products manufactured by rotational molding, indicate whether they would be made with a rock-and-roll or standard rotational molding machine:
    a) A ball
    b) A barrel
    c) A truck bed toolbox
    d) A kayak

# 16    Thermoforming

## 16.1    Introduction

We use thermoforming to convert flat sheets of polymer into three-dimensional two-sided objects. Smaller thermoformed products have thin walls which encompass a large volume relative to their weight, making them ideal for packaging and other single-use applications. Many foods, such as fresh strawberries and individual servings of jam, and hardware items, including screws and nails, are routinely packaged in thermoformed containers. Disposable foamed polystyrene plates and take-out containers are made exclusively by thermoforming. Larger products include automobile bumper covers, pick-up truck bed liners, shower stalls, and hot tubs.

Thermoforming is a secondary process in which we take polymer sheet made by extrusion, warm it to its softening point, then press it into or against a mold. During vacuum forming we clamp the softened sheet over a mold cavity from which we exhaust the air. Atmospheric pressure forces the rubbery polymer into contact with the mold where it solidifies. Common products made by vacuum forming include transparent blister packs used to hold and display items, such as batteries and automotive light bulbs. When we wish to produce items with greater depths or objects with thicker walls, we can augment the vacuum with mechanical assistance from a shaped plunger or positive air pressure opposing the vacuum. Products made by plug assisted thermoforming include disposable drink cups and yogurt containers. Thicker-walled products include refrigerator liners and running boards for sports utility vehicles.

## 16.2    Principles of Thermoforming

We thermoform plastic sheets to create useful items by applying differential pressure to them while they are in a rubbery state. Unlike extrusion, injection molding, and blow molding, the polymer is not shaped in its molten state. In the case of amorphous polymers, we heat them to a temperature just above their glass transition temperature. When working with semicrystalline polymers, we heat them to a temperature that is a little below their peak melting temperature. The principal steps of the process consist of heating the polymer sheet, clamping it over a mold, applying forming pressure, allowing it to cool, and removing it from the mold.

## 16.2.1 Vacuum Forming

Figure 16.1 illustrates the principal stages of vacuum forming. In the first step, shown in Fig. 16.1 a), a thin sheet of polymer is positioned below a radiant heater that warms it until it is rubbery. It is important that we not apply so much heat that the sheet sags unduly, tears under its own weight, or melts entirely. In the next stage, shown in Fig. 16.1 b), we clamp the edges of the sheet against the perimeter of a mold to form an air tight seal. A vacuum is created within the mold by exhausting air through vents located in the bottom of the mold. The resulting pressure differential forces the sheet down into the mold, as shown in Fig. 16.1 c) and d). The polymer sheet is so thin that it solidifies instantaneously when it touches the mold. The polymer is quenched into its final state while still under tension, which locks in molecular orientation. After the product is removed from the mold, we generally have to trim its edges, which yields the final product, shown in Fig. 16.1 f), and generates an appreciable amount of scrap.

The polymer sheet can enter the forming process either as individual cut sheets or as a continuous length. Continuous lengths can either be unwound from a stock roll or come directly from an in-line sheet extruder. Integrated lines that combine sheet extrusion and thermoforming are preferred for high-output production lines. Not only can a manufacturer save on warehousing and handling costs, but scrap from the trimming process can be recycled

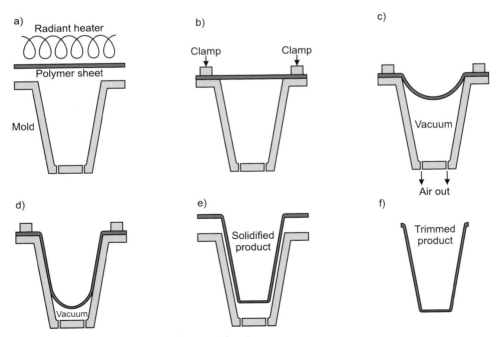

**Figure 16.1**    Schematic diagram of vacuum forming:
a) heating of sheet, b) clamping, c) vacuum drawing, d) cooling, e) removal,
f) trimmed product

directly back to the extruder. The molds for such high-output lines typically contain multiple cavities, so that we can form numerous items simultaneously.

Vacuum forming has limitations due to the non-uniform wall thicknesses of its products. As the sheet is drawn into the mold, its thickness decreases, especially in the corners. For this reason, just as in blow molding, we design vacuum formed products to have rounded corners. If the depth of the cavity is excessive, walls can become locally so thin that they are unacceptably weak. One strategy that we use to alleviate this problem is to pump air into the cavity after the sheet has been clamped. This inflates the sheet, pushing it upwards and expanding its area approximately uniformly. When we subsequently apply a vacuum the expanded sheet is drawn back down into the mold. The finished product has a more uniform wall thickness than if we had applied the vacuum directly.

## 16.2.2 Plug Assisted Thermoforming

We use plug assisted thermoforming when we wish to make thick-walled products or thin-walled, deep draw products. The principal steps of the process are illustrated in Fig. 16.2. The first two stages are similar to those of vacuum forming. Once the sheet has been clamped, a plug made from a material with low heat conductivity is thrust downwards into the cavity,

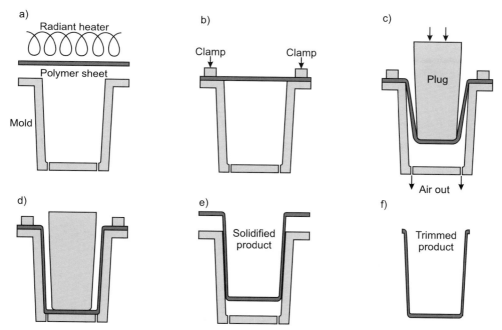

**Figure 16.2**  Schematic diagram of plug assisted vacuum forming:
a) heating of sheet, b) clamping, c) plug insertion, d) vacuum drawing, e) removal,
f) trimmed product

stretching the polymer, as shown in Fig. 16.2 c). The part of the sheet that makes contact with the plug does not stretch, which eliminates the problem of excessive thinning in the base of the product. We simultaneously exhaust air from the die, creating a pressure differential that forces the polymer into the corners of the mold, as shown in Fig. 16.2 d). The solidified item is removed from the mold and trimmed in the same manner that we would employ to finish a vacuum formed product.

### 16.2.3    Mold Considerations

Thermoforming molds typically do not have to withstand loads much greater than atmospheric pressure. We can therefore make them from a range of low-cost materials, which are more easily machined than the high-strength steels that we use for injection molds, or even the stainless steels used for blow molds. Another cost advantage of thermoforming molds is that they generally do not have to be manufactured to such tight machining tolerances, as there are no mating surfaces to be considered.

Due to the low cost of thermoforming molds we can economically make short runs of specialty or niche market products, for which an injection mold would be prohibitively expensive. This is especially important for large moldings, such as whirlpool tubs and large panels for recreational vehicles, for which injection molds and presses would be astronomically expensive.

## 16.3    Products

Packaging is the major outlet for thermoformed products. During a walk down the aisles of a supermarket or hardware store we pass dozens of products in clear thermoformed packages. Most of these packages are made from transparent amorphous polymers that permit a clear view of their contents. Many of the packages are designed as hinged "clam shells" that can be snapped shut to hold their contents and then opened and closed readily. Polyethylene terephthalate and polystyrene are common material choices for this type of packaging. Each of these polymers has a fairly wide processing temperature range in which they are rubbery. Grocery items that we see packaged in this type of container include fresh herbs, individual drink cups, individual syrup, honey, jam or butter servings, microwave ready meals, and pre-made sandwiches. Hardware and automotive items sold in thermoformed packages include just about any item that can be held conveniently within one's hand.

Foamed polystyrene sheet is thermoformed into single use plates, insulated cups, fast food clam shell packages, and take home containers.

An increasing variety of automotive parts is being made by thermoforming processes. Many of these products are made from rubber-toughened polypropylene, which are relatively stiff and can withstand a high level of physical abuse over a wide range of temperatures. When such parts are used in high visibility areas, such as wheel arches, air dams, truck fenders, and bumper covers, they may be painted. Alternatively, we can line a thermoforming mold with a thin polymer film that contains pigments and provides a high gloss surface. When the polymer sheet is pressed into the mold, it makes intimate contact with the surface film to which it permanently adheres. An alternative method for providing surface color economically is to thermoform co-extruded sheets that consist of a primary structural unpigmented polymer and a secondary skin layer of pigmented polymer that is visible when the part is in use. It is possible to produce automotive fuel tanks by heating two sheets, clamping them together around their perimeters, and inflating them within a mold. This method allows us to directly encapsulate the fuel pump and the fuel gauge sender unit, which decreases hydrocarbon emissions by reducing the number of holes that we must subsequently make in the tank. We can also make surf boards by this twin sheet molding process.

Thermoformed products are used in various outdoor and commercial applications which require tough products that have a large surface area. Typically, these products do not have close dimensional tolerances. Such applications include ornamental pool liners, dumpster lids, panels for portable sanitation facilities, and agricultural feed troughs. These products, which are not required in very large numbers, could not be produced economically by other molding processes.

## 16.4  Conclusion

We use thermoforming to make three principal types of product:

- Low cost packaging and single-use products with a high volume-to-weight ratio,
- Automotive parts, and
- High surface area shapes with relatively low dimensional tolerances.

The technique is relatively simple, inexpensive, and can be modified to create a wide range of profiles.

Thermoformed packages have replaced cardboard boxes in many point of sale applications, where a clear view of the product is desirable. In the automotive field, we can expect to see increasing use of thermoformed parts. Reduced weight, reduced solvent emissions from paint lines, and improved dent resistance are the principal concerns that drive the replacement of painted metal parts. As car manufacturers develop vehicles for niche markets and provide more custom options, the economic advantage of thermoforming over traditional production methods will increase. Thermoforming will gradually take a greater market share of products that have traditionally been made from glass fiber reinforced polyester, which release harmful volatile organic compounds during the curing process.

## Review Questions for Chapter 16

1.  Why is thermoforming called a secondary process?

2.  What type of products can be manufactured by thermoforming? Why?

3.  Why is it necessary to use molds that have rounded corners?

4.  If a thermoformed product is placed in an oven set to a temperature just above the softening point of the polymer, the product returns to the form of a flat sheet. Why does this happen?

5.  Why would we use a plug assist process instead of a vacuum forming process?

6.  Why does thermoforming generate a great deal of scrap? How is this scrap recycled?

# 17 Recycling

## 17.1 Introduction

In a world where consumers and manufacturers increasingly focus on dwindling non-renewable resources and cost savings, polymer recycling is a major concern. Recycling efforts associated with polymers focus on two distinct material sources: commercial scrap and post-consumer waste recycling. Both types have their own, specific challenges. This chapter will introduce you to the types of materials that we can recycle, the means by which manufacturers recycle their scrap materials, and how recyclers achieve post-consumer recycling. We will also discuss the limitations and complications associated with polymer recycling and an alternative to recycling into a new polymeric product.

## 17.2 Materials Amenable to Recycling Processes

Thermoplastics are commonly recycled both during the manufacturing process and after consumer use. During manufacturing processes, scrap and off-specification products create material that can be reprocessed. This type of recycling is often referred to as primary recycling. Post consumer recycling, also known as secondary recycling, takes discarded polymeric materials and reprocesses them into a form suitable for use by a processor.

Both manufacturers and post consumer recyclers face many challenges. Of primary concern, scrap and post-consumer waste polymer has already experienced at least one heat and shear history. Because of this, the polymer has already partially degraded and will have lower starting physical properties than virgin resin. Secondly, many polymeric products contain many components. For example, some food packaging films consist of many different polymeric layers, and carbonated beverage bottles not only consist of several layers but also have a label affixed with an adhesive. In both cases, recycling requires either that the components be separated or be miscible with one another to allow reprocessing. Finally, recycling either scrap or post-consumer waste potentially introduces contamination into the processing system. Contamination can lead to reduced physical properties, manufactured parts that fail to meet product specifications, or reduced output.

Thermoset resins, because of their extensively crosslinked structure, cannot be reprocessed. Therefore, recycling of these materials is often limited to grinding or chipping them into small pieces that can be used as fillers in other materials. As you may imagine, this type of recycling is limited to polymer applications that result in large volumes of waste. The most common example is automotive tires. Ground tires are used in asphalt, playground surface covers, and athletic track surfaces. In general, though, most applications of thermoset materials are not recycled.

## 17.3    Recycling Commercial Scrap

From the perspective of a commercial operation, resins often represent the most expensive single component of the cost of production. Their cost can constitute up to 90% of the cost to manufacture a given part. For this reason, in addition to any environmental arguments, processors strive to recycle any scrap that they generate. Many manufacturing facilities operate at a material efficiency of 95 to 98%. The type of scrap generated defines the processes by which it can be recycled. Molding and profile extrusion processes generally rely on grinding the scrap back into a small pellet form and reintroducing the scrap into the resin hopper. Film extrusion processors can shred the scrap film or feed it back into melt processing equipment directly with specialized feeders and crammers.

### 17.3.1    Injection Molding Scrap

Injection molding processes generate scrap in the form of out-of-specification parts, sprues, and runners. The final form of the polymer in the scrap is very different from the form it was fed into the injection molding machine. The randomly sized pieces have to be reprocessed into a form that is compatible with the virgin feed stock. For this reason, it is mechanically chopped into small pieces called regrind. Regrind must be similar in size to the starting feed material and therefore it is often sieved to remove both very small and large pellets. The regrind is then incorporated with the original feed at low concentrations. Since the regrind has experienced one heat history, additional stabilizers might be added to limit further degradation. For water-sensitive resins, the regrind must be dried before reintroduction into the equipment to prevent hydrolysis. If the injection molding process produces parts with different core and skin resins, the regrind is typically introduced to the core material rather than the surface components.

### 17.3.2    Extrusion Scrap

Extrusion processes can generate several types of scrap, including out-of-specification material, material generated when the process is not running at steady state, and edge trim when sheet or film is trimmed to the required width. In extrusion, the scrap can be reintroduced to the material stream and reincorporated into the manufactured products. As with injection molding, the form of the scrap often differs substantially from the starting resin.

### 17.3.2.1  Profile Extrusion Scrap

For profile extruded products, such as pipes or window frames, a chopper or pelletizer processes rejected extrudate into regrind. Screens or centrifuges size the regrind, which is then reintroduced to the extrusion process. Often, regrind constitutes a small proportion of the material in the middle or backing of the extrudate. This is done to prevent visual flaws from appearing on the surface of the part. Manufacturers of rigid polyvinyl chloride house siding introduce regrind to their siding. Typical vinyl siding consists of two or three distinct layers of co-extruded formulated polyvinyl chloride. The regrind is fed into the feeder creating the middle layer or the backing, while the layer that will be seen in use contains only new resin.

### 17.3.2.2  Film Extrusion Scrap

Film extrudate presents specific challenges for recycling. Its high surface area and low bulk density make it both easily contaminated and difficult to feed back into the extrusion equipment. The sources of scrap for these materials include out-of-specification film, film produced prior to the operation achieving steady-state processing conditions, and edge trim when cutting film or sheet to specified widths. There are two general means by which film scrap can be reintroduced to the extrusion process. The method most often used to recycle trim forces the film back into the throat of the extruder with a crammer feeder alongside a preponderance of virgin resin. The crammer feeder reduces the bulk density of the film. Generally, trim recycle is a continuous process. Slitting knives cut the trim off the film, after which it runs directly to the crammer feeder back to the process. Alternatively, the trim can be shredded to form flakes (known colloquially as "fluff") that are returned to the feed hopper. Another means of recycling film involves reprocessing the film to create pellets. Fluff or scrap film is fed to a single-screw extruder via a crammer feeder. The extruder reprocesses the film through a pelletizing die, creating new pellets suitable for introduction to the film manufacturing process.

Film recycling is particularly troublesome because of the high surface area of the scrap. Moisture, air, and dust can easily become entrained in the material, introducing potential problems in the extrudate. To address these issues, vent ports and fine mesh screens are often incorporated to the extrusion system to release volatiles and trap solid contaminants.

# 17.4  Post Consumer Recycling

Once a consumer has purchased a material made of a polymer, the likelihood of it being recycled is dramatically reduced. In the United States there are three resins that are commonly recycled: polyethylene, polypropylene, and polyethylene terephthalate. SPI (The Society of the Plastics Industry) recycling codes, found on the bottom of food containers, provide the consumer with information as to the recyclability of the material. Table 17.1 lists these codes.

**Table 17.1**    SPI's Recycling Codes and Acronyms Found Embossed or Printed on Recyclable Polymeric Materials

| Code | Symbol | Polymer |
|------|--------|---------|
| 1 | PETE | Polyethylene terephthalate |
| 2 | HDPE | High density polyethylene |
| 3 | PVC | Polyvinyl chloride |
| 4 | LDPE or LLDPE | Low or linear low density polyethylene |
| 5 | PP | Polypropylene |
| 6 | PS | Polystyrene |
| 7 | Other | Other |

## 17.4.1    The Recycling Process

Once a waste material enters the recycling stream, it must first be sorted by code and color. For example, a green carbonated beverage bottle would be sorted as a code (1) recyclable and then further sorted into a bin with other bottles of the same color. Sorting can be accomplished by hand or electronically. Hand sorters remove bottles, based on visual inspection, from a conveyor belt and place them in the appropriate bins. Electronic sorting systems monitor chemical signatures in the polymers and separate accordingly. If the melt processing occurs at a separate site than the collection and sorting site, the waste is compacted and strapped together to form a bale. The bale can be easily shipped to the reprocessing site. Regardless of where the reprocessing will occur after sorting, the materials are washed, shredded, and then the shredded components are typically separated by differences in their specific gravity. Taking as an example the same green beverage bottle from earlier illustrates the purpose of shredding and separation. Though the bottle itself is made from polyethylene terephthalate, the label is likely to be polypropylene as is the ring of polymer remaining on the neck that was originally attached to the cap. To separate these components, the bottle is shredded and the shredded pieces dumped in a large vat of water. The polypropylene, with its density slightly lower than water's, floats on the surface while the polyethylene terephthalate sinks to the bottom, separating the two polymers.

Another means of separating polymers from one another relies on the difference in charges generated on the surface of the shredded polymers as they pass over one another due to inductive charging effects. When granulated polypropylene and polyethylene terephthalate mix, the polypropylene becomes positively charged while the polyethylene terephthalate takes on a negative charge. The granulate is then poured through a strong electric field where the individual particles deflect towards the electrode bearing the opposite charge. A series of collection screens gather the material after it passes through the field. The purity of the material on the collection screen is defined by the lateral distance from the electrode. A screen near the negative electrode will have a higher concentration of the positively charged granulated particles – in our example the polypropylene. This method is known as positive/negative sorting.

When the new product to be manufactured is the same as what it started as, for example a new bottle made from bottle scrap, the recycling is referred to as closed-loop. When the new application is different from the starting one, the process is referred to as open-loop recycling, as is the case when the polyethylene terephthalate bottle is used to produce polyester fiber for carpeting.

## 17.4.2    Products Made from Post-Consumer Recycled Polymers

Several different products are made from post-consumer recycled polymers. The predominant recycled polymers include polyethylene terephthalate and high density polyethylene. Recycled polyethylene terephthalate can be made into carpet fiber, sports apparel such as fleece jackets, bottles, and industrial strapping. High density polyethylene is recycled to make non-food containing bottles and pipes. Polyvinyl chloride is recycled to produce carpet backing, industrial flooring, and the non-slip treads on ladder rungs. Nylon from used carpets can be recovered to produce new carpeting.

### 17.4.2.1  Polyethylene Terephthalate Recycling

According to literature from the American Plastics Council, approximately 20% of polyethylene terephthalate beverage bottles are recycled annually in the USA. The dominant use for this recycled material is in the manufacture of fibers to produce carpets. Polyethylene terephthalate and polyvinyl chloride are often used in the same types of applications. This makes these resins difficult to sort. The polymers tend to look the same and possess nearly equivalent bulk densities. Unfortunately, they are not miscible. Their similar densities prohibit their separation by specific gravity and their incompatibilities foul the manufacture of recycled materials. Polyvinyl chloride tends to degrade, releasing highly corrosive hydrogen chloride when reprocessed along with polyethylene terephthalate. For this reason, polyethylene terephthalate bottle designers avoid the use of any components, such as bottle caps, other closures or labels, made of polyvinyl chloride.

Polyethylene terephthalate also has the tendency, because it is produced by a condensation polymerization process, to depolymerize under high pressure and temperatures in the presence of water. Although this is usually a negative attribute, it can be utilized to regenerate pure monomers which can be repolymerized to make fresh polymer. This avoids the issues experienced by reprocessing resins, as the new resin has not experienced a previous heat history. A major drawback to this process is the requirement that the monomers used in polymerization processes must be highly pure. Unfortunately, this process is extremely costly and not performed on a commercial scale.

### 17.4.2.2  Recycling of Electronic Equipment

Electronic equipment, including computers, printers, telephones, televisions, and stereos is ubiquitous in society today. These products often have a very limited lifetime due to the fast pace that innovations are made in their design and production. For this reason, as a society, we generate huge amounts of electronic scrap. Several companies have recognized the need to recycle both the metals and polymers comprising this equipment. Just considering a television set can provide some perspective on what these companies do. In addition to the polymer body of the television, there are polymeric-based electronic components, wires, glass, and metal pieces. There are several different polymers used to produce electronic parts further complicating the process. These include polystyrene, polypropylene, polyvinyl chloride, and polyesters. Electronic equipment recyclers start by separating the polymeric parts from the other materials. They then sort, by color, which provides the first step in isolating the different polymers. After this, the material is shredded, separated by specific gravity, and then melt processed. The recycled parts are then used for manufacturing new electronic components or to make entirely different products such as filler for asphalt and injection molded parts.

### 17.4.2.3  Manufacture of Plastic Lumber

One outlet for polypropylene, polyethylene, and polyvinyl chloride waste is plastic lumber. These materials, often containing more than one polymer and a wide variety of additives, provide superior weather resistance in humid environments when compared to natural wood. To manufacture these materials, the compound incorporates compatibilizers, which allow dissimilar polymers to mix evenly. Additionally, they assist in the incorporation of fillers and additives, such as wood flour, calcium carbonate, and pigments.

# 17.5    Tertiary and Quaternary Recycling

So far we have described recycling efforts in which the material is re-used to form a new product. Another option to recycle polymers is to actually depolymerize the high polymer to manufacture fuels, oils, low molecular weight hydrocarbons, and gases. This type of recycling is called tertiary recycling. The depolymerization of high polymers occurs via different processes depending on the starting polymer. For example, polyethylene can be depolymerized by heating the polymer to high temperatures in a controlled atmosphere containing hydrogen or oxygen, sometimes in the presence of a catalyst. Polyesters can undergo depolymerization by the introduction of water at high temperatures. Crosslinked rubber tires can be recycled this way by gently heating the rubber in the presence of a catalyst system.

Sometimes, it makes more sense to use the energy captured in the polymerization process as a source of heat energy. This can be achieved by burning the polymeric materials under controlled conditions, then using this heat energy to generate electricity. This type of recycling is referred to as quaternary recycling. A major issue with this method of recycling is the release of potentially harmful degradation byproducts or additive components to the atmosphere. For example, burning polyvinyl chloride generates hydrogen chloride. Burning nylons can generate the noxious $NO_x$ gases, which are associated with the creation of acid rain. In both of these examples, these gases must be scrubbed from the effluent gas stream before the water and carbon dioxide can be released to the environment.

## 17.6     Conclusions

Polymer recycling presents considerable challenges either at the manufacturing site or in the recycle of waste. In-plant recycling requires some specialized equipment and care when utilizing the recycled material. This care and equipment, though, generally pays off for the manufacturer as they can avoid financial losses through the use of virtually all of their starting resins. Post-consumer recycling efforts focus on those materials that are most commonly used in the manufacture of non-durable goods. These materials include polyethylene terephthalate, polyethylene, polypropylene, and polyvinyl chloride.

## Review Questions for Chapter 17

1.   Explain the recycling hierarchy indicated by the terms primary, secondary, tertiary and quaternary recycling.

2.   What challenges face manufacturers who recycle their scrap? How do they address these challenges?

3.   What polymers are most commonly recycled from post-consumer sources? Why?

4.   How can different polymers be separated from one another?

5.   Plastic lumber can be manufactured from mixed-polymer post-consumer recycled materials. One example incorporates polyethylene and polypropylene into a final material. How do processors achieve this incorporation in light of the immiscibility of the two polymers?

6.   Why do polyethylene terephthalate bottle manufacturers not use any polyvinyl chloride for closures or labels on the bottle?

# 18 Polyethylene

## 18.1 Introduction

The general structure of polyethylene is shown in Fig. 18.1. A polyethylene molecule consists of a long backbone of covalently bonded carbon atoms, to each of which is attached a pair of hydrogen atoms. The chemical formula of this structure is $C_nH_{2n}$. There are numerous variants on this basic structure that fall under the general umbrella of polyethylenes. Modifications to the general structure include short and long chain branches, crosslinking, and a variety of terminal groups. Branches may consist of simple hydrocarbon chains or other more complex structures.

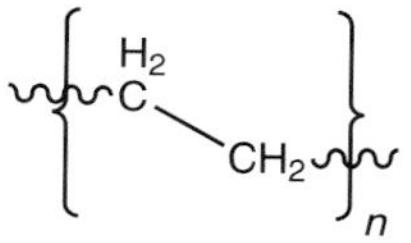

**Figure 18.1** Chemical structure of polyethylene in its simpslest form

Polyethylene exhibits a range of tensile strengths and flexibilities, is generally tough, can be readily extruded or molded, and is relatively inexpensive. These characteristics guarantee that the various families of polyethylene find major use as a commodity polymer. Due to its large number of variants, we use polyethylene in a wider range of applications than any other polymer.

## 18.2 Chemical Structure

There are seven principal variants of polyethylene, which are shown schematically in Fig. 18.2. Within each family there is a range of products that share the same general characteristics.

### 18.2.1 High Density Polyethylene

High density polyethylene, shown in Fig. 18.2 a), consists primarily of linear hydrocarbon chains of the type shown in Fig. 18.1. We commonly abbreviate its name to HDPE. As with all other polymers, high density polyethylenes contain a distribution of molecular weights. The molecules have few, if any, branches.

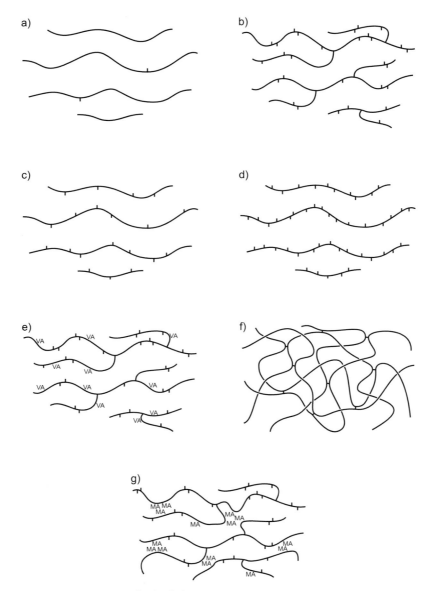

**Figure 18.2**  Principal variants of polyethylene:
a)  high density polyethylene,
b)  low density polyethylene,
c)  linear low density polyethylene,
d)  very low density polyethylene,
e)  ethylene vinyl acetate copolymer,
f)  crosslinked polyethylene,
g)  ionomer

## 18.2.2    Low Density Polyethylene

Low density polyethylene has the general structure shown in Fig. 18.2 b). Its molecules contain many short and long chains. The short branches primarily consist of ethyl and butyl branches, which are often located near each other. We generally refer to it as LDPE. Low density polyethylene is polydisperse in terms of its molecular weight, long chain branching length and placement and short chain branching.

## 18.2.3    Linear Low Density Polyethylene

The general structure of linear low density polyethylene is shown in Fig. 18.2 c). Linear low density resins are copolymers of ethylene and 1-alkenes; principally 1-butene, 1-hexene, and 1-octene. Comonomer levels range from approximately 2 to 8 mole %. This family of polyethylene is widely known as LLDPE. Linear low density polyethylenes are polydisperse with regard to molecular weight and branch distribution.

## 18.2.4    Very Low Density Polyethylene

Figure 18.2 d) illustrates the general structure of very low density polyethylene, which we also call ultra low density polyethylene. In common with linear low density polyethylene, these resins are copolymers of ethylene and 1-alkenes. The comonomer level ranges from approximately 8 to 14 mole %. We normally refer to these polymers as VLDPE or ULDPE. The molecules of very low density polyethylene contain a distribution of lengths and branch placements.

## 18.2.5    Ethylene-Vinyl Ester Copolymer

The most common copolymer of this type is ethylene-vinyl acetate, which we normally refer to as EVA. This variety of polyethylene is illustrated in Fig. 18.2 e), in which the ester branches are indicated by the symbol "VA". This family of copolymers is commercially available containing vinyl acetate concentrations of up to approximately 25 mole %. In addition to the randomly distributed ester branches, these resins also contain the short and long chain branches that are characteristic of low density polyethylene.

## 18.2.6    Crosslinked Polyethylene

Crosslinked polyethylene consists of molecular chains that are linked at random points to form a network, as shown schematically in Fig. 18.2 f). The crosslinks can consist of carbon-carbon bonds, which directly link adjacent chains, or short bridging species, such as siloxanes, which may link two, three, or four chains. We often refer to these materials as XLPE.

### 18.2.7    Ethylene Ionomers

Ethylene ionomers consist of copolymers of ethylene and an organic acid, such as methacrylic acid, the acid moieties of which have been neutralized to form a metal salt. The metal salts from neighboring chains tend to form clusters, such as the one shown schematically in Fig. 18.3. The net result is the overall structure shown in Fig. 18.2 g), in which the ionic clusters form weak crosslinks between adjacent chains. Ionomers also contain short and long chain branches, which are similar to those found in low density polyethylene.

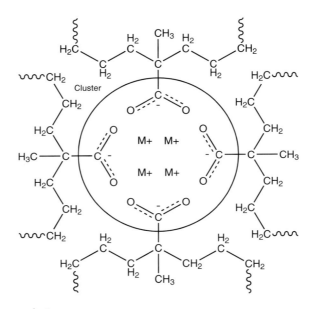

**Figure 18.3**    Ionomer cluster

## 18.3    Manufacture

We make polyethylene resins using two basic types of chain growth reaction: free radical polymerization and coordination catalysis. We use free radical polymerization to make low density polyethylene, ethylene-vinyl ester copolymers, and the ethylene-acrylic acid copolymer precursors for ethylene ionomers. We employ coordination catalysts to make high density polyethylene, linear low density polyethylene, and very low density polyethylene.

Ethylene-acrylic acid copolymers are converted to ethylene ionomers in a separate, post-polymerization reaction.

## 18.3.1    Free Radical Polymerization

Free radical polymerization consists of the four basic steps that we introduced in Chapter 2:

a)  Initiation,
b)  Propagation,
c)  Chain transfer, and
d)  Termination.

These steps are illustrated in Fig. 2.3 and were described in Section 2.3. When chain transfer involves intermolecular or intramolecular reactions, long and short chain branches are created.

The polymerization process is initiated by the decomposition of molecules, such as trace amounts of oxygen or organic peroxides, to form free radicals. Spontaneous decomposition of initiators occurs at reaction temperatures of approximately 180 to 200 °C at pressures typically in the range of 1,000 to 1,500 atm. The free radicals attach themselves to ethylene molecules, transferring their unpaired electron to the opposite end of the monomer, as shown in Fig. 2.3 a), which initiates the formation of a polyethylene molecule. Propagation, shown in Fig. 2.3 b), proceeds as ethylene monomers are attached sequentially to the growing polyethylene chain. Chain transfer, shown in Fig. 2.3 c), occurs when the free radical is transferred to another ethylene molecule, thereby terminating the growth of one chain and beginning that of another. Termination, shown in Fig. 2.3 d), involves the quenching of two growing chain ends, either to form a single molecule or by disproportionation to form two separate chains.

Long chain branches form by the process illustrated in Fig. 18.4. This type of branching occurs when an unpaired electron at the end of a growing chain attacks a pre-existing polyethylene molecule, abstracting a hydrogen atom to create an unpaired electron that is remote from the polymer's ends, as shown in Fig. 18.4 a). Chain growth proceeds from this non-terminal carbon to form a long chain branch, which is shown in Fig. 18.4 b).

An example of a "backbiting" reaction that creates the short chain branches is shown in Fig. 18.5. In this example the growing end of a polyethylene chain turns back on itself and abstracts a hydrogen atom from the carbon atom located four bonds away from the chain end, as shown in Fig. 18.5 a). Chain growth proceeds from the newly formed unpaired electron, leaving a pendant butyl group, as shown in Fig. 18.5 b). There are many variants of backbiting, which create a variety of short chain branches.

We can also incorporate branches by copolymerizing ethylene with vinyl esters and vinyl acids. In addition to their ester or acid side groups, these copolymers also contain the long and short chain branches, which are characteristic of low density polyethylene.

Low density polyethylene is made at high pressures in one of two types of continuous reactor. Autoclave reactors are large stirred pressure vessels, which rely on chilled incoming monomer to remove the heat of polymerization. Tubular reactors consist of long tubes with diameters of approximately 2.5 cm and lengths of up to 600 m. Tubular reactors have a very high surface-to-volume ratio, which permits external cooling to remove the heat of polymerization.

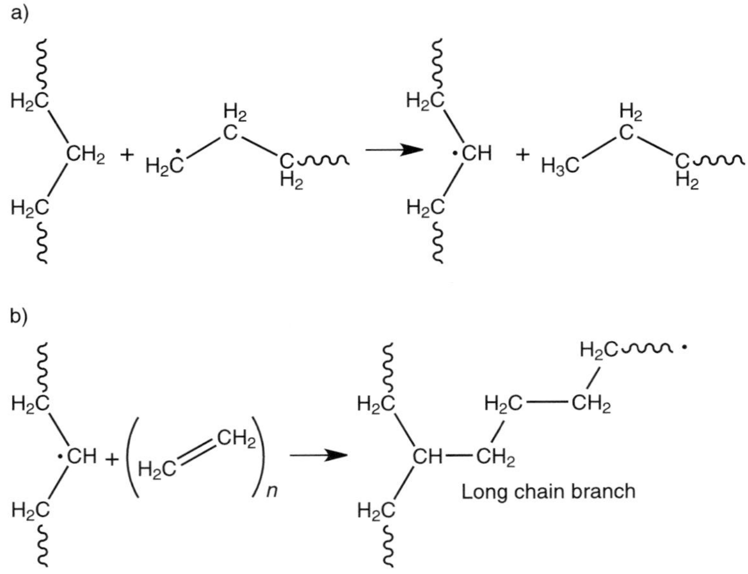

**Figure 18.4**  Long chain branching mechanism:
a) abstraction of hydrogen from adjacent chain,
b) chain growth from non-terminal carbon atom

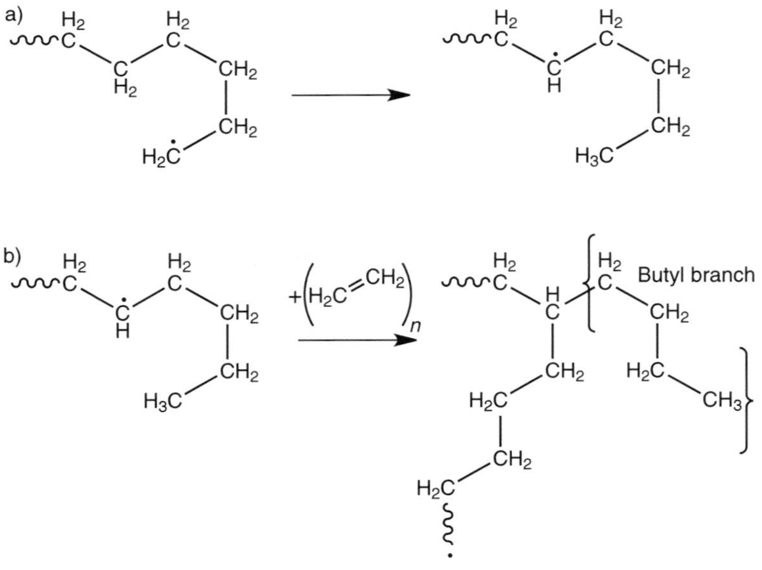

**Figure 18.5**  Short chain branching mechanism:
a) "backbiting",
b) chain growth from non-terminal carbon atom

## 18.3.2   Coordination Catalysis

Coordination catalysts contain a metal atom that is activated by a variety of electron-withdrawing species that are attached to it. Coordination catalysts for polyethylene fall into three main categories:

- Ziegler-Natta catalysts,
- Single site catalysts, and
- Metal oxide catalysts.

Within each category there are many variants. There are also hybrids, which combine chemical structures borrowed from more than one family of catalysts.

Ziegler-Natta catalysts, which we introduced in Chapter 2, consist of a group I, II, or III base metal alkyl or halide complexed with a salt of a transition metal from group IV to VIII. Figure 2.9 illustrates the coordination polymerization of ethylene catalyzed by a Ziegler-Natta type catalyst. The active site consists of a titanium atom coordinated with four chlorine atoms and an alkyl group, which are arranged octahedrally to leave a vacant site. An ethylene molecule attaches itself to the vacant site and is inserted between the metal atom and the alkyl group via a transition complex. In the process a new vacancy is created, which is receptive to another monomer.

One of the drawbacks of Ziegler-Natta catalysts is that we cannot prepare them to contain only one type of active site. Catalysts invariably contain more than one type of active site. Each type of active site has a different reactivity and receptivity to incoming monomer types. The net result is that Ziegler-Natta catalyzed polyethylenes have relatively broad molecular weight distributions. Additionally, when we copolymerize ethylene with 1-alkenes, we produce copolymers with a bimodal range of comonomer incorporation. The product is effectively a blend consisting of higher molecular weight molecules that have few short chain branches and shorter molecules that have a higher concentration of short chain branches.

We use Ziegler-Natta catalysts to make high density polyethylene and linear low density polyethylene. The lowest density of commercial polyethylene made with this type of catalyst is approximately 0.91 g/cm$^3$. If we were to make polymers of lower density by incorporating higher concentrations of comonomer, we would end up with a sticky product. The stickiness is due to an unacceptably high level of low molecular weight, highly branched molecules.

Single site catalysts differ from Ziegler-Natta catalysts in that they contain only one type of active site and thus produce polymers with a more uniform distribution of molecular weight and composition. Some examples of single site catalysts are shown in Fig. 2.10. As with Ziegler-Natta catalysts, there are vast numbers of single site catalysts.

We use single site catalysts primarily to make copolymers of ethylene and 1-alkenes, with densities ranging from approximately 0.87 to 0.93 g/cm$^3$. More recently some higher density grades have been introduced, but these are relatively uncommon at the moment.

A small number of companies use metal oxide catalysts, such as the example shown in Fig. 18.6, to make high density polyethylene. The polyethylene made with this catalyst generally has a narrower molecular weight distribution than high density polyethylene made with Ziegler-Natta catalysts.

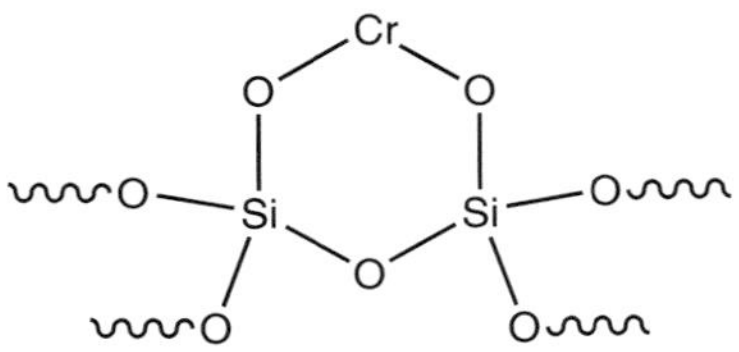

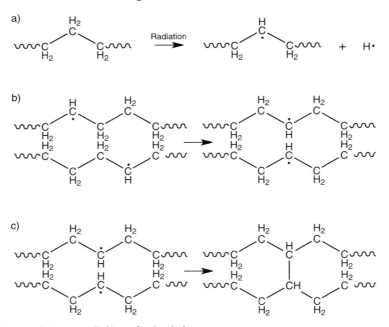

**Figure 18.6**  Example of a metal oxide catalyst

### 18.3.3  Crosslinking

We typically crosslink polyethylene using one of two different strategies; we can create direct crosslinks between chains via a free radical mechanism or we can create short chemical bridges between chains.

Direct crosslinking can be brought about by irradiation of a solid product with gamma radiation or high energy electrons. The radiation splits carbon hydrogen bonds to produce free radicals, as shown in Fig. 18.7 a). The free radicals can migrate along the length of a chain by the process of hydrogen exchange shown in Fig. 18.7 b). When two free radicals meet, they combine to form a covalent bond between the carbon atoms creating a crosslink between adjacent chains, as illustrated in Fig. 18.7 c).

**Figure 18.7**  Radiation crosslinking of polyethylene:
a)  scission of C-H bond,
b)  migration of radicals and
c)  formation of covalent C-C crosslink

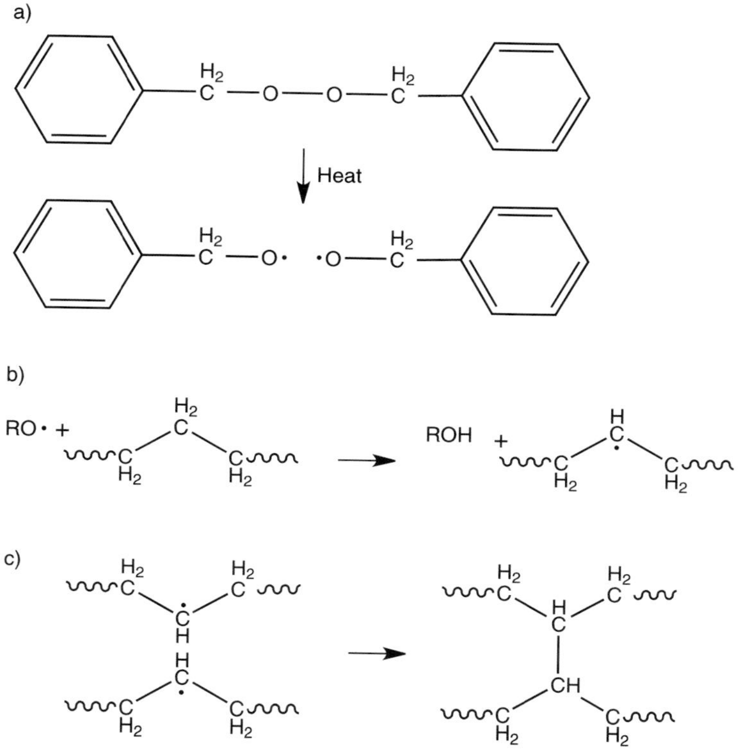

a)

Heat

b)

c)

**Figure 18.8**  Peroxide crosslinking of polyethylene:
a)  decomposition of dicumyl peroxide,
b)  abstraction of hydrogen from polyethylene chain, and
c)  formation of covalent C-C crosslink

We can also produce direct crosslinks by the action of peroxy radicals, as shown in Fig. 18.8. In this process, we blend an organic peroxide, such as dicumyl peroxide, into molten polyethylene at a temperature below that at which the peroxide decomposes. Once we have formed the molten blend into the required shape, we increase its temperature until the peroxide decomposes into peroxy radicals, as shown in Fig. 18.8 a). The peroxy radicals abstract hydrogen atoms from the polyethylene chains to create free radicals, as shown in Fig. 18.8 b). Crosslinking takes place when two radicals react to form a covalent bond, which is shown in Fig. 18.8 c).

The most common method of crosslinking via short bridges is a two-step process involving trimethoxysilane, which is shown in Fig. 18.9. In the first step, we graft siloxane branches onto polyethylene with the aid of peroxy radicals. The second step consists of a condensation reaction, which occurs in the presence of hot water or steam. A siloxane bridge is created and methanol is released.

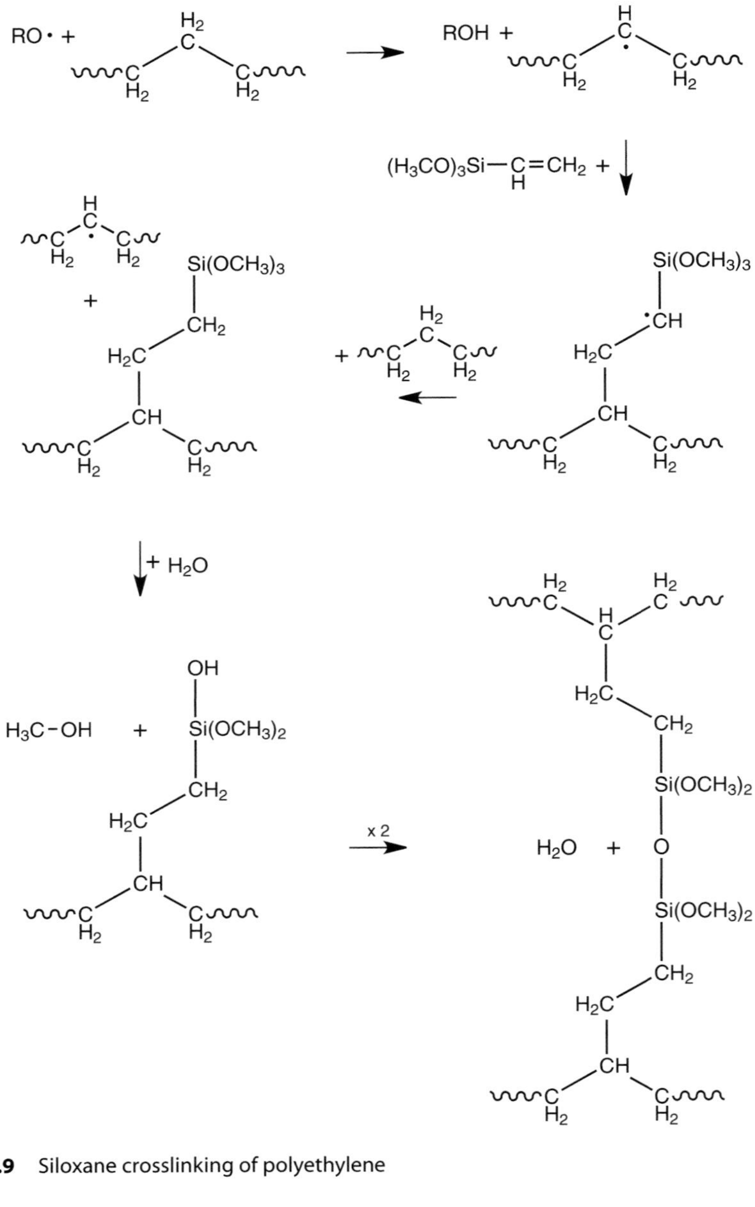

**Figure 18.9**   Siloxane crosslinking of polyethylene

### 18.3.4    Production of Ethylene Ionomers

Ionomers are made in a two-stage process. In the first step, we copolymerize ethylene with small amounts of an organic acid containing a vinyl group, such as acrylic or methacrylic acid, in a high pressure reactor. In the second step, we neutralize the acid comonomers to form metal salts. We can create ionomers with a variety of metal salts, including sodium, calcium, and zinc.

## 18.4    Morphology

Solid polyethylene is semicrystalline in nature. As far as commercial products are concerned, the overwhelming majority contains only orthorhombic crystallinity of the type illustrated in Figs. 7.2 and 18.10. Figure 7.2 a) shows a perspective view of a section from a crystallite with a unit cell outlined. Each orthorhombic unit cell contains a complete chain segment running down its center and parts of four others, each of which is located at a corner of the unit cell. Figure 18.10 illustrates the same structure using the space filling configuration, looking down the chain axis.

We observe a continuum of ordering in the semicrystalline morphology of polyethylene, which improves as the degree of crystallinity increases. Table 18.1 summarizes the semicrystalline ordering of polyethylene as a function of density and crystallinity level. At the lowest levels of crystallinity, represented by very low density polyethylene, crystallites are small and isolated, with dimensions of approximately 2 to 5 nm in any direction. At modest levels of crystallinity, typical of low density and linear low density polyethylene, crystallites are lamellar in form, with thicknesses of approximately 5 to 10 nm and lateral dimensions of several tens of nanometers. These lamellae tend to be aligned with their nearest neighbors, the degree of alignment increasing with the degree of crystallinity. On a larger scale, lamellae are organized

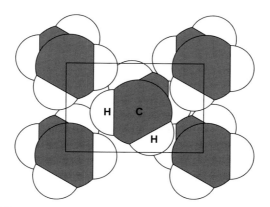

**Figure 18.10**    Space filling view of the orthorhombic unit cell of polyethylene

**Table 18.1**    Summary of semicrystalline ordering in polyethylene

| Class | Approximate density (g/cm³) | Approximate degree of crystallinity (%) | Approximate crystallite size (nm) | | Semicrystalline organization |
| --- | --- | --- | --- | --- | --- |
| | | | Thickness | Lateral dimensions | |
| Very low density | 0.86–0.91 | 5–30 | 2–5 | 2–5 | Isolated crystallites |
| Low density Linear low density | 0.91–0.94 | 30–55 | 5–10 | 10–100 | Small lamellar stacks Poorly defined spherulites |
| High density | 0.94–0.97 | 55–75 | 10–20 | 100–1000 | Extensive lamellar alignment Well defined spherulites |

to form spherulites, the perfection of which increases with the degree of crystallinity. At the highest levels of crystallinity, exhibited by high density polyethylene, crystalline lamellae range from 10–20 nm in thickness with lateral dimensions of up to several hundred nanometers. Lamellae are typically parallel with their neighbors and spherulites are well organized, with diameters of up to several tens of micrometers. Oriented high density polyethylene consists of stacks of lamellae with their planes approximately perpendicular to the principal orientation direction, i.e., with their chain axes aligned with the orienting force.

As density and lamellar thickness increase, the peak melting temperature increases. Very low density polyethylene can begin to soften and melt at temperatures not much above room temperature. Low density and linear low density polyethylenes typically exhibit peak melting temperatures in a range from approximately 95 to 120 °C. The peak melting temperature of high density polyethylene generally occurs between 125 and 130 °C. Morphologies and melting temperatures are strongly influenced by crystallization conditions. Rapid quenching results in smaller crystallites with less well organized semicrystalline morphologies than those created by slow cooling. Slow cooling permits the development of thicker crystallites that are better organized and melt at higher temperatures.

## 18.5    Structure/Property Relationships

The physical characteristics of polyethylene resins vary widely as a function of their density. The density of polyethylene is highest when it has very few branches to impede the crystallization process. A 3 mm thick plaque of high density polyethylene is an opaque white solid that can be flexed moderately by hand. In its lowest crystallinity form, very low density polyethylene is clear and flexible. The forms of polyethylene that are made entirely from hydrocarbons, i.e., the ones with no oxygenated branches, have very little dipole moment and are thus largely chemically inert. Most types of polyethylene have a somewhat waxy feel and are odorless and tasteless.

## 18.5.1 High Density Polyethylene

Due to the low level of branching, there is little to hinder the crystallization of high density polyethylene. We routinely observe crystallinity levels in excess of 60%, which translate into densities ranging from approximately 0.94 to 0.97 g/cm$^3$. High density polyethylene is the stiffest of all the polyethylene types. In some cases we incorporate small amounts of a comonomer, such as 1-hexene, which reduce the crystallinity level. This improves toughness at the expense of stiffness.

High density polyethylene is widely used in films, pipes and in injection molded and blow molded applications. High density polyethylene can be cast or blown into very thin films, such as those used in grocery sacks and institutional trash can liners. They have a characteristic crisp sound when handled. Due to their high crystallinity, pipes and molded items are stiff, with excellent resistance to permeability by aqueous fluids. High density polyethylene deforms by yielding and necking, becoming white in the process, due to light scattering by voids formed within the material.

## 18.5.2 Low Density Polyethylene

The high levels of branching found in low density polyethylene disrupt its ability to crystallize, which limits its density in the solid state to approximately 0.91 to 0.94 g/cm$^3$, which equates to crystallinity levels ranging from approximately 30 to 55%. The highly branched nature of low density polyethylene and its broad molecular weight distribution give it high melt strength and good shear thinning properties, which are desirable processing characteristics. We frequently blend low density polyethylene into other types of polyethylene that are more difficult to process, thereby improving their processability.

Low density polyethylene has modest stiffness and moderately high tear and impact resistance. Films made from low density polyethylene are relatively transparent, because their spherulites are generally significantly smaller than the wavelength of light. Thicker samples are translucent, but text can still be read through a 3 mm thick plaque when it is placed in contact with a printed page. Molded items are flexible and resilient, requiring strains of 20% or more to initiate yielding.

### 18.5.3    Linear Low Density Polyethylene

We control the density of linear low density polyethylene by adjusting the percentage of 1-alkene. By varying the level of comonomer from approximately 2 to 8 mole % we can create polymers with densities ranging from approximately 0.94 down to 0.90 $g/cm^3$.

Linear low density polyethylene products have similar degrees of crystallinity and stiffness compared to their low density polyethylene counterparts. Transparency, flexibility, and resilience increase as density decreases. The properties of the higher densities of linear low density polyethylene approach those of high density polyethylene, and we sometimes refer to them as medium density polyethylene. Linear low density polyethylene products exhibit outstanding toughness, both as films and molded items. Films have excellent tear and puncture resistance, which suits them for the bulk packaging of heavy or irregularly shaped items. Lower density films made from Ziegler-Natta catalyzed resins feel tacky and tend to stick to each other, which we exploit in cling wrap. Grades made with metallocene catalysts are less tacky than those made with Ziegler-Natta catalysts, because they have narrower molecular weight and composition distributions. Molded items exhibit a desirable balance of flexibility and impact resistance. Higher density grades deform by yielding and necking. Lower density grades deform in a more uniform manner.

### 18.5.4    Very Low Density Polyethylene

We can vary the density of very low density polyethylene from 0.90 down to 0.86 $g/cm^3$ by varying the comonomer level from approximately 8 to 14 mole %. At the highest comonomer levels, crystallization is impeded by the branches to such an extent that only about 5% of the material crystallizes. The crystallites of very low density polyethylene are small and poorly organized. We polymerize these resins using single-site catalysts, which give us relatively narrow molecular weight and composition distributions

Very low density polyethylenes are relatively soft and transparent, because they are largely non-crystalline in nature. Films made from these materials are resilient with a moderate level of elastic recovery, as long as they are not stretched beyond strain levels of approximately 100%. At their lower densities, they feel tacky and tend to stick to each other. We often blend very low density polyethylene into higher density grades of polyethylene or isotactic polypropylene to boost their impact resistance.

### 18.5.5    Ethylene-Vinyl Acetate Copolymer

The properties of ethylene-vinyl acetate copolymers vary widely with their ester content. At the lowest levels of vinyl acetate, they have physical properties that are similar to those of low density polyethylene. As the comonomer content increases, the material becomes less crystalline and more elastic. Copolymers made with the highest comonomer levels contain no measurable crystallinity. The resulting products are tough, flexible, and clear. The ester moiety increases the chemical activity of these polymers relative to the purely hydrocarbon variants of polyethylene. In addition to their uses as packaging films, such as in ice bags, and coatings for cardboard and aluminum foil, we make extensive use of ethylene-vinyl acetate copolymers in adhesives.

### 18.5.6 Crosslinked Polyethylene

The physical properties of crosslinked polyethylenes vary widely depending on the resin on which they are based and their level of crosslinking. The crystallinity level of a crosslinked polyethylene is lower than that of its precursor resin, because the crosslinks impede molecular re-organization during crystallization. As a general rule, crosslinked polyethylenes are somewhat more flexible than their precursor resins, due to their lower degrees of crystallinity. Crosslinked polyethylenes are tougher than their precursors, because their chains are bound together to form a network. Unlike other types of polyethylene, crosslinked polyethylenes do not flow when their crystallites melt. We use crosslinked polyethylene in applications that may experience high temperatures or where exceptional toughness is required. Applications include insulation surrounding high voltage electricity distribution cables, chemical storage tanks, and whitewater kayaks.

### 18.5.7 Ethylene Ionomers

The crystallinity levels in ethylene ionomers are generally low due to their high levels of branching and the clustering of the metal salts. At high temperatures, the clusters dissociate and the individual chains can move independently in the molten state, permitting them to be molded. When the ionomer is cooled clusters reform, crosslinking the chains

Due to their low levels of crystallinity, ionomers are transparent and flexible. The ionomeric crosslinks prevent the chains from sliding past each other when a sample is stretched. When the stretching force is released, the material recovers elastically to its original dimensions. The crosslinks impart high toughness, abrasion resistance and resistance to being cut. The combination of high levels of toughness, elasticity, and cut resistance make these polymers the material of choice for the outer layer of golf balls, which are subject to tremendous physical abuse. Other uses include meat packaging, where their resistance to puncture by bones and high resistance to fat permeation are valued.

## 18.6 Products

As cast or blown films we encounter polyethylene in packaging films, bags, cling film, greenhouse skins, diaper back sheets, and high altitude scientific balloons. In these applications we value polyethylene's high toughness, impermeability to aqueous liquids, transparency, and the fact that we can draw it down to thicknesses as low as 10 μm. We use thicker sheets as geomembranes to line effluent collection ponds and landfills, where its impermeability, puncture resistance, and ready weldability are valuable assets. Other extruded items include rigid pipes and flexible tubing, which are used to distribute natural gas and transport aqueous fluids in agricultural, mining, and industrial settings. Polyethylene is used to insulate electrical conductors, ranging from the low voltage wire found in electronic devices to the high voltage electrical cables used to distribute electricity nationwide. In the latter application we crosslink the polyethylene in order to better withstand temporary overheating that could melt regular polyethylene. Ethylene-vinyl acetate copolymer is widely used as an extrusion coated layer on the cardboard used to make milk and fruit juice drink cartons. The thin layer of polymer adheres strongly to the cardboard, forming an impermeable lining that can be melted to form-liquid tight seals.

Polyethylene is widely used in blow molded bottles for foods, household products, and industrial chemicals, where its toughness, impermeability, and light weight are exploited. Injection molded products include crates, outdoor furniture, pails, toys, and food storage containers, where good, but not outstanding strength and stiffness are required. In general, the greater the load that the product will experience, the higher the density of polyethylene needed.

Rotationally molded objects include kayaks and agricultural chemical storage tanks, which are lightweight and tough, especially when crosslinked. Other large items, such as pick up truck bed liners and panels for portable sanitation facilities, are made by thermoforming.

Ultrahigh molecular weight ultra-oriented fibers have outstanding tensile strength and modulus, which suits them for use in bullet proof vests.

## 18.7    Conclusions

Given the wide range of properties that we can obtain from polyethylene, its relatively low cost and ease of conversion, it should be no surprise that it is the most commonly used of all synthetic polymers. It is ubiquitous, entering our lives and making them easier and more convenient in all manner of ways. In its various forms it has supplanted a host of traditional materials, particularly in the realm of packaging. Many household and industrial chemicals that were formerly packaged in metal cans or glass bottles and jars are now shipped in blown polyethylene bottles or cardboard cartons lined with polyethylene. Pallets and crates that used to be made of wood can now made from high density polyethylene, which is generally more versatile and durable. Paper bags have been largely supplanted by polyethylene bags, which are lighter in weight and stronger when wet. A brief inspection of our surroundings will confirm that polyethylene is one of the most useful and widely used materials that contributes to a comfortable and convenient lifestyle.

# Review Questions for Chapter 18

1.  Describe the seven different classes of polyethylene and explain how the structural differences affect the tensile strength, crystallinity and flexural modulus of each of the materials.

2.  What affect would a $Na^+$ cation have on the stiffness of an ionomer relative to a $Ca^{2+}$ cation? Why?

3.  What differences would be observed between linear low density polyethylenes manufactured with a Ziegler-Natta catalyst versus a single site catalyst, such as a metallocene?

4.  How do chain transfer events generate long chain branches in polyethylene?

5.  How is crosslinking achieved in polyethylene?

6.  How do long and short chain branching influence the crystallinity of polyethylene?

7.  What characteristics of high density polyethylene make it suitable for the manufacture of rigid pipe? What limitations would there be in the use of this pipe?

8.  Why is Ziegler-Natta linear low density polyethylene used to manufacture cling films?

9.  How does the vinyl acetate content in ethylene vinyl acetate copolymer affect the properties of the polymer?

10. What properties of polyethylenes make them excellent choices for packaging films?

# 19 Polypropylene

## 19.1 Introduction

Polypropylene, the general chemical structure of which is shown in Fig. 19.1, makes up approximately 30% of the polymer produced in the world today. We find polypropylene in a wide array of disposable and durable products. For example, dishwasher-safe food containers, plastic labels on soda bottles, indoor/outdoor carpet fibers, the non-woven fabric backing on disposable diapers, cleaning wipes, dishwasher liners, thermally insulating fabrics, and children's toys are all manufactured from polypropylene. This wide variety of applications stems from the resin's strength, toughness, and high melting point relative to other inexpensive polymers, as well as the ease with which it is processed. In this chapter, we will explore the chemical nature of polypropylene and will relate this to its properties, applications, and processing.

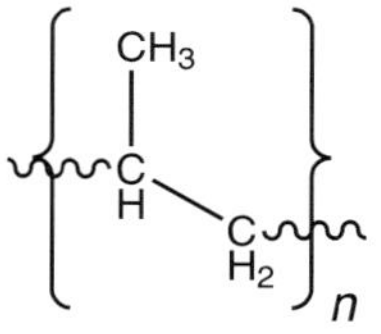

**Figure 19.1**    General chemical structure of polypropylene

## 19.2 Chemical Structure

The pendant methyl groups attached to one half of the carbon atoms of the polypropylene backbone play a major role in defining the characteristics of this polymer. The stereochemical arrangement of polypropylene's methyl groups gives rise to three distinct stereoisomers: isotactic, syndiotactic, and atactic, which are illustrated in Fig. 1.8. The general physical properties of solid polypropylene are defined by its stereoregularity (or lack of it) as well as by the other properties we have discussed throughout the book: average molecular weight, molecular weight distribution, and the degree of crystallinity. Propylene is routinely copolymerized with other monomers including ethylene, tetrafluoroethylene, and vinyl chloride.

### 19.2.1    Isotactic Polypropylene

Polymer scientists cannot create a 100% isotactic polypropylene on the commercial scale. Current catalyst technology, though, can produce materials that are more than 95% isotactic as determined by the percent of isotactic pentads (that is, the percentage of five adjacent repeat units that are isotactic). The advantages of the isotactic form of the polymer arise from its ability to crystallize. The degree of crystallinity of typical isotactic polypropylenes ranges from 40 to 70%, creating a polymer with a high melting point (160 to 180 °C) and a high density relative to the atactic form.

### 19.2.2    Atactic Polypropylene

Historically, the first polypropylene manufactured was an atactic, low molecular weight, soft and tacky material. The random placement of the pendant methyl group arises from sterically uncontrolled addition of the monomer. When monomers approach the reactive site, they randomly orient relative to the growing chain, creating the atactic form. Atactic polypropylene's irregular structure inhibits the regular packing required to create crystallites. The polymer exhibits poor strength, high tackiness and is, in general, not terribly useful as a thermoplastic resin in its pure state. It does, though, find a wide range of applications as a component of some adhesives and as a filler in asphalt, for some of the same reasons that make it ineffective in polymer applications.

### 19.2.3    Syndiotactic Polypropylene

Syndiotactic polypropylene became commercially available about ten years ago with the advent of single-site catalysts. Unlike its atactic and isotactic counterparts, its manufacture presented serious challenges to polymer scientists and engineers. Even under the best conditions, its syndiotacticity rarely exceeds 75%, based on pentad sequences. It typically has both a lower melting point (approximately 138 °C relative to approximately 155 to 160 °C) and density ($0.89 \text{ g/cm}^3$ relative to $0.93 \text{ g/cm}^3$) than isotactic polypropylene. Syndiotactic polypropylene crystallites have a much more complex structure than the isotactic form, which impedes its crystallization. Therefore, in general, the syndiotactic form of polypropylene crystallizes very slowly.

## 19.3    Structure and Morphology: Isotactic Polypropylene

Isotactic polypropylene exhibits four crystalline structures: alpha, beta, gamma and mesomorphic. Each of these structures forms under specific processing conditions and defines the properties of the polypropylene. In polypropylene containing the alpha, beta, and gamma structures, the final material is typically opaque due to the scattering of light by spherulites with sizes similar to the wavelength of visible light.

## 19.3.1  Alpha Crystallinity

The most common polypropylene crystalline form is the alpha structure, in which the polypropylene chain exhibits either a left or right handed helical conformation. This type of crystallinity forms monoclinic unit cells created by the alpha helices, which are shown in Fig. 19.2. Crystallites grow to form sheet-like lamellae due to the preferential growth of the crystalline regions perpendicular to the principle axis of the helices. Secondary lamellae grow tangentially from pre-existing lamellae, resulting in a characteristic crosshatched microstructure, as shown in Fig. 19.3. These tangential growth structures can interact with neighboring lamellae to create a far-spreading supermolecular structure.

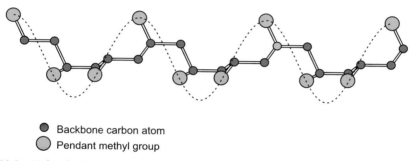

● Backbone carbon atom
◯ Pendant methyl group

**Figure 19.2**  Molecular helix of alpha crystalline form of isotactic polypropylene. (For simplicity only the backbone carbon atoms and pendant methyl groups are shown.)

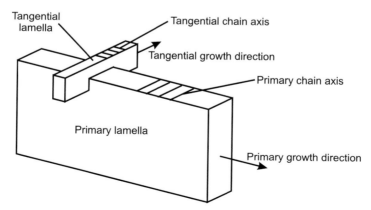

**Figure 19.3**  Crosshatched microstructure of alpha crystalline isotactic polypropylene

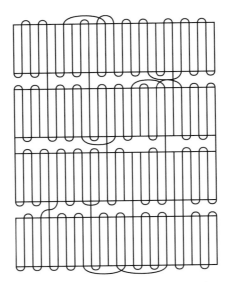

**Figure 19.4**    Parallel stacks of lamellae in beta crystalline isotactic polypropylene

## 19.3.2    Beta Crystallinity

The beta crystalline form of isotactic polypropylene differs from the alpha form by having a lower crystalline density and lower melting point. The beta form is metastable to the alpha form and will rearrange to the alpha structure when heated to approximately 100 °C or placed under strain. Beta lamellae form parallel stacks as shown in Fig. 19.4.

## 19.3.3    Gamma Crystallinity

Isotactic polypropylene's gamma structure rarely forms under standard processing conditions. It is believed that this form arises when an alpha crystalline material is sheared in the growth direction as crystals form. Gamma crystallites form a crosshatched structure that is similar to that of the alpha form. It differs, though, in that the crystallites grow in two directions simultaneously, as shown in Fig. 19.5. The resulting structure is more uniform than that seen in alpha crystalline isotactic polypropylene. The density of the gamma crystallites exceeds that of both the alpha and beta crystallites.

## 19.3.4    Smectic or Mesomorphic Crystallinity

We find an additional crystalline structure in rapidly quenched isotactic polypropylene products. In these materials, the polymer chains do not have the necessary time to orient,

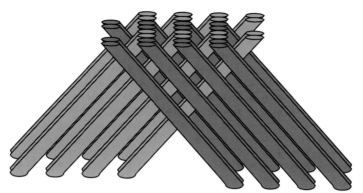

**Figure 19.5**  Microstructure of gamma crystalline isotactic polypropylene

preventing them from forming the large crystalline domains seen in the other three forms. This means that the crystalline regions are small and that there is poor alignment among the individual crystallites. The small crystallites are surrounded by amorphous regions. The behavior of smectic isotactic polypropylene falls between that of alpha crystalline polypropylene and the atactic form. The poorly organized crystallites do not effectively scatter light resulting in a transparent material. This form of isotactic polypropylene is metastable, reorganizing to form alpha crystallinity when heated.

## 19.4    Manufacture

The method we choose to manufacture polypropylene defines its tacticity, average molecular weight, and molecular weight distribution. Early attempts to make polypropylene involved the pressurization of propylene in the presence of a radical generator, akin to the manufacture of low density polyethylene described in Chapter 18. This technique yielded oligomers with little commercial value. It was not until the invention of Ziegler-Natta catalysts that the production of isotactic polypropylene became commercially feasible. Continued catalyst research led to the discovery of metallocene catalysts. Both Ziegler-Natta and metallocene catalysts are used commercially, the choice between the two depends largely on the desired molecular weight distribution.

### 19.4.1    Ziegler-Natta Catalyzed Polymerization of Polypropylene

Polypropylene grades manufactured using Ziegler-Natta catalysts are predominantly isotactic, with a comparatively broad molecular weight distribution ($M_w/M_n \approx 3.5$). Ziegler-Natta catalyzed polypropylene comprises the majority of commodity grade resins.

### 19.4.1.1  Ziegler-Natta Catalysts

The Ziegler-Natta catalysts developed for the polymerization of propylene are based on the same chemistry used to produce polyethylene. The key difference is that $TiCl_3$ and aluminum diethyl chloride replace the $TiCl_3$ and trialkyl aluminum system used for polyethylene. Polypropylene made with this catalyst system has a high isotactic content and can reach high molecular weights (over $4 \times 10^6$ g/mol on a weight basis) making it viable for the fabrication of polymeric parts. Since the development of the initial Ziegler-Natta catalyst system, further work focused on ways to increase the efficiency of the catalysts as well as increase the tacticity of the final polymer. Several means of achieving this were discovered, including modifying the crystal structure of the $TiCl_3$, increasing the surface area of the inorganic catalyst support particles, purifying the catalyst systems, using different transition metal compounds as the co-catalyst, and introducing a Lewis base, such as ethyl benzoate, as an electron-donating species.

### 19.4.1.2  Manufacturing Polypropylene with Ziegler-Natta catalysts

We have a choice of four major polymerization techniques by which to manufacture polypropylene using Ziegler-Natta catalysts: slurry, liquid propylene, solution, and gas phase. Regardless of which technique is employed, all polymerization plants must accomplish the same basic goals: they must

a) introduce and mix the monomer with the catalyst system,

b) achieve a high conversion rate from monomer to polymer,

c) recycle unreacted monomer, and

d) purify the final product.

*Slurry Process Polymerization of Polypropylene*

In the slurry process, propylene monomer is dissolved in a hydrocarbon diluent in which the polymerization process occurs. The polymerization products are either soluble (the highly atactic components) or insoluble. Both the insoluble and soluble components are collected and form separate product streams. The insoluble species form a slurry in the solvent, from which they are removed by centrifugation. The soluble, atactic component is removed with the solvent as another product stream. To separate the atactic polymer from the solvent, the solution is heated allowing the solvent to flash off, leaving the atactic polymer behind. Any unreacted monomer is degassed from the solution and recycled to the start of the polymerization process.

*Liquid Propylene Polymerization of Polypropylene*

Another method of manufacturing polypropylene employs the liquid monomer as the polymerization solvent. This process, known as the liquid propylene or bulk-phase process, has a major advantage over the slurry method in that the concentration of the monomer is extremely high. The high concentration increases the rate of the reaction relative to that seen in the slurry process. In addition, the heat of polymerization can be removed from the process by the vaporization of the monomer. The gaseous monomer is then recycled to the reactor, after liquefaction, as condensed monomer. Just as in the slurry process, the polypropylene forms an insoluble phase in the propylene diluent. The insoluble phase is isolated from the liquid propylene. Unlike the slurry process, the insoluble phase contains both the atactic and isotactic fractions. Separation of these two components requires an additional step in which the soluble portion is dissolved into an organic solvent.

### Solution Polymerization of Polypropylene

During solution polymerization the monomer, catalyst, and diluent are introduced to a reactor maintained at a temperature between 175 and 250 °C. The resulting polymer forms a viscous solution in the solvent, which is pumped out of the reactor. If necessary, the solution can be filtered to remove the catalyst residue. Solvent is removed from the solution and recycled, leaving behind a mixture of isotactic and a small amount of atactic polypropylene.

### Gas Phase Polymerization of Polypropylene

In gas phase reactors, the monomer is introduced to the bottom of reactor where it percolates up through a fluidized bed of polymer granules and inert-media supported catalyst. A fraction of the monomer reacts to form more polymer granules, the remaining monomer being drawn from the top of the reactor, cooled, and recycled. Polymer granules are continuously withdrawn from the bottom of the fluidized bed and the catalyst is replenished.

## 19.4.2   Metallocene Catalyzed Polymerization of Polypropylene

Ziegler-Natta catalysts were essential to the development of the polypropylene industry, but there were limitations preventing the manufacture of low polydispersity or controlled tacticity polymer with these catalyst systems. Additionally there was a need for improved catalyst efficiency. Controlling tacticity and polydispersity allows more specialized products to be manufactured, while improved catalyst performance reduces the catalyst residue in the final product. The most effective means of achieving these goals found to date is with the use of metallocene catalysts. As described in Chapter 2, metallocene catalysts control the polymerization by permitting only those monomers that approach with a specific orientation to attach to the growing chain. This results in increased control of the tacticity in the final polymer. In addition, the metallocene catalysts have only one active polymerization site, which results in a narrower molecular weight distribution than achieved by Ziegler-Natta catalysts ($M_w/M_n \approx 2.0$). These catalysts introduce their own type of defect which actually results in a slightly lower melting temperature than can be achieved with the best Ziegler-Natta catalysts.

Metallocenes are homogeneous catalysts that are often soluble in organic solvents. Therefore, polymerization can occur via a solution process with a non-polar diluent dissolving the propylene gas, the catalyst, and the co-catalyst system. They can also be adsorbed onto an inert substrate which acts as part of the fluidized bed for gas phase polymerization processes.

## 19.5    Structure and Property Relationships

In general, isotactic homopolymer polypropylene has a high degree of crystallinity thereby creating a material that is strong, with low permeability to vapor or solvents, and high chemical resistance. Isotactic polypropylene – with the exception of the mesomorphic crystalline form – is typically opaque, due to the high concentrations of crystalline regions. Syndiotactic polypropylene, which has lower crystallinity, is transparent and not as strong. The atactic form, due to its lack of crystallinity, has poor physical strength with lower resistance to dissolution in solvents and greater permeability to low molecular weight gases such as oxygen and water vapor.

Polypropylene's melting temperature is high, though nowhere near as high as that seen in the engineering polymers, making it a useful material for many high-temperature applications. It also has a high glass transition temperature of approximately 0 °C. The result of this is that polypropylene is prone to brittle failure on impact, which is especially noticeable at low temperatures. To address the engineering issues created by the high propensity for brittle failure, polypropylene is often compounded with an impact modifying agent.

### 19.5.1    Oriented Polypropylene

Sometimes the hazy optical properties of polypropylene are a detriment to its use in thin films. To address this problem, as well as to improve physical properties and reduce material costs, we can orient the molecular chains by stretching the polymer film after it has left the extrusion die. There are two methods of orienting films, film blowing, shown in Fig. 11.6, and tentering, shown in Fig. 19.6.

During blown film processing, orientation is imparted to the film by both pulling the melt into the nip rollers and by increasing the circumference of the bubble with air. By managing

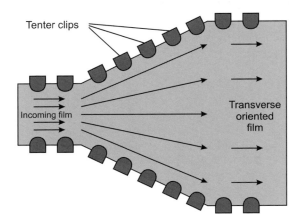

**Figure 19.6**    Tentering of cast Film

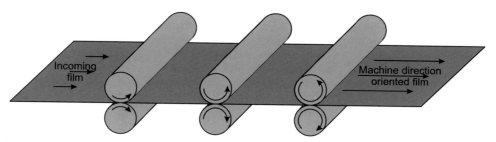

**Figure 19.7**    Machine directional orientation using pairs of nip rollers

the rate at which we pull the film from the die and the amount of air we use to expand the film bubble we can control the relative levels of machine and transverse orientation that we introduce.

The tentering process takes a warmed sheet of polymer from a flat die and stretches it by pulling it using clips that clamp to the edges of the film sheet. This creates orientation in the transverse direction, perpendicular to the machine direction. Tentering can be coupled with machine directional orientation by passing the sheet through a series of nip rollers which rotate at increasing speeds, as shown in Fig. 19.7. A typical ratio of the final to initial roll speeds is approximately 2 : 5. Orientation is induced since the sheet leaves the final set of nip-rolls at a higher speed than it enters the first set of nip-rolls. Another means to create the machine directional orientation is to allow the tenter clips to pull the film in both directions simultaneously. This happens by accelerating the clips as they carry the film through the oven.

In both methods used, the orientation interferes with the formation of spherulites. Beyond the aesthetic reasons for orienting polypropylene, there are significant changes in the properties of oriented films relative to the unoriented material. The polymer chains are oriented during stretching, thereby improving the tensile properties. This improvement results from the alignment of the crystallites and polymer molecules, which reduces the free volume of the material. Failure then occurs either through breaking a large number of intermolecular interactions or by rupturing the chains of the individual polymer molecules. Both of these processes require high energy input. In addition, oriented materials have higher impact strength because the alignment of the chains allows dissipation of the impact forces over a wider area. Finally, vapor permeability drops as a result of the alignment of the crystallites which reduces the free volume and increases the tortuosity of the path traveled by a permeant.

## 19.5.2  Impact Modified Isotactic Polypropylene

Even though isotactic polypropylene has poor impact strength, its other properties and low cost attract engineers. For this reason, impact modifying agents have become a common ingredient in polypropylene products that require high impact strength, especially at low

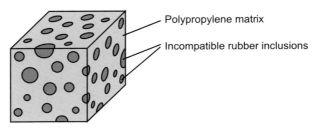

**Figure 19.8**   Two-phase structure of impact modified polypropylene

temperatures. Soft rubbers that are incompatible with the polymer matrix are effective impact modifiers. In the solid state, a two-phase structure forms, as shown in Fig. 19.8. Small discrete rubber particles are surrounded by a continuous matrix of polypropylene. The rubber particles enhance the material's impact strength by blunting the cracks that propagate during the impact event. The rubber also absorbs and helps dissipate the impact energy.

## 19.6    End Uses

Polymer scientists and engineers have developed a wide variety of processes by which to convert polypropylene into useable articles. These processes include fiber-spinning, production of melt blown and spun bond non-woven fabrics, film production, blow molding, and injection molding. The following sections describe these processes and the resulting properties of the product as a function of the process.

### 19.6.1    Fibers

Fibers created from polypropylene are used in carpets, thermal underwear, filters, disposable medical garments, and many other applications because of the ease of their production and their high strength. Regardless of the manufacturing process, all fibers are comprised of oriented molecules. For carpet fibers and yarns, we extrude a very high melt flow polypropylene through a die with many tiny holes, called a spinneret. The number of holes in the spinneret determines the number of filaments in the final yarn. The yarn is rapidly drawn down as it exits the spinneret and simultaneously cooled. Orientation of the polypropylene molecules occurs as the melt is drawn from the die orifice to the spindle that collects the fiber and locked in when the material cools. Further orientation can be imparted by the use of godet rolls, as shown in Fig. 11.7.

## 19.6.2     Melt-Blown and Spun-Bond Non-Woven Fabrics

Spun-bond and melt-blown non-woven fabrics are manufactured from very fine filaments that are collected together and bonded to one another. In both cases, a low viscosity melt is pumped through the small orifice of the spinnerets under pressure and the material is collected on a passing screen which forms the final fabric as shown in Fig. 11.8. The advantage of these materials over traditional woven fabrics, beyond the reduced cost, is seen in the random way that the individual filaments overlay one another. The randomness creates a tortuous pathway for liquids or solids with pore sizes of 1 to 100 μm, depending on the processing method used. The smaller the pore size, the more effective the barrier properties of the material are. In addition, the overall density can be as low as 5 g/m$^2$ which is considerably less than for lightweight woven fabrics (90 to 160 g/m$^2$). This combination makes non-woven fabrics ideal for use in medical drapes, medical garments, and disposable pillow covers as used on airplanes. Non-woven fabrics are found in a wide variety of applications. Spun-bond fabrics are often used on cloth-feel baby diapers and mattress covers. Melt-blown fabrics are often used in conjunction with spun-bond fabrics in medical drapes and disposable hygienic garments. The melt-blown fabric provides the highly impermeable layer while the spun-bond creates the strength necessary for the article.

In the spin-bonding process, many filaments are extruded simultaneously from a die. The filaments are stretched and pulled into a collector by a high pressure air stream. The filaments cool as they leave the extruder and pass onto a collection screen. A vacuum creates a negative pressure at the surface of the screen, thereby holding the extrudate tightly to the screen. The extrudate of a spun bond process has a larger diameter than a melt-blown process. Because of the thicker diameter, the process is less demanding, creating a stronger fabric than that made by the melt-blown process.

Melt-blown filaments are very fine and are, therefore, suitable for the manufacture of lightweight fabrics. During the melt-blown process, the low viscosity resin is pulled swiftly from the die spinnerets by high temperature, high velocity air. The material is blown onto a forming screen which moves under the curtain of filaments. The molecular weight of the resins used in this process must be very low to reduce the viscosity of the melt. The resins must also be very clean, as small impurities can block the orifices in the spinnerets. Finally, maintenance of the equipment is also quite demanding. Any charred resin or contaminant particles can plug the spinnerets resulting in off specification product.

## 19.6.3     Films

Polypropylene film is found in a variety of applications. Films are thin plastic sheets with thicknesses less than 250 μm. Applications for these films include: candy wrappers, plastic label material, electrical capacitors (which requires very pure polymer), and fiber glass insulation backing. There are two general categories of polypropylene films: unoriented and oriented. Films are manufactured through either a cast or a blown film extrusion process. Oriented films exhibit improved strength relative to unoriented films. This means that they can be made much thinner than an unoriented film, while maintaining the same tensile properties of the thicker film. Oriented polypropylene is used as the overwrap for cigarette packages and electronic storage media such as compact disk and digital video disk cases.

### 19.6.4  Extrusion Coating

Extrusion coating is the process by which a very thin layer of polymer is extruded onto a substrate, such as cardboard, another polymer, or metal sheet. We find polypropylene in extrusion coated packaging applications that need high grease resistance or high thermal stability. Due to the acceleration of the molten curtain between the die and the rapidly moving substrate, the polymer becomes oriented. Extrusion coated polypropylene is found on non-woven fabrics used as geomembranes, on the peel-off sheet of microwavable meals, and on other packaging materials.

### 19.6.5  Injection Molded Parts

Polypropylene is often formed into useful articles by injection molding. For example, disposable cutlery and cups, plastic outdoor furniture, children's toys, electronic housings, dishwasher linings, washing machine tubs, pallets and crates can all be manufactured from polypropylene. The high strength and low cost of the polymer explains the wide variety of applications. During processing, some orientation of the molecules occurs as the polymer passes through narrow entry gates into the die. Cooling occurs rapidly at the surface of the part, leading to small oriented crystalline regions in the outer skin. The interior, due to the longer cooling time, contains larger, less oriented crystalline regions. This mismatch in crystalline morphology can lead to part warpage since the density of the interior is higher than that of the exterior. To counteract this, we spend a great deal of time and effort on constructing dies that will prevent preferred orientation during mold filling and adjust the molding conditions such that there is more uniform cooling throughout the part. The choice of the grade of polypropylene for injection molding depends on the needs of the final part as well as the processing requirements. The balance between part strength, which would be increased with high molecular weight resins, and ease of processing, which requires low viscosity resins, must always be considered.

## 19.7  Conclusions

Polypropylene is a versatile polymer used in applications ranging from carpet fibers to packaging films and medical garments to washing machine tubs. The wide variety of applications results from the relatively high temperature resistance compared to other inexpensive polyolefins and the high crystallinity which imparts strength to the final parts. It is also, generally, easy to process and is therefore amenable to a wide range of processing methods.

# Review Questions for Chapter 19

1.  How does the type and level of tacticity of the polypropylene chain affect the degree of crystallinity of the solid polymer?

2.  Which crystalline form of polypropylene exhibits the lowest free energy of crystallization? Which form exhibits the highest?

3.  Compare and contrast the slurry, gas phase, solution and liquid propylene polymerization methods.

4.  How do metallocene catalysts define the molecular weight distribution and tacticity of polypropylene?

5.  Oriented polypropylene is used for clear films used as food wrap. What advantageous properties do the oriented films have over standard films? How can a polymer that is typically semicrystalline be transparent?

6.  Atactic polypropylene exhibits a greater vapor permeability relative to either syndiotactic or isotactic polypropylene of the same molecular weight distribution. Why is this so?

7.  How do impact modifiers improve the toughness of polypropylene?

8.  Oxidative degradation of polypropylene chemically incorporates oxygen to the polymeric chains. Why would an isotactic polypropylene be less susceptible to this type of degradation than one that is atactic?

9.  What causes polypropylene injection molded parts to be susceptible to warpage upon cooling? How can this issue be addressed during the manufacture of these parts?

10. Rank the following polypropylene conversion processes in terms of their relative sensitivity to particulate contamination: film casting, injection molding or fiber spinning.

# 20    Polycarbonates

## 20.1    Introduction

Polycarbonates are condensation polymers with many desirable properties. They exhibit very high thermal stability, have a high heat distortion temperature and, despite their hardness, display ductile rather than brittle failure on impact, making them very tough. The combination of these physical properties places polycarbonates in a select class known as engineering polymers. Engineering polymers display properties that meet the needs of challenging applications such as high temperatures, or endure considerable physical abuse during the lifetime of the product. In addition, polycarbonates have high clarity and process easily on standard extrusion, injection, and blow molding equipment. By far the most commonly used polycarbonate, making up over 90% of commercial usage, is produced from the condensation reaction between bisphenol A and a difunctional, proton-accepting species such as diphenyl carbonate or phosgene. The general chemical structure of this type of polycarbonate is illustrated in Fig. 20.1.

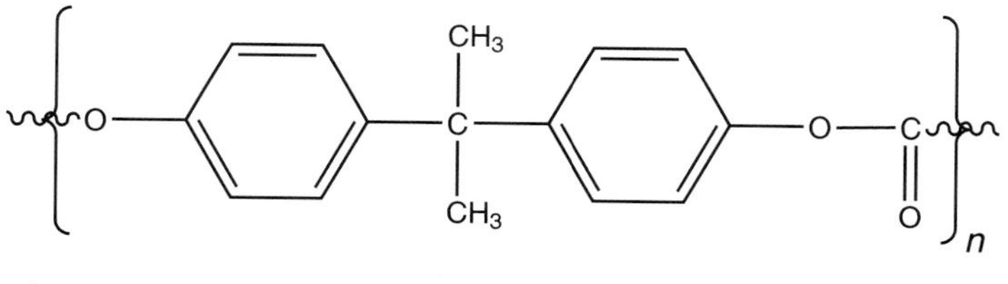

**Figure 20.1**   Chemical structure of bisphenol A based polycarbonate

## 20.2    Polymer Chemistry

Bisphenol A, whose official chemical name is 2,2-bis(4-hydroxyphenyl)propane, is a difunctional monomer with two reactive hydroxyl groups, as shown in Fig. 20.2. It polymerizes with dicarbonyl organic monomers, such as phosgene or diphenyl carbonate, which are illustrated in Fig. 20.3. During polymerization, shown in Fig. 20.4, the hydroxyl groups of the bisphenol A deprotonate in the presence of a base. After deprotonation, the oxygen atoms on the bisphenol A residue form ester bonds with the dicarbonyl compounds. The polymerization process terminates when a monohydric phenol reacts with the growing chain end.

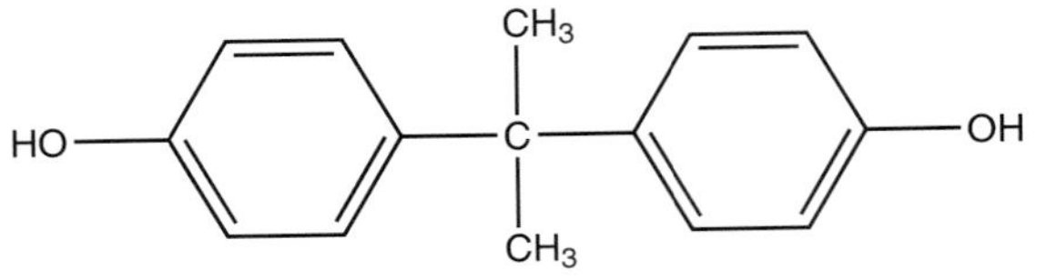

**Figure 20.2**    Chemical structure of bisphenol A

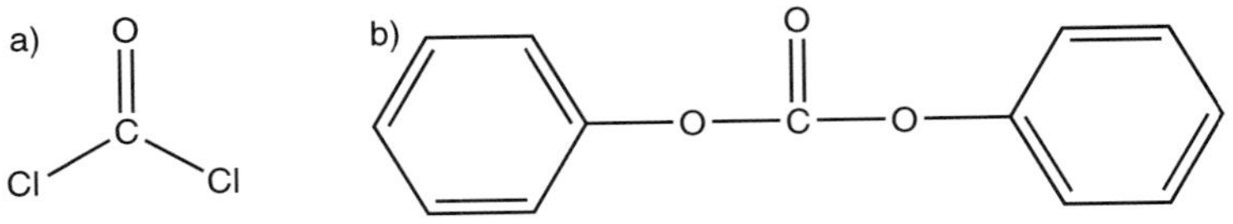

**Figure 20.3**    Chemical structures of:
                   a) phosgene and b) diphenyl carbonate

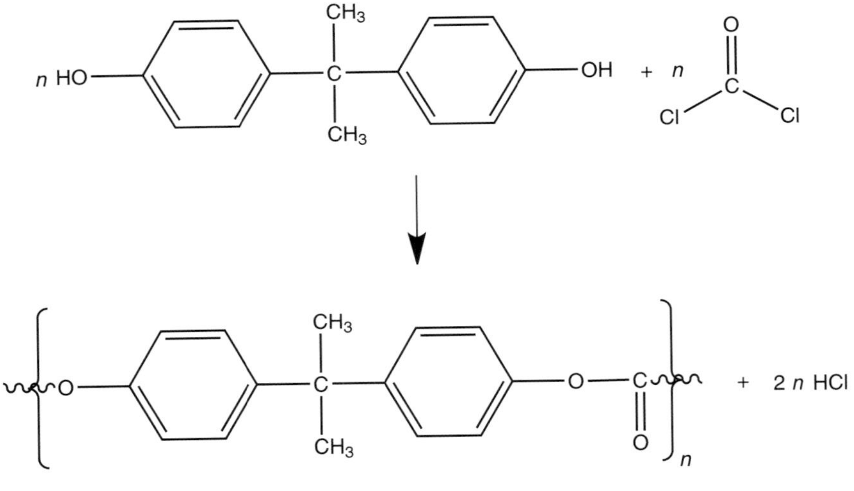

**Figure 20.4**    Polymerization of polycarbonate from bisphenol A and phosgene

In addition to bisphenol A, there are several other bisphenol-based monomers that can be used to create polycarbonates. The choice of a different monomer depends on what properties we wish to alter. We consider using a modified bisphenol when we wish to improve or alter specific properties, including heat resistance, ductility, and/or glass transition temperature. A variety of functional groups or bridging groups can modify the bisphenol structure creating different properties in the final polymer. Figure 20.5 shows two such functionalized bisphenols. In addition to altering the bisphenol structure, we also can choose different co-monomers other than phosgene or diphenyl carbonate. The two most common other condensation reactants with bisphenol A are ortho- and para- terephthaloyl chloride, the chemical structures of which are shown in Fig. 20.6.

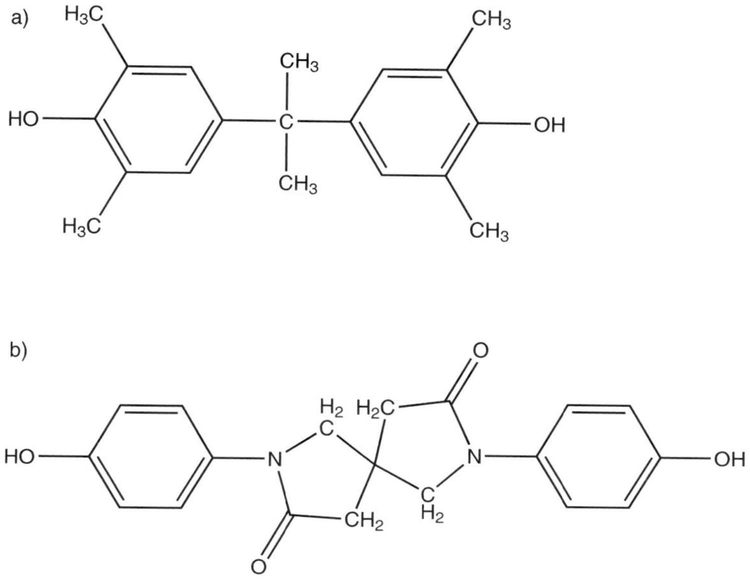

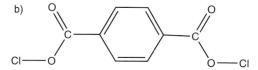

**Figure 20.5**  Chemical structures of non-bisphenol A monomers:
a) 3,3′,5,5′-tetramethyl bisphenol and
b) spirodilactam bisphenol

**Figure 20.6**  Chemical structures of alternative condensation reactants with bisphenol A:
a) ortho-terephthaloyl chloride and
b) para-terephthaloyl chloride

## 20.3    Structure and Morphology

Polycarbonates exhibit no crystalline structure when used in common manufacturing processes. At the molecular level, there is some evidence of localized ordering along a chain. It is believed that the polycarbonate repeat units can fold back onto the chain in a structure that resembles the letter Z. These individual units do not associate on a large enough scale to create a regular crystalline material. Under specific conditions, such as forcing the polymer to cool very slowly, we can create small crystalline domains. Since these conditions are not met in commercial processes, it is safe to say that the polycarbonates that we encounter are universally amorphous.

## 20.4    Manufacture

Polycarbonates are manufactured via interfacial polymerization or through a melt esterification process. The properties of polycarbonate can differ greatly based on the method of polymerization. Specifically, the molecular weight distributions created by the two methods differ because of kinetic effects. Polycarbonates manufactured via interfacial polymerization tend to be less stable at high temperatures and less stiff than those produced via melt esterification, unless proper manufacturing precautions are taken. Therefore, when choosing a polycarbonate resin grade for a specific application, it is important to know the method by which it was produced. Either polymerization method can be performed as a continuous or batch process.

### 20.4.1    Interfacial Polymerization

At the start of interfacial polymerization, bisphenol A is dissolved in methylene chloride, then introduced into a reactor. Phosgene is injected into the reactor as a liquefied gas together with an aqueous solution of sodium hydroxide. The methylene chloride and the aqueous solutions are immiscible; polymerization occurs at the interface between them. The reactants are combined in a rapidly stirred reactor as shown in Fig. 20.7. The sodium hydroxide neutralizes the hydrochloric acid that is generated by polymerization, while the organic phase serves as a solvent for the polymer. The organic phase is separated and washed to remove traces of the base or salts after which the solvent is removed.

Polymerization occurs very quickly and the process is controlled via kinetic effects rather than thermodynamic ones. The net result is that the molecular weight distribution of the product does not match the thermodynamically stable one. If the chains were not capped with monofunctional phenols, the polymer chains would depolymerize, allowing the monomers to rearrange themselves at elevated temperature to approach the thermodynamically stable

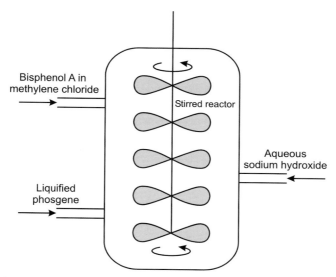

**Figure 20.7**   Schematic diagram of interfacial polymerization process

molecular weight distribution. In order to reduce the likelihood of reorganization during processing it is standard practice to minimize the time that the polymer is molten. It is important to remove as much of the methylene chloride as possible to avoid it plasticizing the final material, thereby reducing its stiffness.

## 20.4.2   Melt Esterification Polymerization

Melt esterification occurs when bisphenol A and diphenyl carbonate are maintained at a temperature in excess of 180 °C. At this temperature, the condensation reactants, oligomers, and final polymer exist in the molten state. The pressures in the reactor are initially very high, in the range of 150 to 200 atmospheres, and then reduced to sub-atmospheric pressures immediately preceding the outlet of the reactor. The large pressure drop allows the devolatilization of the generated phenol. The most common catalyst used in this process is sodium hydroxide; other catalysts that have been investigated include fluoride and phosphate salts, amines, and guanidine. The esterification reaction occurs under thermodynamic rather than kinetic control. This means that the final product is less likely to suffer the molecular weight redistribution problems seen in the polycarbonates produced by the interfacial reaction. Typically, the purity of the monomer defines the purity of the final product. Therefore, it is possible to achieve extremely pure polycarbonates via this method. Also, since no solvent is introduced, there is no concern with residual solvents plasticizing the final product. Despite the many advantages of melt esterification, the higher cost associated with the process makes it the lesser used method of the two manufacturing processes.

## 20.5    Structure and Property Relationships

Polycarbonates typically exhibit exceptionally high impact resistance, good thermal stability, excellent clarity, and high modulus at a wide range of temperatures. For these reasons, they have found use in many applications with challenging performance requirements, such as fire fighter's helmets, power tool and appliance housings, face shields, and automotive and aircraft instrument panels. Negative attributes of polycarbonates include low solvent resistance, photodegradation when exposed to ultraviolet and gamma radiation, and a high susceptibility to crazing, especially when exposed to solvents, mechanical stresses, or high temperature conditions.

### 20.5.1    Impact Properties

The most desirable property of polycarbonates is their high ductility on impact, relative to other engineering polymers in the unmodified state. There is no consensus on the mechanism of ductility; researchers continue to explore this behavior through molecular dynamics studies of chain segment motion during the formation of crazes and propagation of the failure.

### 20.5.2    Heat Resistance and Modulus

The high modulus of polycarbonates, even under high temperature conditions, arises from the high glass transition temperature (141 to 150 °C) of the polymer. Below this temperature, the material is rigid with little distortion under load. Improved modulus at high temperatures can be attained by raising the glass transition temperature, which can be accomplished by judicious selection of the bisphenol.

To raise the glass transition temperatures, substituents can be added to the phenol groups, thereby hindering the motion of the polymer chain segments. If only one functional group is added to the phenol ring on the bisphenol the glass transition temperature can be depressed. Rotation along the polymer backbone will not be hindered much while steric interference prevents the formation of stable intermolecular interactions. If a second functional group is added to the ring, the glass transition temperature increases. This happens because the presence of the second group reduces segmental associations. The bisphenol structure can also be modified to affect the rigidity of the final polymer. For example, the structure shown in Fig. 20.5b inflexibly joins the two phenol rings to one another. The inflexible molecular architecture creates a rigid polycarbonate backbone, thereby creating a stiffer polymer. In general, the stiffness of the polymer increases as the rigidity of the functionalized bisphenol moiety increases.

### 20.5.3    Optical Properties

The very high clarity of polycarbonates is a result of their amorphous nature. There are no crystalline/amorphous interfaces that can scatter light, which would lead to opacity. Polycarbonate's refractive index differs very little from glass, making the material an excellent choice for glass replacement applications. For example, clear produce drawers in refrigerators are often made of polycarbonate. Polycarbonate mimics the appearance of glass by creating a sparkling, clear drawer with low weight and high structural integrity. Polycarbonate has unique birefringence behavior as a result of small-scale orientation of the polymer chains in injection molded parts. You see this as a rainbow effect where a spectrum of colors is observed at specific angles of observation. Injection molded materials exhibit birefringence due to a highly ordered, anisotropic skin.

## 20.6    Manufacturing Methods Using Polycarbonate

We routinely manufacture polycarbonate products by injection molding, blow molding, extrusion, and thermoforming. Injection molding is the most common processing technique.

### 20.6.1    Injection Molding of Polycarbonates

Polycarbonate can be readily injection molded. Polycarbonates typically require only a short injection molding cycle time, because the polymer flows into the mold easily and solidifies rapidly. We injection mold polycarbonate to produce a wide variety of commercial goods, including compact disks, jewel cases, aircraft windows, kitchen utensils, and clear refrigerators drawers. Polycarbonates are also found in a wide range of disposable medical devices, such as the flow locks on intravenous tubes and the hard, disposable components of dialysis machines. Impact resistant polycarbonate is used to manufacture sports and other safety helmets. Glass fiber reinforced polycarbonate is used in the housings for power tools.

### 20.6.2    Blow Molding

We frequently blow mold polycarbonate to make non-disposable water bottles and water cooler storage tanks. These applications take advantage of the polymer's high impact resistance and low levels of extractable molecules. The low extractable levels mean that water will not become tainted or taste funny after long term storage in polycarbonate containers. Long-chain branched polycarbonates are used in these applications, because blow molding requires the formation of a parison with sufficiently high melt strength to be self supportive. Long-chain branching increases the entanglements to provide the required high melt strength.

### 20.6.3    Sheet Extrusion and Thermoforming

We extrude polycarbonate sheets in a wide range of thicknesses that are subsequently thermoformed. Thermoforming is used to make large parts, such as automotive instrument panels and the large light-up advertising signs in front of fast food restaurants. Polycarbonates are chosen for automotive instrument panels because of their high impact resistance and their ability to withstand the temperatures of up to 70 °C that can be encountered in automotive interiors. Light-up signs take advantage of polycarbonate's high heat distortion temperature and excellent impact resistance. An additional advantage is that it accepts colored pigments and dyes well as it is naturally both transparent and colorless.

## 20.7    Conclusions

Polycarbonate is an extremely versatile polymer with a wide range of applications. Its stiffness, high heat distortion temperature, impact resistance, and clarity make it a favorite choice for durable goods with demanding application requirements. Its high modulus allows it to be used in many automotive, medical and domestic applications.

## Review Questions for Chapter 20

1.    How does the molecular architecture of the bisphenol molecule affect the physical properties of the final polycarbonate polymer?

2.    How does the molecular architecture of the bisphenol molecule affect the glass transition temperature of polycarbonate?

3.    Why is it necessary to cap off the polycarbonate chains with a monofunctional molecule when producing the polymer via interfacial polymerization?

4.    What advantages does melt esterification have over interfacial polymerization in manufacturing polycarbonates?

5.    What two phases create the interface at which polymerization occurs in interfacial polymerization of polycarbonate?

5.    What properties of polycarbonate allow it to be classified as an engineering resin?

6.    Why are polycarbonate resins easy to process via injection molding?

7.    Why are polycarbonates used to manufacture drinking water bottles?

8.  What molecular process creates the birefringence effect seen in some injection molded polycarbonate parts?

9.  What is the difference between brittle and ductile failure upon impact? Why is ductile failure preferable for most applications?

# 21    Polystyrene

## 21.1    Introduction

The majority of commercial polystyrene molecules consist of a backbone of carbon atoms with phenyl groups attached to half of the carbon atoms, as shown in Fig. 21.1. Free radical initiator residues terminate each end of the chain. Minor variants include chains terminated by anionic or cationic initiator residues. All commercial polystyrene products are atactic; that is, the placement of the phenyl groups on either side of the chain is essentially random, as illustrated in Fig. 21.2.

Pure polystyrene is stiff, transparent, reasonably strong, brittle, chemically inert, and a good electrical insulator. It can be readily converted into useful items by extrusion, injection molding, blow molding, and thermoforming. In its foamed state it is an excellent thermal insulator, which can be formed into sheet, tubes, and more complex shapes. Polystyrene can be chemically modified by grafting it to rubbery polymers, which greatly increases its toughness. These characteristics, coupled with its relatively low cost, guarantee that polystyrene finds use in a broad range of applications.

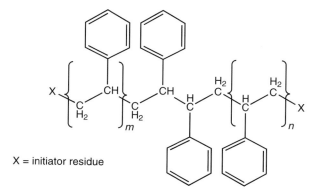

X = initiator residue

**Figure 21.1**    Chemical structure of most commercial polystyrene molecules

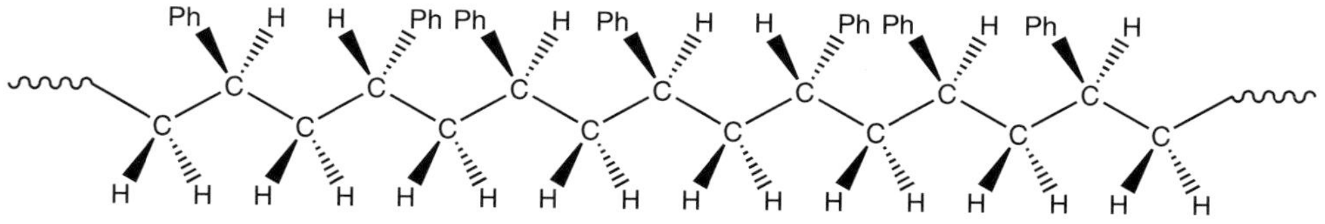

**Figure 21.2**    Atactic polystyrene molecule

## 21.2     Chemical Structure

There are three principal families of styrene containing polymers, which are used to make commercial plastic products. The first family is pure polystyrene, the second family comprises random copolymers, and the final family consists of polystyrene chains grafted to blocks of rubbery polymers. There are also synthetic rubbers that contain significant concentrations of styrene, but these are outside the scope of this book.

### 21.2.1     Pure Polystyrene

All commercial polystyrene is atactic. The random placement of polystyrene's aromatic rings prevents it from crystallizing. At room temperature, pure polystyrene is a glass; that is, it is an amorphous solid that is below its glass transition temperature of approximately 100 °C. Commercially, the term "crystal polystyrene" is often used to describe pure polystyrene, because objects made from it are transparent, stiff, and shatter when dropped. This term is a misnomer because it incorrectly suggests the presence of crystallinity. The density of atactic polystyrene at room temperature is approximately 1.04 to 1.05 g/cm$^3$. Pure polystyrene is essentially non-polar, which makes it chemically inert and a good thermal and electrical insulator. Polystyrene is soluble in a number of common organic solvents, including toluene, benzene, and chloroform.

Small amounts of isotactic polystyrene have been synthesized in the laboratory using non-commercial polymerization techniques. These polymers are capable of partially crystallizing, albeit at a very slow rate. Syndiotactic polystyrene was available commercially for several years, but its continued production proved unprofitable.

### 21.2.2     Random Copolymers of Polystyrene

We can readily copolymerize styrene with a variety of comonomers. Commercially, the two most important random styrene copolymers are styrene co-acrylonitrile and styrene co-butadiene, the general chemical structures of which are shown in Fig. 21.3.

By incorporating acrylonitrile into polystyrene we can depress the copolymer's glass transition temperature below that of pure polystyrene. When sufficient acrylonitrile is present, the copolymer's glass transition temperature falls below room temperature. The resulting copolymer is tough at room temperature and at higher temperatures.

Styrene co-butadiene is a rubbery amorphous polymer with a glass transition temperature well below room temperature. Polystyrene co-butadiene is an important component of several commercial families of plastic that contain polystyrene blocks.

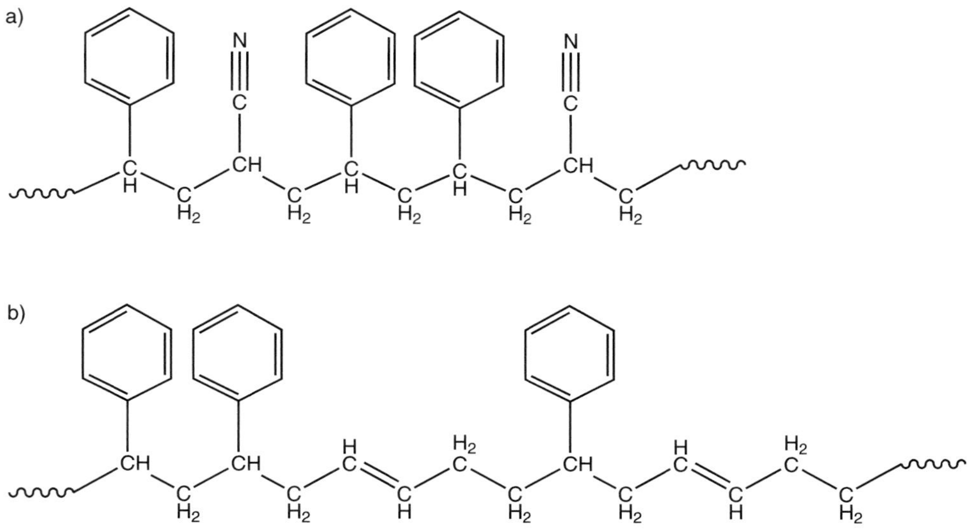

a)

b)

**Figure 21.3**   General chemical structures of:
a) polystyrene-co-acrylonitrile and b) polystyrene-co-butadiene

## 21.2.3   Block Copolymers of Polystyrene

Block copolymers of polystyrene with rubbery polymers are made by polymerizing styrene in the presence of an unsaturated rubber such as 1,4 polybutadiene or polystyrene co-butadiene. Some of the growing polystyrene chains incorporate vinyl groups from the rubbers to create block copolymers of the type shown in Fig. 21.4. The combination of incompatible hard polystyrene blocks and soft rubber blocks creates a material in which the different molecular blocks segregate into discrete phases. The chemical composition and lengths of the block controls the phase morphology. When polystyrene dominates, the rubber particles form

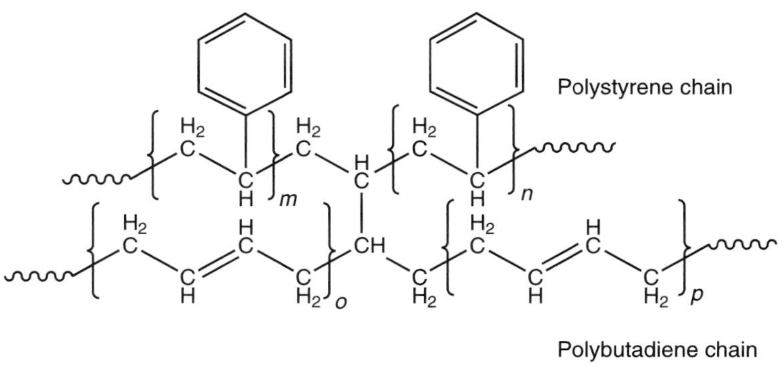

Polystyrene chain

Polybutadiene chain

**Figure 21.4**   General chemical structure of styrene block co-butadiene

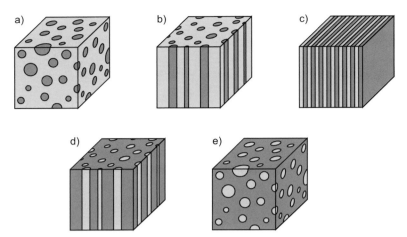

**Figure 21.5**   Schematic examples of two-phase block morphologies:
a)  high styrene content, isolated domains of rubber,
b)  moderately high styrene content, rods of rubber,
c)  equal parts of rubber and styrene, alternating platelets of rubber and polystyrene,
d)  moderately high rubber content, rods of polystyrene, and
e)  high rubber content, isolated domains of polystyrene

discrete approximately spherical phases within a matrix of polystyrene. Conversely, if the rubber dominates, the polystyrene blocks form discrete phases. Between these extremes we can create morphologies consisting of rod-like phases of one polymer surrounded by the other or alternating layers of the two blocks. The general types of morphology available are illustrated in Fig. 21.5. By adjusting the molecular structure and concentrations of the two blocks, we can create materials that exhibit specific mechanical properties. At high rubber contents the products are soft with good elastic recovery. At the other end of the spectrum we can create rubber toughened polystyrene that is stiff, but tough.

## 21.3    Synthesis

Polystyrene is unusual among commodity polymers in that we can prepare it in a variety of forms by a diversity of polymerization methods in several types of reaction vessel. Polystyrene may be atactic, isotactic, or syndiotactic. Polymerization methods include free radical, cationic, anionic, and coordination catalysis. Manufacturing processes include bulk, solution, suspension, and emulsion polymerization. We manufacture random copolymers by copolymerizing styrene directly with comonomers containing vinyl groups. In addition, we can polymerize styrene in the presence of polymer chains containing unsaturation in order to create block copolymers. Crosslinked varieties of polystyrene can be produced by copolymerizing styrene with difunctional monomers, such as divinyl benzene.

## 21.3.1 Polymerization Methods

The unsaturated phenyl ring of the styrene monomer plays a key role in its versatility. The ring can behave as either an electrophile or nucleophile, stabilizing active chain ends in both anionic and cationic states. In addition, polymerization can proceed by a free radical mechanism, during which the terminal radical is stabilized by resonance with the unsaturation of the ring. Coordination catalysts can guide the insertion of incoming monomers to the growing chain end to create stereoregular structures.

### 21.3.1.1 Free Radical Polymerization

Free radical polymerization is the predominant commercial technique for producing polystyrene. It follows the general process described in Chapter 2. The principal reactions of initiation, propagation, and termination are illustrated in Fig. 21.6. Polymerization can be initiated by a variety of molecules that decompose to form free radicals. One of the most common initiators is benzoyl peroxide, which decomposes to form a pair of peroxy radicals when heated. Azo compounds are another popular family of initiators. The majority of propagation reactions are head-to-tail because the benzylic radical is stabilized by resonance with the unsaturated ring, as shown in Fig. 21.7. Due to steric interference between the incoming monomer and the ring on the end of the growing chain, there is a tendency towards the formation of syndiotactic sequences. Sufficient random placements occur that the resulting polymer is considered to be atactic. Chain transfer can occur in the presence of various aliphatic and aromatic molecules, such as chloroform and ethylbenzene. The chain transfer rate to pre-existing polystyrene chains is very low, which virtually eliminates the formation of long chain branches.

### 21.3.1.2 Anionic Polymerization

Anionic polymerization of polystyrene takes place very rapidly – much faster than free radical polymerization. When practiced on a large scale, this gives rise to heat transfer problems and limits its commercial practice to special cases, such as block copolymerization by living reactions. We employ anionic polymerization to make tri-block copolymer rubbers such as polystyrene-polybutadiene-polystyrene. This type of synthetic rubber is widely used in the handles of power tools, the soft grips of pens, and the elastic side panels of disposable diapers.

### 21.3.1.3 Cationic Polymerization

The cationic polymerization of polystyrene occurs very fast. We perform this type of reaction at low temperature in order to obtain small scale samples with very high molecular weights. Cationic polymerization is not widely practiced outside the laboratory.

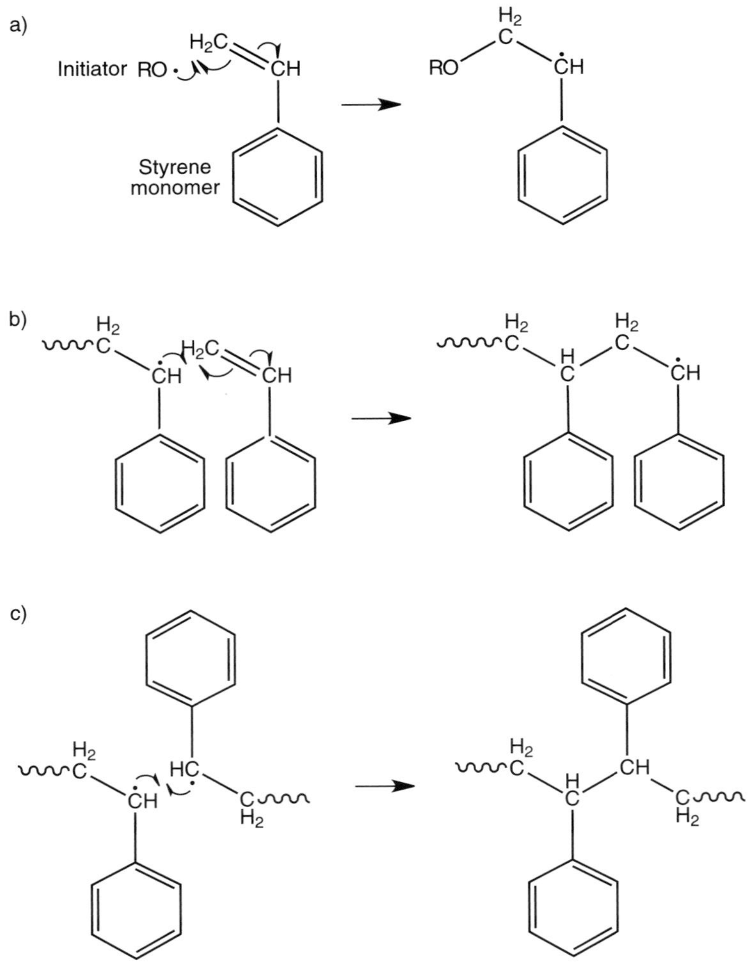

**Figure 21.6** Free radical polymerization of polystyrene:
a) initiation, b) propagation, and c) termination

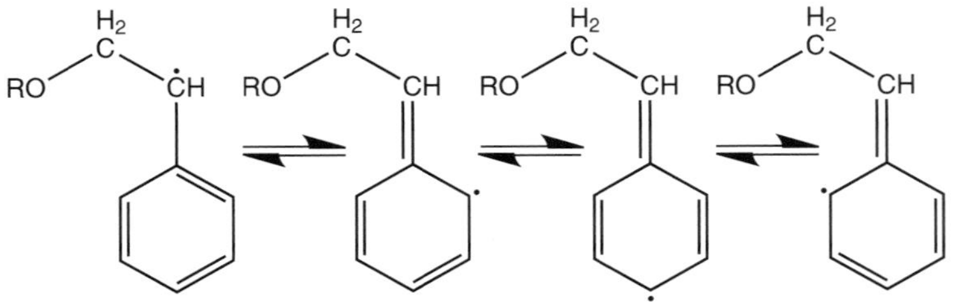

**Figure 21.7** Stabilization of a benzylic radical by resonance with the unsaturated ring

### 21.3.1.4  Coordination Polymerization

We can employ coordination polymerization to produce stereoregular polystyrene. By performing this type of reaction at low temperatures, using Ziegler-Natta or single-site catalysts, we can prepare isotactic and syndiotactic versions of polystyrene.

## 21.3.2  Manufacturing Processes

We can readily polymerize styrene by a variety of methods including solution, emulsion, suspension, and bulk processes. Historically, bulk polymerization was the first commercial process, but it has now largely been superseded by solution and suspension polymerization.

### 21.3.2.1  Bulk Polymerization

Pure styrene polymerizes to form polystyrene due to the spontaneous generation of styrene radicals at elevated temperatures. When transporting styrene monomer, we must add an inhibitor, such as t-butylcatechol at levels of approximately 12–50 ppm, to retard thermal polymerization. Continuous bulk polymerization is performed in tall reaction vessels charged with styrene from the top. As the monomer enters the column, it is heated and starts to polymerize. The resulting polymer is soluble in the monomer and starts to descend through the column, increasing in molecular weight as it does so. The temperature of the contents of the reaction vessel increases towards the bottom of the column. As the product descends, the styrene monomer boils away, increasing the concentration of the polystyrene solution. By the time the material reaches the bottom of the column, it consists largely of polystyrene with traces of styrene monomer. Molten polymer is drawn off from the bottom of the column and extrusion pelletized. The final product contains traces of residual monomer that impart a characteristic odor and lower the polymer's glass transition temperature.

### 21.3.2.2  Solution Polymerization

Solution polymerization is the preferred process by which we make general purpose polystyrene. It yields high purity polymer with low residual monomer content. Typically, we employ a train of about five reactors in series, through which the reaction mixture is pumped. At the start of the reaction styrene is dissolved in an aromatic solvent. Free radical initiators, such as azo compounds and peroxides, can be injected at various stages. By adding extra initiator at different times during the polymerization process, we can generate bimodal molecular weight distributions and other specialized grades. The temperature of each reactor is higher than the preceding one, which broadens the molecular weight distribution of the final product. (In general, a broader molecular weight distribution improves the polymer's melt flow characteristics, making it easier to mold and thermoform.) The molecular weight of the polymer increases as the reaction temperature and rate are reduced. The most common polymerization solvent is ethylbenzene, which also acts as a chain transfer agent that moderates the molecular weight of the product. Continued polymerization at high temperature in the final reactor reduces the residual monomer content in the product.

### 21.3.2.3  Suspension Polymerization

Commercially we perform suspension polymerization of polystyrene batchwise in stirred tanks with volumes in excess of 10,000 L. Droplets of styrene are suspended in water by vigorous stirring, aided by suspension agents. We initiate free radical polymerization by the action of heat on azo compounds or peroxides. Polymerization proceeds to high conversion within the styrene droplets. The surrounding water acts as a heat sink and conductor to remove the heat of polymerization. Suspension polymerization yields small spherical beads of polystyrene, with the consistency of coarse sand. The beads are centrifuged to separate them from the solvent, washed to remove traces of the suspension agent, and then dried. After drying the beads we can extrusion pelletize them or ship them as made.

### 21.3.2.4  Emulsion Polymerization

The process of emulsion polymerization is similar to that of suspension polymerization. The chief difference is that the emulsified droplets of styrene are significantly smaller than the droplets found in suspension polymerization, due to the presence of an emulsifying agent. Polystyrene made by emulsion polymerization contains significant amounts of the emulsifying agent, which is expensive to remove. For this reason, emulsion polymerization is an uneconomic process for producing pure polystyrene. It is, however, a viable commercial route to make higher value polystyrene copolymers, such as acrylonitrile-butadiene-styrene. Emulsion polymerization of styrene is readily demonstrated in the laboratory as a teaching tool.

## 21.3.3    Styrene Copolymer Production

We commonly copolymerize styrene to produce random and block copolymers. The most common random copolymers are styrene-co-acrylonitrile and styrene-co-butadiene, which is a synthetic rubber. Block copolymerization yields tough or rubbery products.

### 21.3.3.1  Random Copolymerization of Styrene-co-Acrylonitrile

The polymerization rates of styrene and acrylonitrile monomer are not equal. If we were to initiate polymerization in an equimolar solution of the two monomers, the styrene monomer would initially be depleted at a faster rate than the acrylonitrile. Thus, the copolymer molecules initially produced would contain a higher concentration of styrene than acrylonitrile. As the reaction progressed, the styrene would be depleted from the solution and the comonomer ratio in the copolymer would gradually shift towards a higher acrylonitrile content. The final product would consist of polymer chains with a range of comonomer compositions, not all of which would be compatible. Phase segregation would inevitably occur, resulting in a hazy product. To avoid this situation it is common to copolymerize styrene and acrylonitrile at a molar ratio of 62 parts of styrene to 38 parts of acrylonitrile. At this composition the relative polymerization rates of the monomers match and they are consumed at the same rate. At this "sweet spot" the resulting polymer has a uniform molecular composition from start to finish of the polymerization process. Styrene-co-acrylonitrile of this composition is made commercially by bulk or suspension processes.

### 21.3.3.2 Block Copolymerization

As a homopolymer, polystyrene is relative brittle. We can improve its toughness by polymerizing it in the presence of an unsaturated rubber. This is achieved by dissolving the rubber in styrene, then initiating polymerization. The general process is illustrated in Fig. 21.8. The resulting product is known as elastomer modified, rubber toughened, or high impact polystyrene. As the reaction progresses, the polystyrene and rubber tend to phase-segregate, but cannot do so completely because they are grafted to each other. Thus, the grafting limits the

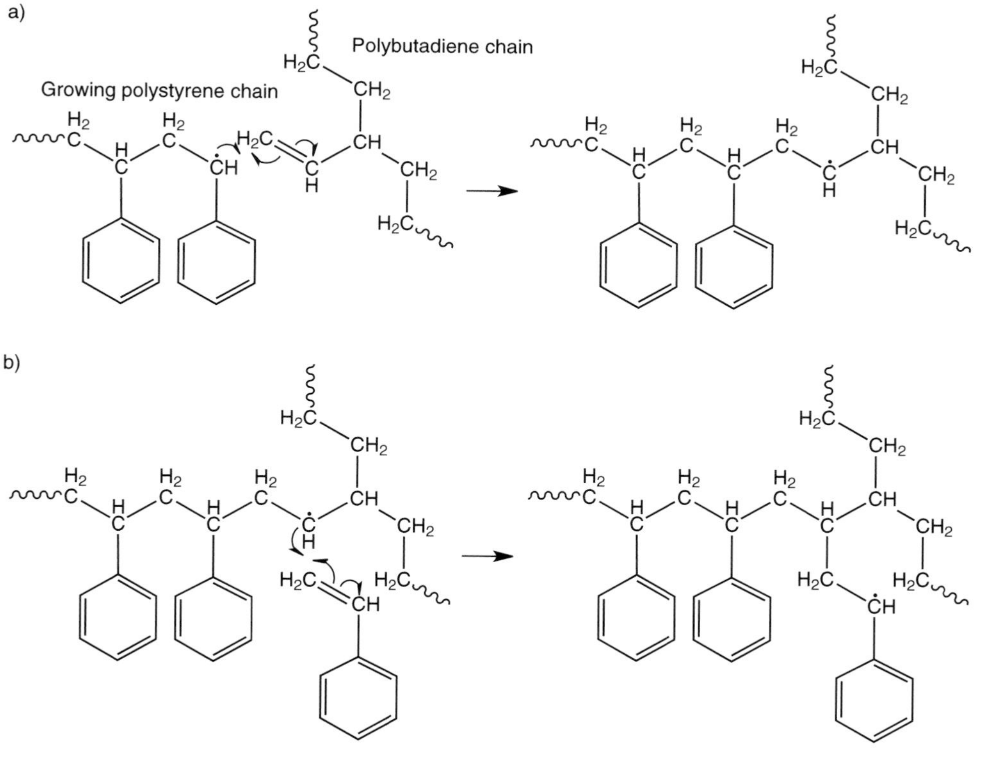

**Figure 21.8**  Incorporation of polybutadiene blocks into growing polystyrene chain: a) addition of polybutadiene chain, b) continuation of styrene polymerization

size of the rubber particles in the matrix of polystyrene. Grafted chains act as "compatibilizers" at the interface between the phases, which improves the adhesion between the phases and hence the effectiveness of the toughening agent. When manufacturing rubber toughened polystyrene, we carefully control the relative concentrations of the styrene and rubber and the agitation level, thereby controlling the size of the rubber particles. If the rubber particles were allowed to become too large, we would not obtain the full toughening effect. We normally target rubber particle sizes in the range of 1 to 10 μm at a level of about 5 to 20%.

We have considerable latitude when it comes to choosing the chemical composition of rubber toughened polystyrene. Suitable unsaturated rubbers include styrene-butadiene copolymers, cis 1,4 polybutadiene, and ethylene-propylene-diene copolymers. Acrylonitrile-butadiene-styrene is a more complex type of block copolymer. It is made by swelling polybutadiene with styrene and acrylonitrile, then initiating copolymerization. This typically takes place in an emulsion polymerization process.

## 21.4    Foam Manufacture

Foamed polystyrene – which is also known as expanded polystyrene – is used extensively in a variety of applications, ranging from packaging "peanuts" to insulation board and single-use cups and plates. We produce it by two processes: foam extrusion and bead expansion. Both types of expanded polystyrene consist of closed cells, i.e., bubbles with continuous walls. We can visually distinguish the two types of foam by the fact that products made by the expanded bead process consist of discrete beads that are welded together

### 21.4.1    Foam Extrusion

We typically produce extruded polystyrene foam using a pair of extruders arranged in series. The feed and homogenization zones of the first extruder heat, melt, and mix pure polystyrene. Once the molten polystyrene is homogenized, we inject a blowing agent, such as n-butane or n-pentane. The final section of the first extruder uniformly blends the blowing agent into the polymer. The first extruder directly feeds the second, where the temperature of the mixture is reduced to control its viscosity. The second extruder pumps the solution into an extrusion die. Expansion of the blowing agent occurs just before and after the die lip. The most common types of extrusion dies are flat or annular. Annular dies produce tubes, which may be stretched over a mandrel and split to make sheet. The mandrel imparts bi-axial orientation, much as inflation does in blown film. Boards of expanded polystyrene with thicknesses of up to 15 cm can be extruded from flat dies with widths of over 1 m. Expansion of the blowing agent within the board cools the product, which is then trimmed to the desired size and shape. Foaming with hydrocarbon blowing agents requires meticulous attention to safety to minimize the risk of fire.

Expanded sheet is wound up on rolls in which state it is allowed to age for two to five days. During this time, the blowing agent diffuses out and air diffuses in until equilibrium is reached. After aging, the sheets are unwound and passed through an oven where secondary expansion of up to 100% occurs, which yields the final product. The whole process consumes a relatively high amount of energy due to heating then cooling in the first and second extruders, followed by reheating for secondary expansion. The process typically generates a fair amount of scrap due to trimming, which requires additional energy input to recycle.

In order to control the cell size in expanded polystyrene, we add nucleating agents, such as citric acid or sodium bicarbonate. The foaming process requires a relatively high molecular weight in order to provide sufficient viscosity to sustain the cell walls during expansion. We typically use polystyrene with a weight average molecular weight of up to 320,000 g/mol. Using this process, we can produce foam with densities ranging from 27 to 160 kg/m$^3$.

## 21.4.2   Bead Expansion

We use the bead expansion process to create foamed polystyrene from solution polymerization beads into which an aliphatic hydrocarbon blowing agent has been dissolved. The blowing agent can be incorporated either during polymerization or afterwards under the influence of heat and pressure. The first step in the manufacture of polystyrene foam is pre-expansion of the beads, which takes place in a steam bath. Here, the beads soften and the blowing agent vaporizes, creating bubbles within the polymer. Steam diffuses into the cells due to osmotic pressure, further expanding the beads up to forty times their original volume. We must carefully control the pre-expansion conditions. If the pre-expansion time is too long, the natural elasticity of the cell membranes will cause the cells to collapse as the blowing agent diffuses out. In addition, the pre-expansion temperature must be sufficient to soften, but not melt the polystyrene. After pre-expansion, the polymer beads are conditioned in air for 24 hours or more. During this time, air diffuses into the beads to equalize their internal pressure with their surroundings. In the final stage, we load pre-expanded beads into a mold with perforated walls through which steam is injected into the cavity. The steam initially displaces air, then pressure is increased. The blowing agent revolatilizes and steam diffuses into the cells, further inflating them. Expansion of the hot beads causes them to impinge upon and fuse with their neighbors to form a solid block. The mold may determine the final shape of the product or we can produce slabstock from which final products of the desired shape are cut. The expanded bead process can produce foams with densities ranging from 16 to 160 kg/m$^3$.

## 21.5    Structure/Property Relationships

We most often encounter polystyrene in one of three forms, each of which displays characteristic properties. In its pure solid state, polystyrene is a hard, brittle material. When toughened with rubber particles, it can absorb significant mechanical energy prior to failure. Lastly, in its foamed state, it is versatile, light weight thermal insulator.

### 21.5.1    Pure Solid Polystyrene

At room temperature, atactic polystyrene is well below its glass transition temperature of approximately 100 °C. In this state, it is an amorphous glassy material that is brittle, stiff, and transparent. Due to its relatively low glass transition temperature, low heat capacity, and lack of crystallites we can readily raise its temperature until it softens. In its molten state, it is quite thermally stable so we can mold it into useful items by most of the standard conversion processes. It is particularly well suited to thermoforming due to its high melt viscosity. As it has no significant polarity, it is a good electrical insulator.

### 21.5.2    Rubber Toughened Polystyrene

One of the principal weaknesses of pure polystyrene is its low impact resistance. To counteract this problem, we toughen it with various types of rubber. This is most effective when a portion of the rubber is chemically grafted to the polystyrene. The rubber forms small inclusions within a matrix of polystyrene. The presence of rubber also improves polystyrene's extensibility, ductility, and resistance to environmental stress cracking.

Rubber toughened polystyrene absorbs energy when it is struck, by the formation of numerous microcrazes. The microcrazes are prevented from growing to become catastrophic fractures by the rubber particles. When a growing craze encounters a rubber particle, its growth is impeded by blunting of its tip. Each of the microcrazes scatters light, which explains why rubber toughened polystyrene turns white at the point of impact. Naturally, a price is paid when polystyrene is toughened with rubber; the higher the rubber content, the better its toughness, but the lower is its stiffness and tensile strength. To obtain the highest toughening using the least amount of rubber it is important to keep the rubber particles small. In most applications the rubber particles are less than 10 μm in diameter. The two-phase structure of rubber toughened polystyrene tends to scatter light, which can be minimized by limiting the rubber particle dimensions to less than approximately 1 μm.

### 21.5.3  Foamed Polystyrene

We can prepare foamed polystyrene at densities ranging from approximately 16 to 160 kg/m$^3$. The closed cells of foamed polystyrene effectively trap air, making it an excellent thermal insulator. In general, the lower the foam's density, the higher is its insulating capacity. The closed cells of foamed polystyrene effectively dissipate mechanical energy, which, combined with its low cost per unit volume, explains why foamed polystyrene finds such extensive use in packaging applications. In its higher densities, foamed polystyrene has an excellent balance of stiffness to weight, which suits it for certain types of load bearing packaging applications. The low density of the foamed polystyrene, coupled with its moderate stiffness and hydrophobic nature, suit it for many applications where buoyancy is required.

## 21.6  Products

We can readily process polystyrene into a wide variety of products. Its relatively low heat capacity and lack of crystallinity allow us to raise its temperature to its softening point without having to input large amounts of energy. Polystyrene is quite thermally stable in the molten state, permitting a wide range of processing temperatures. By selecting appropriate molecular weight characteristics, we can convert polystyrene into useful items by many standard processing techniques.

### 21.6.1  Homopolystyrene and Random Copolymers

Polystyrene's combination of ease of processing, relatively low cost, stiffness, and clarity suit it for many packaging and household applications. Injection molded items include disposable eating utensils, drinking tumblers, toys, and medical devices. Extruded sheet can be further processed by thermoforming to manufacture disposable drinking cups, yogurt containers, and packaging clam shells for foodstuffs and small hardware items. When additional toughness is required, we often select random copolymers, such as styrene-co-acrylonitrile, which may be vacuum formed to make items such as refrigerator door liners. We also take advantage of the toughness of copolymers to make injection molded toys. The relatively low glass transition temperature of polystyrene generally limits its use in load bearing items to room temperature applications.

### 21.6.2    Rubber Toughened Polystyrene

Rubber toughened polystyrene is widely used in electronic and kitchen appliances. This type of application requires a good balance of stiffness, impact resistance, and ready coloration. Telephones, which are frequently dropped, are an excellent example of the benefits of rubber toughened polystyrene. The high surface gloss that we desire is obtained by minimizing the size of the rubber particles. Larger items, such as canoes, can be thermoformed from extruded sheet.

### 21.6.3    Foamed Polystyrene

Expanded polystyrene articles are typically made by extrusion, thermoforming, or foaming within a cavity mold. We extrude foamed polystyrene to form various profiles, planks, sheet, and packaging peanuts. Profile extrusion includes thick-walled tubes that are slit and placed over pipes to insulate them. We most commonly encounter planks and boards in construction, where their light weight and insulating properties make them valuable. Extruded sheet is used directly for packaging, or it may be further processed by thermoforming. Vacuum-formed applications include disposable plates and bowls, egg cartons, and meat trays. Foamed polystyrene items made by expansion within a mold include disposable cups, coolers and the shaped blocks that support goods within cardboard boxes. By foaming polystyrene beads within large cavity molds we can make very large billets or blocks that can be cut or otherwise shaped as required for construction, maritime flotation, and packaging purposes.

## 21.7    Conclusions

Polystyrene's ease of processing, relatively low cost, and wide range of product densities suits it for use in many disparate applications. As a packaging material we can encapsulate items in transparent clamshells or protect delicate goods with foam blocks or peanuts. In its rubber toughened forms, we use it in appliance and consumer electronics housings, where it withstands the rigors and abuse of modern life. On a larger scale, we can vacuum-form it to make smooth and easily cleaned liners for refrigerators and coolers. Other rubber toughened applications include thermoformed recreational items and storage containers that must withstand repeated impact. Within the walls of many buildings we find foamed polystyrene board that reduces our energy costs to heat or cool our indoor environment. In general, polystyrene contributes to our comfort and convenience of our lives in a wide variety of areas.

# Review Questions for Chapter 21

1.  Why is commercial polystyrene an amorphous polymer? Why is it sometimes referred to as "crystal" polystyrene?

2.  Why is it possible to manufacture polystyrene by radical, anionic, cationic and coordination polymerization methods?

3.  What is it about the monomer that promotes a syndiotactic arrangement of the pendant styrene group? Why is the polymer still considered atactic if the syndiotactic addition is more energetically favorable?

4.  Describe the bulk polymerization process for the manufacture of polystyrene. What limitations exist with this method?

5.  Why, in the production of styrene-co-acrylonitrile, do we introduce a 62 : 38 ratio of styrene to acrylonitrile in the reaction vessel?

6.  Compare and contrast the bead expansion and foam extrusion methods of creating foamed polystyrene.

7.  How can the toughness of polystyrene be improved? How do these alterations increase the toughness?

8.  What properties of polystyrene are lowered by the introduction of an impact modifier?

9.  What properties of foamed polystyrene make it amenable for the manufacture of packing peanuts?

10. What properties of polystyrene make it amenable to thermoforming processes to produce food packaging items such as clam shell containers?

# 22    Polyvinyl Chloride

## 22.1    Introduction

Polyvinyl chloride is one of the most versatile and maligned polymers in production today. Its versatility arises from its wide range of properties and the varied processes by which it is manufactured into finished goods. We find polyvinyl chloride in intravenous fluid and blood bags, children's toys, vinyl siding, vinyl flooring, wallpaper, shower curtains, packaging materials, and shoes, to name just a few applications. Its poor reputation arises for two totally different reasons. The first, and most often heard of outside of the polymer industry, comes from environmental activists. Many environmental activists believe that we should ban the global use polyvinyl chloride. One valid concern lies in the combustion products released when burning the polymer which include halogenated, low molecular weight organic volatiles. Though this concern has merit, the claims tend to be overblown to create irrational hysteria rather than calm conversations of benefits and risks. The other reason for polyvinyl chloride's infamy arises from its rather finicky nature during manufacturing. When not properly stabilized or processed, polyvinyl chloride can cause considerable harm to both equipment and personnel. For this reason, processing polyvinyl chloride requires a cautious and careful approach. In this chapter we will describe the chemistry of this polymer, as well as provide some basics on formulating recipes to yield the desired properties in a final product.

## 22.2    Chemical Structure

The vinyl chloride monomer polymerizes via addition polymerization to form polyvinyl chloride. The final polymer has the chemical composition shown in Fig. 22.1. The polymer exhibits limited crystallinity, though this property is not often considered as important in defining its performance. It tends to be atactic or regionally syndiotactic, surrounded by extended atactic runs. When exposed to temperatures above 100 °C, polyvinyl chloride decomposes, creating free radicals that further attack the polymer chain, as we shall discuss in more detail later. For this reason, the degradation of polyvinyl chloride is autocatalytic

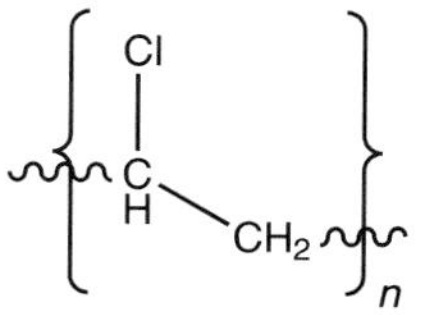

**Figure 22.1**    Repeat unit of polyvinyl chloride

and can occur very quickly. In addition to destroying the properties of the original polymer, degradation releases hydrogen chloride which can attack the internal metal surfaces of processing equipment. For these reasons, we will spend some time describing the proposed mechanisms for degradation and ways to prevent this process from occurring.

## 22.3  Manufacture

Polymerization of vinyl chloride occurs through a radical chain addition mechanism, which can be achieved through bulk, suspension, or emulsion polymerization processes. Radical initiators used in vinyl chloride polymerization fall into two classes: water-soluble or monomer-soluble. The water-soluble initiators, such as hydrogen peroxide and alkali metal persulfates, are used in emulsion polymerization processes where polymerization begins in the aqueous phase. Monomer-soluble initiators include peroxides, such as dilauryl and benzoyl peroxide, and azo species, such as 1,1′-azobisisobutyrate, which are shown in Fig. 22.2. These initiators are used in emulsion and bulk polymerization processes.

Chain transfer occurs frequently during the polymerization of vinyl chloride. The ratio of propagation events to chain transfer events in a given time period determines the average molecular weight of the final polymer. This finding can be summarized by the empirical formula:

$$M_w = 62.5 \frac{k_p}{k_{tr}} \tag{22.1}$$

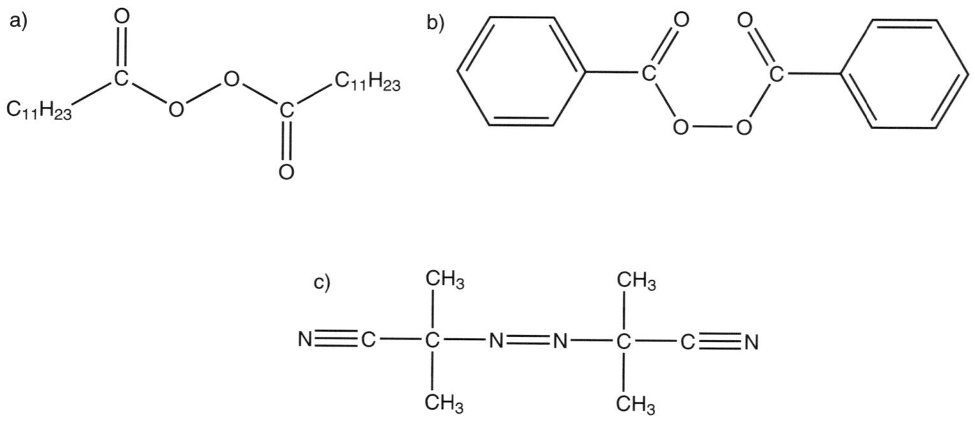

**Figure 22.2**  Chemical structure of monomer soluble initiators for polyvinyl chloride:
a) dilauryl peroxide, b) benzoyl peroxide and c) azobisisobutyronitrile

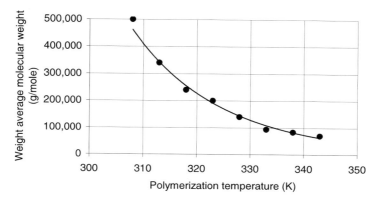

**Figure 22.3**    Example of weight average molecular weight of polyvinyl chloride as a function of polymerization temperature

where $k_p$ is the rate constant defining the propagation reaction and $k_{tr}$ represents the rate constant for chain transfer reactions.

The greater the rate constant for chain transfer, the lower the molecular weight of the polymer. One way to affect the rate constants is by changing the temperature. In general, the chain transfer rate constant is much more sensitive to temperature effects, increasing dramatically as the temperature is increased. For these reasons, there is an inverse correlation between temperature and molecular weight of polyvinyl chloride as shown in Fig. 22.3.

Vinyl chloride polymerization occurs via an exothermic radical reaction. In fact, the reaction is approximately 25% more exothermic than polyethylene polymerization. The highly exothermic nature of the reaction and the strong molecular weight dependence on temperature make heat transfer, and its control, critical to the manufacture of polyvinyl chloride.

## 22.3.1    Bulk Polymerization

Vinyl chloride monomer, when exposed to high energy light or radical initiators, bulk polymerizes to form high polymer. Unfortunately, this simple process does not produce a consistent, commercially useful polymer. One reason for this is that the conversion levels reached are very low: only approximately 50 to 60% of the monomer converts to polymer. The other major issue arises from thermal considerations. Localized heat generation around the initiator and low thermal transfer rates create hot spots resulting in an inconsistent product. Also, in extreme circumstances, the heat can build up to such high levels that the material catastrophically degrades, causing an explosion. Considering the degradation products, this is a highly undesirable event. Another issue is the fact that the unconverted monomer is entrained in the polymer and is difficult to remove. Since vinyl chloride monomer has been associated with many health problems it must be fully removed before use, this renders bulk polymerization commercially unacceptable.

### 22.3.2   Suspension Polymerization

To avoid issues of variable molecular weight, heat build up, low conversion levels, and entrained monomer, polyvinyl chloride manufacturers typically rely on either suspension or emulsion polymerization methods. The most commonly used of the two, suspension polymerization, suspends tiny droplets of the vinyl chloride monomer in an aqueous medium. The aqueous medium contains organic soluble initiators and a very small amount of polyvinyl alcohol. The polyvinyl alcohol reduces the surface tension between the monomer droplets and the water, thereby stabilizing the suspension. The initiators partition into the monomer droplets where polymerization occurs. In essence, this process is bulk polymerization at a very small scale; each droplet can be thought of as a micro-reactor. The aqueous medium dissipates the heat of the reaction, allowing good control of the molecular weight. The small size of the polymer droplets makes monomer removal more efficient. Typical monomer conversion levels range from 75 to 90%.

### 22.3.3   Emulsion Polymerization

Emulsion polymerization also uses an aqueous solution as the reaction medium. In this process, though, the aqueous solution contains anionic surfactants which micellize to create a stable emulsion of the monomer and polymer. Water-soluble free radical sources, such as alkali persulfates, interact with the monomer to initiate polymerization. The growing chain is stabilized by the surfactant molecules, allowing further growth. The final product is a latex that has a 35 to 45% solids level, which can then be dried by flashing off unreacted monomer and water. Conversion levels for this process range from 85 to 95% and any residual entrained monomer is easily removed during the drying process.

## 22.4   Morphology

Polyvinyl chloride exhibits low crystallinity, in the vicinity of 5 to 10% crystalline structure. The crystallinity arises from syndiotactic portions of the chain in which there is little branching. Typically, there are two to four long chain branches per 1000 carbon atoms. The combination of branching and non-stereospecific polymerization processes ensures that commercial polymers have no significant crystallinity. It is possible to reduce branching, which thereby increases crystallinity, but this effect is only appreciably seen in very low temperature polymerization processes (which are not commercially viable). In academic circles there is some discussion of manufacturing highly crystalline polyvinyl chloride using stereoselective catalyst systems.

## 22.5    Degradation of Polyvinyl Chloride

Unstabilized polyvinyl chloride suffers from very low thermal stability. Degradation arises from the dehydrochlorination of the polymer, creating unsaturation in the polymer chain. The process is autocatalytic; the hydrogen chloride evolved during degradation will attack and further degrade adjacent polymer chains if it is not quickly removed. To combat this, all commercial polyvinyl chloride is stabilized. Thermal stabilizers come in a variety of different forms, but all act to remove the hydrogen chloride generated, thereby eliminating catastrophic failure.

### 22.5.1    Observations of Degradation

Initially, degradation of polyvinyl chloride leads to a discoloration of the polymer as double bonds form along the backbone of the polymer. The delocalized electrons, along regions of seven or more adjacent double bonds, create colors such as tan, yellow, and pink. The deeper the color is, the greater the extent of degradation. Severe degradation lowers the tensile and impact strength of the polymer.

### 22.5.2    Mechanisms of Degradation

Several mechanisms are proposed to explain the degradation of polyvinyl chloride. These include unimolecular expulsion, radical chain decomposition, and ionic decomposition. The unimolecular expulsion theory states that hydrogen chloride is simply released from the chain at high temperatures. The site of release then promotes the next monomer unit to release hydrogen chloride because of the destabilizing effect of the newly formed double bond. This is often referred to as a zipper mechanism, as the degradation "zips" along the chain. This zipping leads to consecutive double bonds along the chain, which creates the color indicative of degradation. This mechanism is not strongly supported by kinetic data but has yet to be disproved. The radical chain mechanism, shown in Fig. 22.4, begins with the release of free radical chlorine from the polymer backbone. Chlorine atoms propagate the degradation process by abstracting protons, thereby creating unsaturation. At the site of unsaturation, the stability of the next chlorine on the polymer backbone is reduced, allowing it to be released to form another free chlorine radical. This mechanism also provides an explanation for the zipper effect and the origin of changing coloration during degradation. Ionic theories of decomposition propose that a base reacts with a hydrogen on the chlorine-bound terminal backbone carbon, releasing both the hydrogen and the chlorine and creating a double bond. Again, this mechanism supports a zipper effect as the adjacent hydrogen becomes susceptible to basic abstraction, thus further permitting degradation.

These mechanisms describe how degradation occurs, but do not adequately account for the fact that, theoretically, the polymer should be stable at temperatures of up to 300 °C, based

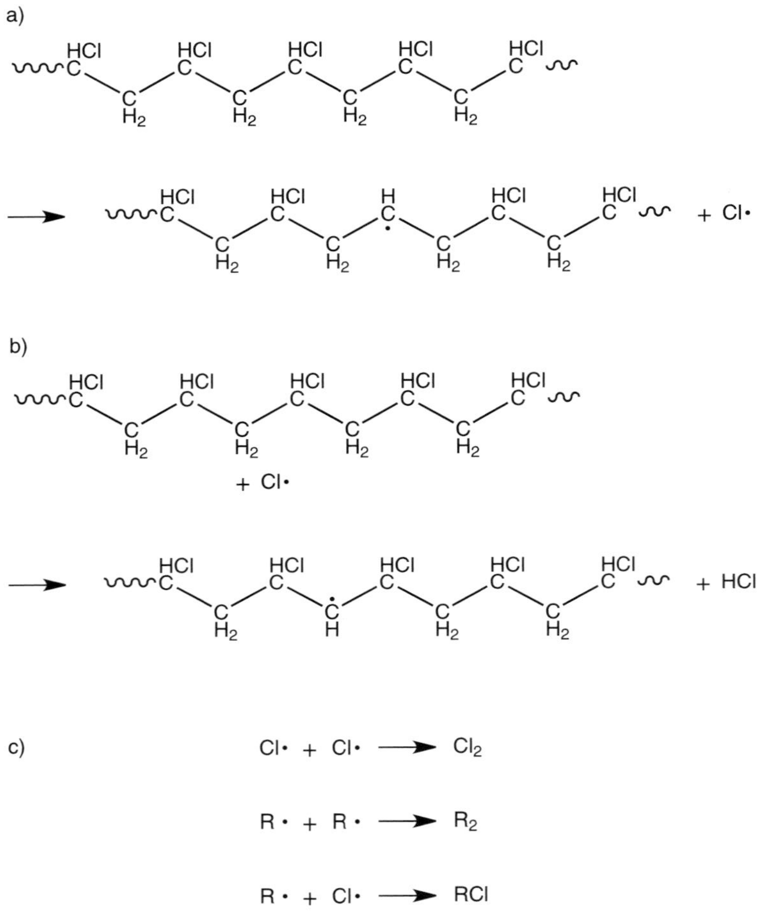

**Figure 22.4** Radical chain mechanism for the degradation of polyvinyl chloride: a) initiation, b) propagation and c) termination

on the ideal chemical formula. For this reason, it is believed that initiation of decomposition originates at defects in the chain structure, such as branching points, pre-existing unsaturation, and sites where there was localized incorporation of the polymerization initiators.

## 22.5.3  Preventing Degradation

Thermal stabilizers combat degradation by removing the hydrogen chloride that is generated. Additionally, we treat polyvinyl chloride more gently than we do polyolefins. We use milder processing conditions (lower temperatures and lower shear rates) and add lubricants to ease the polymer through the process in the hopes of minimizing chain scission and radical decomposition.

## 22.6 Additives to Polyvinyl Chloride

Polyvinyl chloride resin, because of its inherent thermal instability and wide range of applications, requires us to develop additive recipes based on specific application and processing requirements. Typical additive packages include stabilizers, plasticizers, waxes, processing aids, pigments, and mineral additives.

### 22.6.1 Thermal Stabilizers

Thermal stabilizers act by removing hydrogen chloride from the thermal degradation process to prevent catastrophic autocatalytic degradation. Stabilizers must be mobile in the polymer at processing temperatures and must be able to react quickly with hydrogen chloride. Generally, we separate the stabilizers into two classes: primary and secondary. Primary stabilizers readily migrate to the site of degradation and react with the hydrogen chloride providing short-term chemical stability. Secondary stabilizers are less mobile but can react with more acid than the primary ones, making them effective at long-term stabilization. The two classes are often co-formulated providing the best stabilization of the material. Typical stabilizers include inorganic lead compounds, such as tribasic lead phosphate, metal soaps, such as calcium, magnesium, barium, or zinc stearates or laurates, metal complexes of barium, cadmium, and zinc, and organotin stabilizers, such as dibutyltin diacetate.

The stabilizer or stabilization system used depends on the heat and shear likely to be experienced by the polymer during processing, the end use application requirements, such as clarity or color, and the health concerns. A major health issue has been identified with the lead salts and soaps, because of their relative solubility and their corresponding potential to leach into water. For this reason, lead stabilizers currently find use only when other stabilizer systems do not provide the necessary stabilization or end use properties. Wire and cable sheathing is the only remaining application where the use of the lead stabilizer systems is widespread. Since most humans do not chew on wires (though mice, rats, and squirrels do) and lead-based stabilizers provide superior electrical properties, lead salts persist in this application.

### 22.6.2 Plasticizers

We find polyvinyl chloride in so many applications because we can formulate it to be rigid, flexible, or any intermediate hardness required. Rigid polyvinyl chloride is used to manufacture siding, pipes, fences, gutters, window and door frames, credit cards, and many other extruded products. We encounter flexible polyvinyl chloride in shoes, imitation leather products, wire coatings, and tool handle grips. The range of products and properties differ so dramatically because of the use of organic plasticizers. Plasticizers consist of relatively low molecular weight oils with bulky side chains, such as dioctyl phthalate (shown in Fig. 22.5) and adipyl stearate. The plasticizer molecules, when compounded with the polymer, disperse among the polymer chains. Within the polymer they lower the intermolecular attraction between the chains, permitting them to easily slide past one another, lowering the stiffness of the part. Figure 25.6 shows a schematic of how these materials work.

Plasticization has been explained by a variety of theories in an attempt to explain how the plasticizer reduces the rigidity of the final part. All theories rely on the premise that the plasticizer reduces the strength of the intermolecular forces between the polymer chains. The theories fall into two broad categories: interference mechanisms and expansion mechanisms. The interference mechanisms state that plasticizer molecules interact only weakly with the polymer chains after separating the chains from one another, thereby reducing the overall cohesion of the material. The expansion mechanisms state that the reduced rigidity arises from an increase in the free volume of the system as the system expands to incorporate bulky,

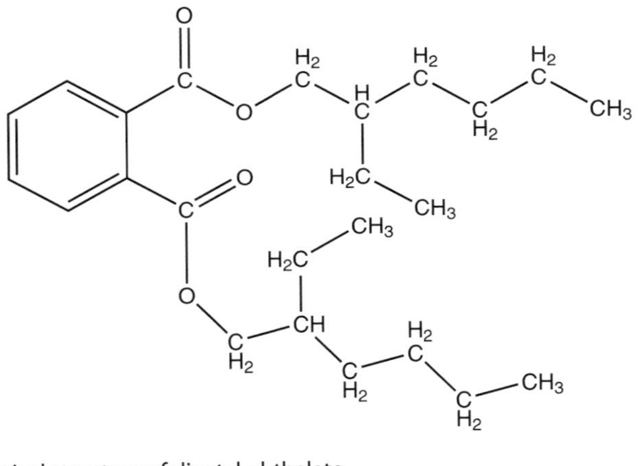

**Figure 22.5**   Chemical structure of dioctyl phthalate

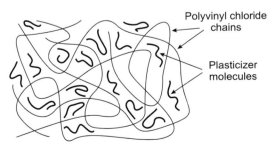

**Figure 22.6**   Schematic diagram showing how plasticizer molecules interbed between polyvinyl chloride chains

low molecular weight species. The best models seem to incorporate aspects of both of these ideas.

### 22.6.3 Waxes and Processing Aids

Processing polyvinyl chloride requires the consideration of likely sources of degradation in order to combat them. Good equipment design is the first line of defense in preventing degradation, but we also employ chemical modifiers. These chemical modifiers, in the form of waxes and processing aids, reduce the viscosity of the melt, thereby reducing the heat generated during melt processing and moderating the processing temperature. In addition, they provide a protective coating on metal surfaces which prevents polymer from sticking and burning as it passes through. Determining the correct levels for these materials can be tricky. If the loading of the migratory components is too high, they can accumulate on the metal surfaces of the processing equipment or volatilize at the die exit, creating a sticky mess that can contaminate the product. In addition, there is a potential to impair the physical properties of the polymer at high loadings. Inadequate addition levels lead to reduced thermal stability and a narrow processing window.

## 22.7   Processing Methods and Applications

Polyvinyl chloride can be processed by a number of different methods. Extrusion processing through profile dies creates pipes, gutters, vinyl siding, and wire sheathing. Injection molding produces shoe soles, flexible handlebar grips, and pipe fittings. Rotational molding can create soft vinyl toys and balls. Blow molding can be used to produce consumer bottles such as spice and cooking oil bottles. It can be calendered to form sheets for use in making credit cards. Polyvinyl chloride is amenable to so many different processing methods due to its wide range of properties when formulated.

### 22.7.1   Compounding

Polyvinyl chloride formulations, because of their many components, typically begin as a dry blend of the materials, which is then either compounded or directly extruded, depending on the final process it will undergo. We achieve dry blending by tumbling the resin granules with the stabilizer, plasticizer (if used), processing aids, and mineral additives. The result is a homogeneously blended powder that can be fed to compounding equipment to produce pellets for the injection molding, blow molding and calendering manufacturers. Extrusion processors can either feed the dry blend directly to the extruder or can use premixed compound, depending on their specific application and equipment.

Compounding is achieved by extrusion, mixing, and melting in large heated mixers or by milling on a two-roll mill. Though the methods differ, the final product comes in the forms of pellets that are fed directly to the processing system.

### 22.7.1.1 Extrusion Compounding

Extrusion compounding of polyvinyl chloride can be accomplished by either single or twin screw machines. Standard single screw extruders (such as those used to extrude polyolefins) do not impose sufficient shearing to adequately mix polyvinyl chloride formulations. Single screw extrusion can be used to compound polyvinyl chloride if several modifications are made to standard single screw extrusion processing. These modifications include the incorporation of mixing elements to the screw, changing the motion of the screw to include longitudinal reciprocation in addition to rotation, and by fitting the barrel with obstacles around which the melt must pass to increase mixing.

Twin screw extruders introduce the necessary shear to thoroughly compound the dry blend into a polymer melt. They achieve this by utilizing aggressive mixing and kneading elements on the screw. Unfortunately, processing speeds are limited by the requirement that temperatures must be kept low to prevent excessive degradation.

### 22.7.1.2 Two Roll Mills

Heated two roll mills consist of two large metal cylinders mounted closely in parallel, which rotate toward one another, creating a narrow gap between the two rolls. A diagram of a two roll mill can be seen in Fig. 22.7. To heat the surfaces of the rolls, steam, hot water, or oil circulates through the rolls. Preblended material is dropped into the nip in small quantities.

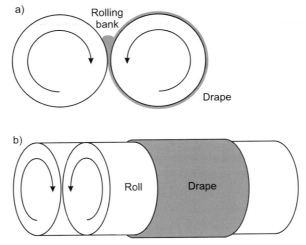

**Figure 22.7**   Schematic diagram of a two roll mill:
a) cross-section and b) perspective view

The preblend melts and forms a drape, a thin layer of polymer wrapped around the roll, over one of the rolls. The gap between the rolls is narrowed until a rolling bank of the melt forms between the two rolls, while the draped material stays on one roll. As the material processes on the mill, it is folded over itself repeatedly to homogenize the ingredients as they mix in the nip's rolling bank. The material can be stripped off the mill once it is fully mixed and can then be further processed by calendering and slitting to make pellets suitable for further processing. This method, though effective, is not particularly efficient. The time it takes for each batch to mix thoroughly and the labor required to monitor the mill makes it a less desirable method for mixing than either extrusion or high shear mixing. Despite this, the method is still used in many laboratories to develop and test compounds on a limited scale.

### 22.7.1.3  Internal Mixers

Internal mixers are large, heated, high shear batch mixers that are used to produce compounds. The design of these mixers arises from the rubber industry's Banbury Mixer which is, itself, derived from two-roll mill technology. Dry blend is forced into the top of the mixer and passes into a high-shear, two-zoned mixing chamber. Each zone of the mixing chamber contains a kneader that rotates to force material into the center of the two zones. The material is then mixed until it is homogenous and dropped into calender rolls to form a sheet. The sheet is subsequently cut to make pellets.

## 22.7.2    Extrusion Processing

Both plasticized and unplasticized polyvinyl chloride can be processed via extrusion methods. Wire and cable sheathing on high voltage wires is manufactured from plasticized polyvinyl chloride, as is flexible vinyl tubing. The unplasticized material is typically found in durable goods such as siding, pipes, gutters, soffits, fences, decking, window frames, electrical conduit, and even as segmented wall panels in modular houses. It can also be found in home goods such as vinyl window blinds.

Both single and twin screw extrusion processes are used to produce polyvinyl chloride goods. It is important to insure that there are no rough spots on the metal processing surfaces or "dead" spots within the flow channels where polymer can stagnate in the extrusion equipment. Material that accumulates and experiences an extended heat history will degrade and spoil the product. The metal used to make components that contact the melt must be hardened, treated, or plated to resist the effects of the corrosive degradation products. Regardless of how mild the processing conditions are, some hydrogen chloride will inevitably be generated, which will attack most metals unless suitable precautions are taken.

The polymer feedstock for extrusion processing can be either compounded pellets or a dry blend. Compounded pellets provide the advantage of being easy to handle. This must be weighed against the disadvantage of higher raw materials costs. Dry blending on site has the advantage that you can economize on raw materials and have the ability to formulate for many different applications or requirements. Major disadvantages include material handling issues when managing multiple additives, the capital costs associated with equipment, and the potential for moisture absorption if the dry blend sits in a humid environment before extrusion. The choice of your extruder feed will define the design of the extruder screws. A pre-compounded material requires a screw design that introduces lower shear forces than that used for a dry blend feed.

Die designs depend on the final product. Profile dies, (such as the pipe extrusion die shown in Fig. 11.4) which have an opening similar to the cross section of the final part, are the most commonly used. Some examples where profile dies are used are in the manufacture of vinyl siding, pipe, sheet, and window frames. For all these parts, the final profile of the material is extremely important so the melt exits the die itself and enters a sizing die. The sizing die supports the extrudate as it cools and forces it into the correct shape. Immediately after the sizing die, the material enters a cooling system within which it freezes into the required shape. Cooling can be achieved by passing the extrudate through a water bath, spraying it with cold water, or passing it under forced air jets.

A wire crosshead die is used to manufacture wire coatings, which is illustrated in Fig. 11.10. This specialized die turns the melt flow 90° before it leaves the die. At this turn, the wire to be coated enters the melt stream and exits the die co-axially with the polymer. This process yields a seamless polymer coating around the wire.

### 22.7.3    Calendering

Calendering processes are used to make vinyl sheets and vinyl coated fabrics. Some examples of calendered polyvinyl chloride include vinyl flooring, vinyl uppers used in athletic shoes, wallpaper, and the covers of three-ring binders. The process consists of several pairs of heated, non-contacting rotating rolls, as shown in Fig. 22.8. Molten polymer drops in between the first two rolls. The material passes between the pair of rolls to another pair by draping on one of the rolls until it hits the next nip. The final sheet comes off the terminal roll with a thickness defined by the nip width between the last two rolls. In laminated processes, the laminating polymer and the substrate pass through the final nip simultaneously.

### 22.7.4    Injection Molding

Injection molding is used to make rigid polyvinyl chloride pipe fittings, toys, and small appliance housings, as well as flexible waterproof boots and the soles of shoes. Great care is taken to minimize shear and dead spots during processing to prevent degradation. Mold designs need to include rounded, rather than sharp internal corners and short flow paths. Vent paths need to be included to remove hydrogen chloride in case of degradation during molding. In addition, the mold surfaces are typically chromium plated and polished to resist degradation and prevent polymer adhering to them.

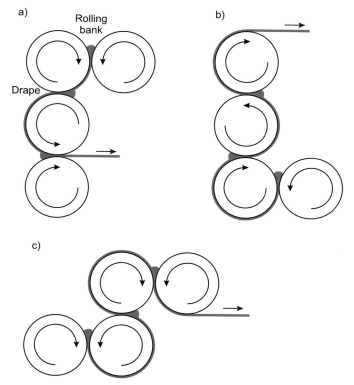

a)

Rolling
bank

Drape

b)

c)

**Figure 22.8**   Various calendar configurations:
a) F-type, b) L-type and c) Z-type

### 22.7.5   Blow Molding

Several types of bottles are manufactured from polyvinyl chloride via blow molding. Examples include clear bottles used for cooking oils, cosmetic bottles, and clear spice bottles.

## 22.8   Conclusions

Simply by changing the components added to polyvinyl chloride, we can create a wide range of properties. The receptiveness of the resin to different additives, the many processing methods available to producing polyvinyl chloride-based products, and its low cost combine to make it one of the most versatile polymers in commercial use today.

## Review Questions for Chapter 22

1.  Polyvinyl chloride is considered to be a "self-extinguishing" polymer – in other words, it does not support combustion well. Why does it exhibit this behavior?

2.  Why is the reaction temperature of polymerization of polyvinyl chloride so important to its manufacture?

3.  How do primary and secondary stabilizers act to reduce degradation in polyvinyl chloride?

4.  Why are suspension and emulsion polymerization processes the primary methods by which polyvinyl chloride is manufactured? How are these processes carried out?

5.  Why are some initiators water soluble and others monomer soluble? Which type is used for suspension polymerization? Emulsion polymerization?

6.  What chemical moiety creates color in degraded polyvinyl chloride?

7.  How do waxes and processing aids work to improve the processability of polyvinyl chloride? What issues can arise if they are used in excess?

8.  The theoretical temperature required for degradation is much higher than that actually observed with polyvinyl chloride. What explanations have been proposed to explain this?

9.  What measures are taken to protect injection molding equipment while processing polyvinyl chloride?

10. What type of extrusion die is used to coat wire with polyvinyl chloride?

# 23    Nylons

## 23.1    Introduction

The family of polymers that we refer to as nylons consists of molecules composed of amide groups alternating with short runs of methylene units. These molecules are also known as polyamides, which may be shortened to PA. The generic chemical structure of a nylon molecule is shown in Fig. 23.1. Variations on this basic structure include the length of the polymethylene sequences and the orientation of the amide groups relative to their neighbors. Figure 23.2 shows the chemical structures of nylon 6 and nylon 66, which are the two most common types of nylon.

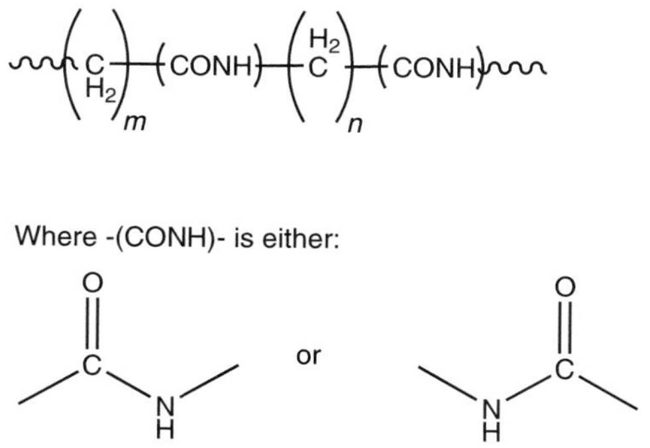

Where -(CONH)- is either:

**Figure 23.1**    Generic chemical structure of a nylon molecule

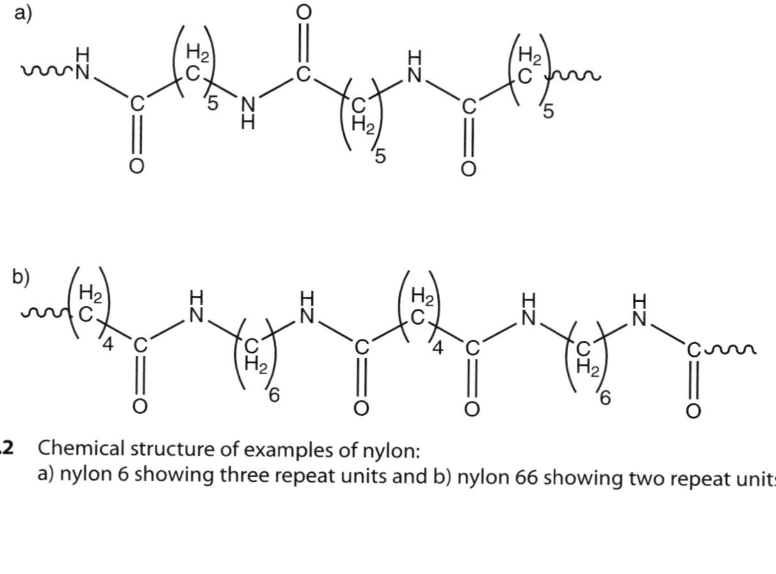

**Figure 23.2**    Chemical structure of examples of nylon:
a) nylon 6 showing three repeat units and b) nylon 66 showing two repeat units

Nylons belong to the class of polymers known as engineering polymers; that is, they are strong, tough, and heat resistant. We can readily extrude and mold nylons to form a wide variety of useful objects, such as tubing, furniture casters, and automotive air intake ducts. Nylons are commonly spun into filaments or fibers. These can be used directly, or braided, or twisted to form threads, yarns, cords, and ropes, which may be further woven to make fabrics. In their fibrous forms, nylons are used in carpets, backpacks, and hosiery.

## 23.2     Chemical Structure

Nylons are classified into two categories: AB and AABB polymers. Molecules of the AB type consist of repeat units that contain a single amide group, such as nylon 6, which is illustrated in Fig. 23.2 a). Molecules of the AABB type consist of repeat units that contain two amide groups, such as nylon 66 (pronounced "nylon six six"), which is illustrated in Fig. 23.2 b). Type AB nylons are made from a single monomer, such as caprolactam, which is shown in Fig. 23.3 a). We make type AABB nylons from diamines and diacids, such as adipic acid and hexamethylene diamine, which are shown in Fig. 23.3 b) and c), respectively.

The nomenclature of nylon is based on the number of carbon atoms found in the monomers. Thus, caprolactam, which contains six carbon atoms per molecule is polymerized to form nylon 6. Nylon 46 (pronounced "nylon four six") is made from 1,4 diaminobutane, which contains four carbon atoms and adipic acid which contains six carbon atoms. The convention that we use to name nylons is summarized in Fig. 23.4.

Nylon molecules contain short, non-polar, polymethylene sequences and polar amide groups. The ratio of the polar to non-polar components strongly influences the properties of the polymer. We can envisage a vast number of polymers that fall into the class of nylons, but

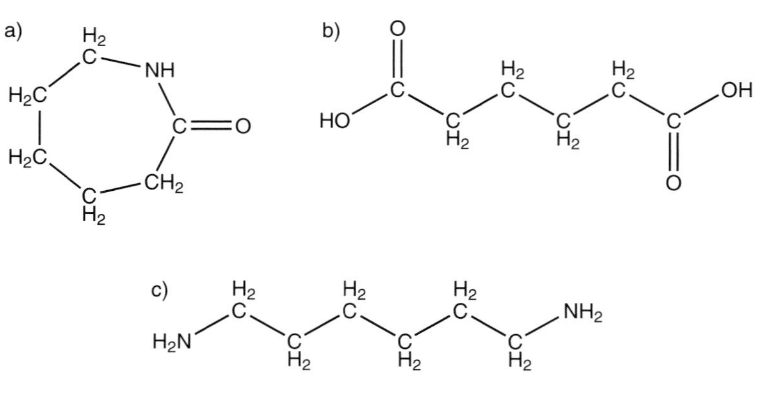

**Figure 23.3**   Examples of monomers used in the polymerization of nylons: a) caprolactam, b) adipic acid and c) hexamethylene diamine

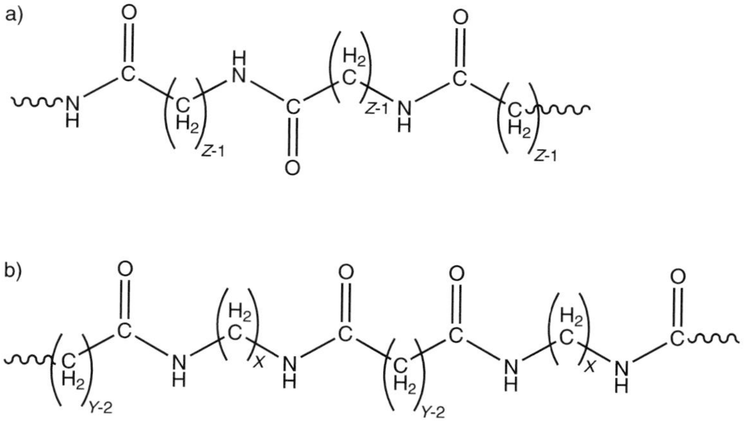

**Figure 23.4**   Summary of nylon nomenclature:
a) type AB, nylon Z and b) type AABB, nylon XY

for our purposes we will concentrate primarily on nylon 6 and nylon 66, which constitute the majority of commercial nylons.

## 23.2.1   Nylon 6

The molecular structure of nylon 6 is shown in Fig. 23.2 a). Its repeat unit consists of an amide group and a pentamethylene sequence, i.e., $-(CH_2)_5CONH-$. The polar amide groups from adjacent chains interact with each other to form hydrogen bonds, which link the chains.

## 23.2.2   Nylon 66

Figure 23.2 b) illustrates the molecular structure of nylon 66. Its repeat unit contains two amide groups and two polymethylene sequences containing four and six carbon atoms respectively. Its repeat unit has the formula $-(CH_2)_4CONH(CH_2)_6CONH-$. In common with other nylons, hydrogen bonds link neighboring amide groups.

## 23.2.3   Other Commercial Nylons

Other commercially important polyamides include nylon 11, nylon 12, nylon 46, nylon 610 ("nylon six ten") and nylon 612 ("nylon six twelve"), the chemical structures of which are shown in Fig. 23.5.

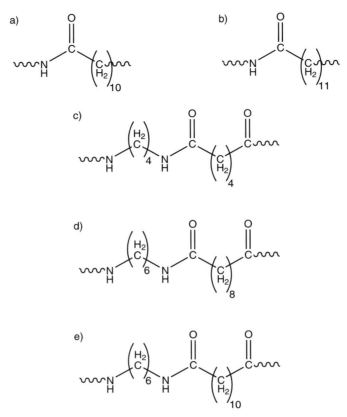

**Figure 23.5**    Chemical structure of repeat units of various commercial nylons:
a) nylon 11, b) nylon 12, c) nylon 46, d) nylon 610, and e) nylon 612

## 23.3    Manufacture

There are three methods by which we commercially manufacture nylons. We make nylons of the AB type by hydrolytic ring opening reactions of lactams and the self condensation of ω-amino acids. Nylons of the AABB type are made from diamines and diacids via an intermediate known as a nylon salt.

### 23.3.1    Ring Opening to Make Type AB Nylons

We make type AB nylons, such as nylon 6 and nylon 12, by the hydrolytic ring opening of lactams in the presence of water at temperatures of about 260–270 °C.

The polymerization process for nylon 6 consists primarily of the three types of reaction illustrated in Fig. 23.6. Each of the reactions is reversible, with the equilibrium of the products being controlled primarily by the concentration of water in the reaction vessel. The reaction is initiated by the hydrolytic ring opening of caprolactam to form 6-aminohexanoic acid, as shown in Fig. 23.6 a). Chain extension of the type shown in Fig. 23.6 b) dominates when water is abundant (10 to 20%) in the reaction mixture. At lower water levels (2 to 5%) chains grow primarily by the mechanism shown in Fig. 23.6 c). In order to limit the average molecular

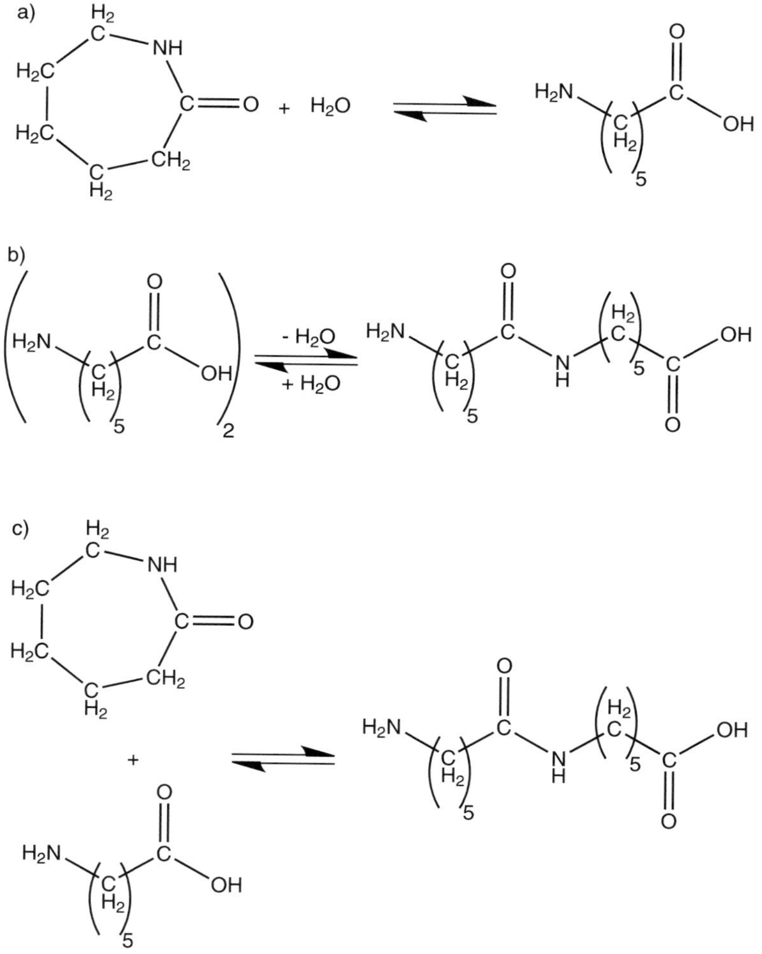

**Figure 23.6**   Principal reactions during the of the polymerization of nylon 6, a type AB nylon, from caprolactam:
   a)  hydrolytic opening of caprolactam to form 6-aminohexanoic acid,
   b)  condensation of two 6-aminohexanoic acid molecules, and
   c)  direct addition of caprolactam to the growing polymer chain

weight of the final product it is standard practice to add small amounts of a monofunctional chain-terminating molecule such as acetic acid.

A typical commercial reactor consists of a vertical tube, up to 10 m in height, into the top of which the monomer is fed continuously. As polymerization proceeds, the increasingly viscous polymer solution travels down the column. Molten polymer is drawn from the bottom of the reaction tube and is subsequently cooled and chopped into pellets. The final manufacturing stage consists of exposing the pellets to a strong vacuum at a slightly elevated temperature to remove residual monomer and water.

### 23.3.2    Self Condensation of ω-Amino Acids to Make Type AB Nylons

The process of self condensation of ω-amino acids involves reactions of the type shown in Fig. 23.6 b). This type of reaction is used to make polymers, such as nylon 11, from molten 11-aminoundecanoic acid, which is stirred at 220 °C. Such specialty nylons are made in much smaller quantities than nylon 6 and are thus more likely to be made in a batch reactor than in a continuous reaction tube.

### 23.3.3    Manufacture of Type AABB Nylons via an Intermediate Nylon Salt

The diamine and diacid monomers used to make type AABB nylons are typically rather difficult to handle in their pure form. Diamines are liquids or semisolids at room temperature, while the diacids are crystalline solids. These monomers become much more manageable when they are combined to form nylon salts, as shown in Fig. 23.7 a). Nylon salts are solids that can be easily handled and ensure a stoichiometric balance between the diacid and diamine, which is necessary to produce high molecular weight polymers. In the case of nylon 66, the precursor salt is made by boiling adipic acid and hexamethylene diamine in methanol, from which the nylon salt precipitates.

An aqueous solution of the nylon salt is charged into a polymerization reactor where it is subjected to high pressure and temperature. These conditions facilitate reversible reactions of the types shown in Fig. 23.7 b). Polymerization can take place either in batch or continuous reactors. As polymerization proceeds, water is driven off, which drives the equilibrium of the reaction to the right, producing high molecular weight polymer. During polymerization, the temperature of the reactor is gradually increased from approximately 220 °C up to 280 °C. We limit the molecular weight by incorporating trace amounts of acetic acid or some other monofunctional molecule. At the conclusion of the polymerization process, the molten polymer contains about 6% water. The molten polymer is extruded, pelletized, and cooled prior to vacuum stripping of the final few percent moisture. As with all nylons, the finished product is packaged and transported in sacks or boxes that exclude moisture.

Copolymers such as nylon 66/610 are made by the polymerization of the salts of nylon 66 and 610.

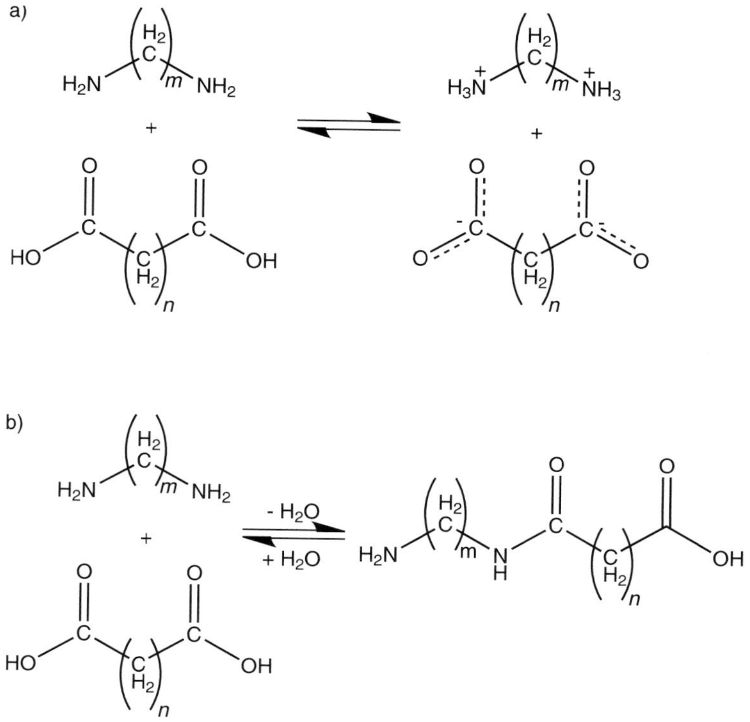

**Figure 23.7**   Reactions occurring during the polymerization of type AABB nylons via a nylon salt: a) production of the nylon salt and b) condensation reaction

## 23.4   Morphology

Nylons are semicrystalline, consisting of strong crystallites linked by flexible amorphous chains. This combination makes them strong and tough, with high melting points. The morphology and properties of nylons in the solid state are strongly influenced by hydrogen bonds linking amide groups from adjacent chains.

### 23.4.1   Crystallite Structure

The molecular arrangement within the crystal units cells of nylon is governed by the need to maximize hydrogen bonding between adjacent chains. Hydrogen bonding within crystallites is facilitated by the fact that nylon chains adopt planar zig-zag conformations with dipoles perpendicular to the chain axis within the plane of the molecule. Examples of nylon crystallite structures are shown in Figs. 23.8 and 23.9 for nylon 6 and nylon 66, respectively. In the

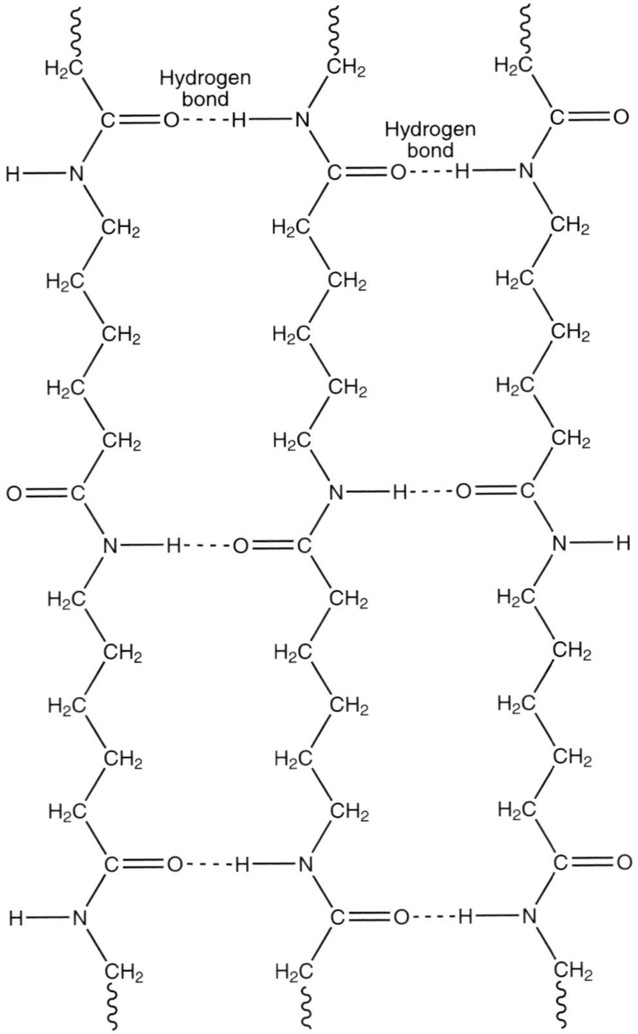

**Figure 23.8**  Planar crystallite structure of nylon 6, showing the anti-parallel alignment of neighboring chains

case of nylon 6, chains are anti-parallel, that is neighboring chains are aligned 180° to each other. In nylon 66, neighboring chains can be arranged in either parallel or anti-parallel configurations.

Nylon crystallites consist of sheets of chains that are hydrogen-bonded to their neighbors. On a supermolecular scale, crystallites have a lamellar structure, that is they are many times longer and broader than they are thick. When nylon crystallizes from an isotropic molten state, it generally forms spherulites, which consist of ribbon-like lamellae radiating in all directions

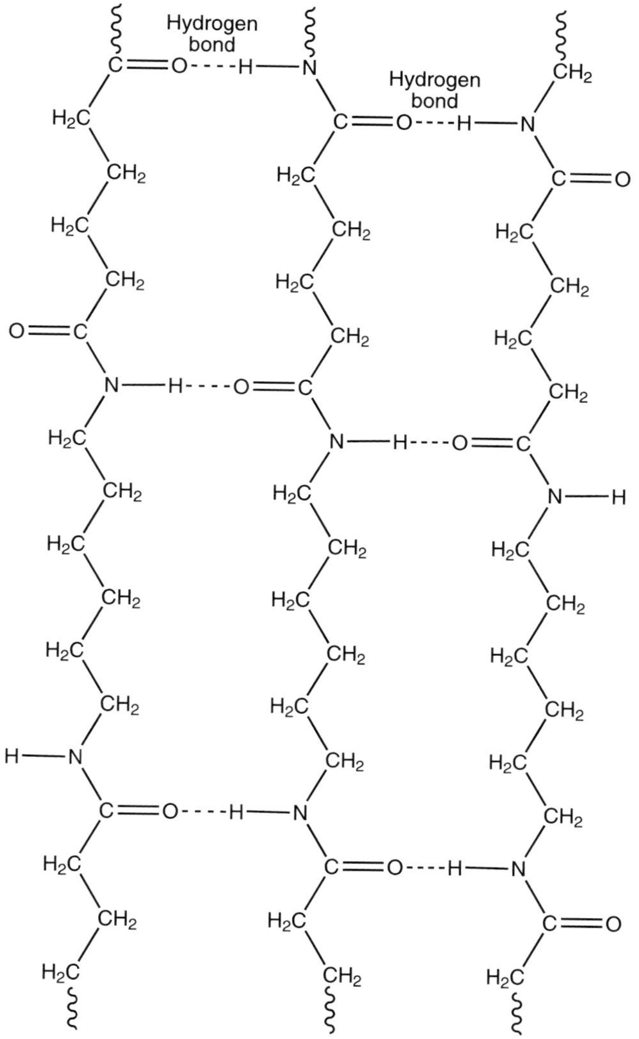

**Figure 23.9** Planar crystallite structure of nylon 66

from nucleation points. The higher the nucleation density, the smaller are the spherulites. In injection molded items supercooling occurs, which allows the orientation induced by flow into the mold to largely relax out. This facilitates the formation of spherulites. The crystallites that form when highly oriented melts cool are aligned parallel to each other in microfibrillar structures. Products with this morphology have high tensile strength when stretched parallel with their chain axes. We take advantage of this morphology when we spin fibers from highly oriented molten strands.

We can strongly influence the degree of crystallinity in nylon articles by changing their crystallization conditions. Rapid quenching results in lower degrees of crystallinity than observed with slow cooling. The crystallinity level in quenched items can be increased by annealing the material at a temperature approaching the crystalline melting point. Under these conditions the chain segments in the non-crystalline regions rearrange to increase the dimensions of existing crystallites. Crystallization from highly oriented melts results in high levels of crystallinity. The higher the level of crystallinity, the stiffer will be the product.

Hydrogen bonding stabilizes crystallites and contributes to their high melting points. Extensive hydrogen bonding exists even in the molten state, which results in a relatively low difference in entropy between the crystal and molten states. This promotes higher crystallization temperatures than found in other polymers with highly flexible chains, such as polyethylene or polyesters.

To a first approximation, the melting point of a nylon crystallite increases linearly with the concentration of its amide groups. Thus, nylon 6 has a melting point of 223 °C versus a melting point of 180 °C for nylon 12. In addition, nylons of the AABB type, such as nylon 66, that can accommodate both parallel and anti-parallel chains have higher melting points than nylons with identical amide concentrations in which neighboring chains are anti-parallel. It is for this reason that nylon 66 has a melting point of 264 °C versus 223 °C for nylon 6.

The crystallites within pure nylon samples melt over a narrow range of temperatures. The resulting melts have low viscosities, reflecting the highly flexible nature of the nylon molecules.

## 23.5    Structure/Property Relationships

Nylons are semicrystalline polymers whose properties are controlled primarily by their amide concentration, molecular orientation, crystallization conditions, and the level of absorbed water. As discussed earlier, the level of crystallinity and hence product stiffness, is maximized by high concentrations of amide groups, high orientation, slow cooling, and the absence of absorbed water.

Highly crystalline nylon articles are hard, tough, and stiff with good heat resistance, which qualifies them as engineering polymers. A 3 mm thick plaque of high crystallinity nylon can be flexed slightly by hand. We can further stiffen nylon products by incorporating fillers, such as glass fibers and inorganic particles.

Nylons are inert to most organic and inorganic liquids, but they will dissolve in highly polar solvents, such as phenol and glacial acetic acid. Nylons degrade hydrolytically when they are exposed to strong acids or when they are immersed in hot water or steam for extended periods of time.

### 23.5.1    Effect of Molecular Composition

The properties of nylon articles are strongly linked to their molecular composition. Nylons with a high concentration of amide groups exhibit strong inter- and intramolecular hydrogen bonding, which stiffens both the crystalline and non-crystalline regions. The degree of crystallinity, and hence the material's stiffness and yield strength, is improved by regular arrangements of amide groups, such as those found in nylon 6 and nylon 66. Nylon copolymers, such as nylon 66/610, exhibit less molecular regularity and hence lower crystallinity than either nylon 66 or nylon 610.

The presence of the amide groups in nylon molecules introduces local dipole moments. When nylons are subjected to alternating electric fields, the molecular dipoles try to align with the field, resulting in high electrical dissipation factors. When the field is strong enough, this can lead to internal heating, which renders nylons unsuitable for use as insulators in strong alternating electrical fields. Nylons can be employed as moderately good insulators in electric fields associated with direct currents.

### 23.5.2    Effect of Crystallization Conditions

We can manipulate the properties of nylon products by changing the conditions under which we crystallize them. The degree of crystallinity is increased by slow cooling, annealing, and by crystallization from highly oriented melts. As we increase the crystallinity level, stiffness and yield strength increase at the expense of impact strength.

We can nucleate crystallization from the melt by incorporating finely ground inorganic crystalline compounds such as silica. Nucleation of injection molded nylons has three primary effects: it raises the crystallization temperature, increases the crystallization rate, and reduces the average spherulite size. The net effect on morphology is increased crystallinity. This translates into improved abrasion resistance and hardness, at the expense of lower impact resistance and reduced elongation at break.

### 23.5.3    Effect of Absorbed Water

Water has a strong effect on the properties of nylons. Under humid conditions, water is readily absorbed into nylons where its presence is stabilized by the formation of hydrogen bonds with the amide groups in the non-crystalline regions. The amount of water that can be absorbed increases as the degree of crystallinity is reduced and the amide content increases. Nylon 6 and nylon 66 can absorb as much as 6% of their own weight in water. Absorbed water plasticizes the non-crystalline regions by interrupting inter-chain hydrogen bonds and providing internal lubrication. The glass transition temperature of nylon 66 falls from approximately 80 °C in its dry state to approximately −15 °C at 100% humidity. As the level of absorbed water increases, impact strength and elongation at break increase at the expense of stiffness, strength, and dimensional stability.

As nylons absorb water, their electrical properties deteriorate. Their performance as insulators declines and they become more susceptible to energy dissipation in alternating fields.

An undesirable effect of the hygroscopic nature of nylons is that we must take care that they are dry before trying to process them in the molten state. If we were to attempt to extrude wet nylon, the water would vaporize within the extruder, creating bubbles in the melt. The result would be a non-uniform extrudate containing voids and exhibiting an uneven surface texture.

### 23.5.4    Effect of Fillers

In engineering applications we routinely incorporate inorganic fillers, such as glass fibers and mineral particles, into nylons. In general, fillers perform three functions: they reduce the cost of the product, they increase its stiffness, and they reduce shrinkage and warpage in molded products.

## 23.6    Products

We encounter nylons principally in four categories: molded items, fibers, films, and coatings. Molded nylon products, such as the buckles on back packs, automotive air intake ducts, and roller blade boots take advantage of nylon's high stiffness, strength, toughness, and abrasion resistance. Nylon is spun into fibers that can be twisted into yarns for use in carpets and woven fabrics, or braided into rope. Thicker filaments are used directly as fishing line and weed trimmer line. To a lesser extent, nylon is used as a barrier film or layer in packaging, where its excellent oxygen resistance, strength, and toughness are valued. The majority of molded items are made from nylon 66, while nylon 6 is used predominantly for fibers and films. Coatings of nylon 11 are used to protect steel tubing and wire in situations where they are routinely exposed to water. Nylon 11 is chosen over nylon 6 or nylon 66 because its longer polymethylene sequences result in lower water absorption.

Molded nylon items exhibit good strength, stiffness, toughness, high heat distortion temperature, abrasion resistance, a low coefficient of friction, and generally good chemical resistance. We take advantage of these properties in a variety of load bearing applications such as furniture casters, screwdriver handles, and gear wheels, bushings and rollers in electronic appliances. Nylon's stiffness, strength, abrasion resistance, and modest electrical characteristics suit it for many electrical applications, such as connectors, light switches, wall plates, and wire and cable insulation that must be pulled through conduits. If we need enhanced stiffness and high temperature performance, we incorporate particulate or fibrous fillers into injection molded products. Glass reinforced nylons are extensively used in automotive applications, such as radiator fan blades, brake fluid reservoirs, wheel covers, and mirror housings, where their resistance to gasoline, oil, other solvents, and temperature extremes are strong assets. We routinely use reinforced nylons at temperatures up to 125 °C, and even 150 °C, for short periods of time. Other applications of glass reinforced nylons include molded bicycle and wheelchair wheels, power tool housings, and pump housings in domestic appliances. Glass reinforced products make up approximately half of all molded nylon articles.

We use nylon in a wide range of fiber and filament applications. Fine fibers can be twisted into yarns that are extensively used in residential carpet, where fiber strength, abrasion resistance, and dyeability are valued attributes. Nylon fibers are woven into fabric as light as that used for parachutes and underwear, and as heavy as the ballistic cloth used for back packs and luggage. The strong and abrasion resistant webbing used in automobile safety belts and climbing harnesses is woven from nylon fibers. Climbing and rappelling ropes consist of a braided nylon fiber sheath surrounding a core of continuous filaments. In this application, strength, abrasion resistance, and shock resistance are valued by rock climbers and cavers (who incidentally refer to their rappelling ropes as their "nylon highway"). Fine nylon filaments are knitted into women's hose, while thicker filaments are used for paintbrushes and toothbrushes, or are twisted into twine for use in fishing nets. Filaments that are exposed to water are often made from nylon 612, which absorbs less water than nylon 6, due its lower amide concentration.

Nylon barrier layers are used in packaging films and blow molded bottles. In these applications, their excellent resistance to oxygen permeation are valued when packaging greasy foods, such as potato crisps (chips), that rapidly turn stale when exposed to oxygen.

To protect steel piping and wire we can powder coat it with a layer of nylon 11. The powder coating process consists of immersing metal heated to above the melting point of nylon 11 (approximately 190 °C) in a fluidized bed of the polymer powder. When the polymer powder comes into contact with the heated metal, it melts and fuses to form a continuous coating. Powder coated products include hospital bed frames, shopping trolleys, and dishwasher racks.

## 23.7    Conclusions

Nylons, with their high modulus, high strength, and high heat distortion temperature, are found in a wide variety of engineering and fiber applications. As engineering polymers, we mold nylons into a tremendous variety of products where they replace metal in an increasing number of applications. Typically, a nylon molding can be made in a single process, which may replace traditional products that were fabricated from several individual pieces or involved multiple processing steps. Nowhere is this more apparent than in our automobiles, in which nylon parts are found inside the passenger compartment, under the hood, and around the exterior. The use of nylon and other polymers permits designers to reduce the overall weight of vehicles and thus improve their fuel efficiency. Nylon fibers have displaced natural fibers in many applications where long-term strength and durability are at a premium. The most common uses of nylon fibers are in residential carpet and in ballistic cloth. The use of this durable and versatile family of polymers in their molded and fiber forms contributes greatly to our comfort and convenience.

## Review Questions for Chapter 23

1.  What do the numbers following the name nylon indicate about the polymeric structure? Why is this nomenclature used?

2.  Why does water content affect the chain extension step of the hydrolytic ring opening of caprolactam?

3.  What advantages are realized by the intermediate nylon salt method over direct polymerization of the diamines with the diacids?

4.  Why is acetic acid used to control the molecular weight of nylons during polymerization?

5.  What factors control the formation of crystalline structure within nylons?

6.  How does annealing affect the properties of rapidly quenched nylons?

7.  Why would the number of amide groups in a nylon chain affect the melting point of the crystallites?

8.  Why does nylon 66 have a melting point that is 40 °C higher than that of nylon 6?

9.  Why does nylon dissolve in highly polar solvents such as glacial acetic acid?

10. How is absorbed water stabilized within nylon? Is its presence necessarily all bad?

# 24 Polyesters

## 24.1 Introduction

Polyesters, which are a class of engineering thermoplastics, are found in a wide variety of applications including carbonated drink bottles, fibers for synthetic fabrics, thin films for photographic films and food packaging, injection molded automotive parts, and housings for small appliances. In this chapter, we will explore the synthesis of this class of polymers. We will also look at the typical properties and end uses for the most common of these resins, polyethylene terephthalate and polybutylene terephthalate, which are commonly known as PET and PBT, respectively.

## 24.2 Chemistry of Polyesters

Polyesters form via a condensation reaction between a dicarboxylic acid and a dialcohol to create an ester linkage, as shown in Fig. 24.1. By far, the two most common polyesters are polyethylene terephthalate and polybutylene terephthalate, the chemical structures of which are shown in Fig. 24.2. These two polymers differ from one another by the length

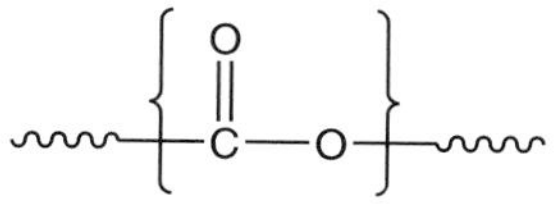

**Figure 24.1** Ester linkage in polyesters

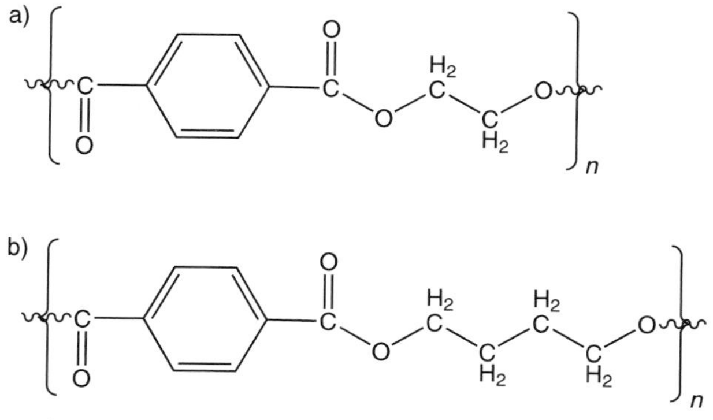

**Figure 24.2** Chemical structures of:
a) polyethylene terephthalate and b) polybutylene terephthalate

of the aliphatic chain introduced through the dialcohol. In polyethylene terephthalate, the chain contains two carbons, while in polybutylene terephthalate the chain consists of four carbons.

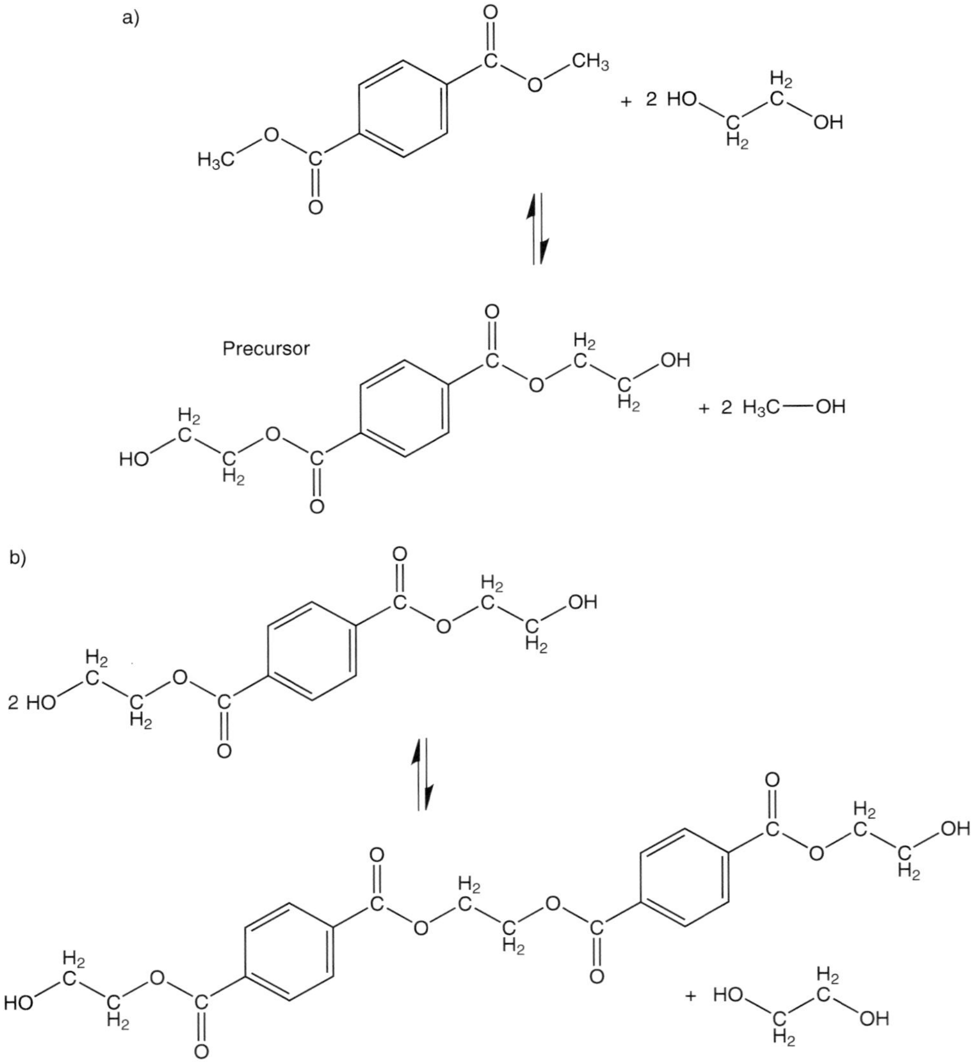

**Figure 24.3**  Two-step polymerization process for the manufacture of polyethylene terephthalate:
a) formation of the precursor via the condensation reaction of dimethyl terephthalate and a dialcohol,
b) polymerization of the precursors to form the polymer

A unique characteristic of polyesters is their ability to undergo additional condensation reactions during processing or when in the solid state. These reactions redistribute the molecular weight of the polymer until a dynamic equilibrium is established. Water, when present at high temperatures in polyester melts, can depolymerize polyesters via a hydrolysis reaction. For this reason, manufacturers must carefully dry the polymer before processing.

## 24.2.1  Synthesis of Polyesters

The simplest way to create polyesters is by the equilibrium condensation reaction between a dicarboxylic acid, such as terephthalic acid, and a dialcohol, such as ethylene glycol or 1,4 butanediol. The chemical structures resulting from these reactions are illustrated in Fig. 24.2, which shows the ester linkage and the location of the carbon chain introduced by the dialcohol. Commercially, polyethylene terephthalate is produced via a two-step process as shown in Fig. 24.3 a) and b). In the first step, dimethyl terephthalate and the dialcohol react, in the presence of a catalyst at high temperature, to create a dihydroxy alkyl terephthalate precursor and an alcohol byproduct. After removing the alcohol byproduct, the precursors enter a second stage reactor. In this reactor, condensation linkages form between the precursors at high temperatures to create a viscous, polymeric melt. The dialcohols produced during this step volatilize and are removed from the melt.

Failure to remove the alcohols generated in either of the equilibrium condensation steps will reduce the efficiency of the polymerization process. This effect can be explained by Le Chatelier's principle, which was discussed in Chapter 3. The volatile alcohols produced during polymerization act as a chemical stress on the product side of the reaction, which inhibits polymerization. Another implication of the equilibrium nature of this polymerization process is seen in the molecular weight distribution of the final polymer. All polyesters contain a few percent of low molecular weight oligomers, regardless of the polymerization process.

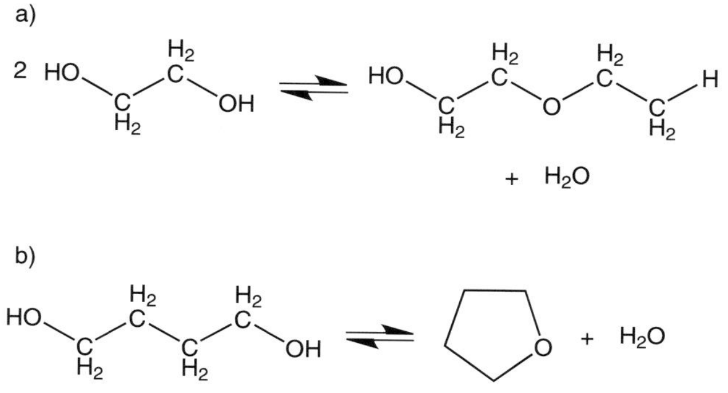

**Figure 24.4**  Side reactions of:
a) ethylene glycol to form a dialcohol ether and
b) 1,4 butanediol to form tetrahydrofuran

Removing these oligomers from the polymer accomplishes nothing as they will eventually reform from the high polymer. The reformation of the oligomers occurs because the system responds to their absence by creating more, in accord with Le Chatelier's principle.

### 24.2.1.1  Side Reactions During Polymerization

In addition to the desired polymerization reaction, the dialcohol reactants can participate in deleterious side reactions. Ethylene glycol, used in the manufacture of polyethylene terephthalate, can react with itself to form a dialcohol ether and water as shown in Fig. 24.4a). This dialcohol ether can incorporate into the growing polymer chain because it contains terminal alcohol units. Unfortunately, this incorporation lowers the crystallinity of the polyester on cooling which alters the polymer's physical properties. 1,4 butanediol, the dialcohol used to manufacture polybutylene terephthalate, can form tetrahydrofuran and water as shown in Fig. 24.4b). Both the tetrahydrofuran and water can be easily removed from the melt but this reaction reduces the efficiency of the process since reactants are lost.

### 24.2.1.2  Solid State Reactions

Polyesters can undergo additional reactions after cooling to the solid state. These reactions include condensation and exchange reactions. During condensation, two separate polymer chains react through their terminal groups to create one long chain, as illustrated in Fig. 24.5. During exchange reactions, two proximate polyester chains react to form two new chains with different relative lengths, as illustrated in Fig. 24.6. In the exchange reaction, one chain's

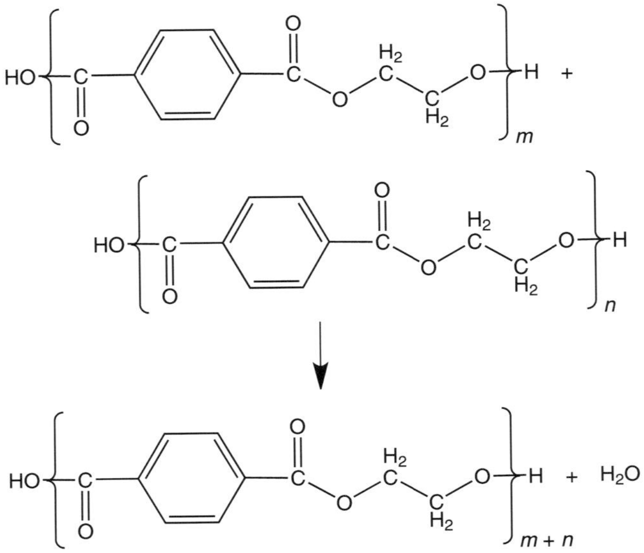

**Figure 24.5**    Solid state condensation of two separate polyester chains

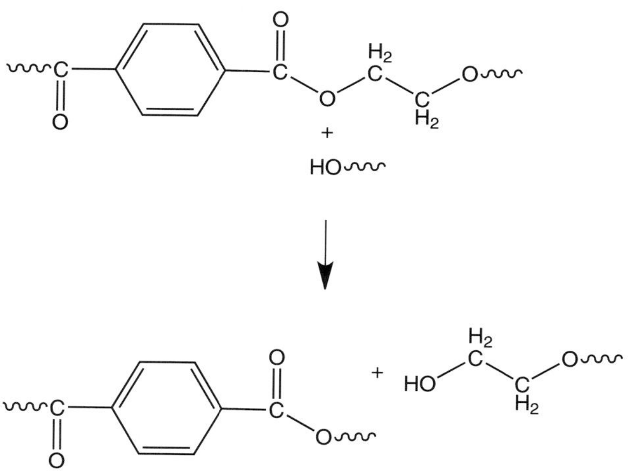

**Figure 24.6**   Exchange reaction between two polyester chains

terminal group interacts with an ester linkage of a nearby chain. During this reaction, the terminal group breaks the chain and adds to one section of the chain while releasing the other, which now has a new terminal alcohol group. Both of these processes affect the average molecular weight of the polymer. Over time, these processes develop dynamic equilibria, so the net result to the polymer's molecular weight distribution will be undetectable.

The solid state condensation reaction of polyesters is exploited in the manufacture of high molecular weight polymer. Polymer produced through standard means is held at a temperature just below its melting point for a period of time. The elevated temperature promotes the solid state condensation reaction by increasing the molecular motion of the chains. These condensation reactions increase the average molecular weight of the material relative to the starting polymer.

## 24.2.2   Degradation of Polyesters

Polyesters can undergo several different degradative reactions. The most important reaction, from a processing perspective, is hydrolysis. Additionally, polyesters can undergo thermal and thermooxidative degradation. Hydrolysis occurs approximately three orders of magnitude faster than thermooxidative degradation and four orders of magnitude faster than thermal degradation. For this reason, we will focus on hydrolysis.

During hydrolysis, water adds across the ester linkage at high temperatures and pressure. This addition breaks one of the bonds that connects the polyester chain together, as shown in Fig. 24.7. The result of this reaction is that the average molecular weight decreases, reducing the polymer's melt viscosity. If hydrolysis goes unchecked, the melt will become so fluid that it becomes impossible to pump and will, instead, dribble out of the end of the processing

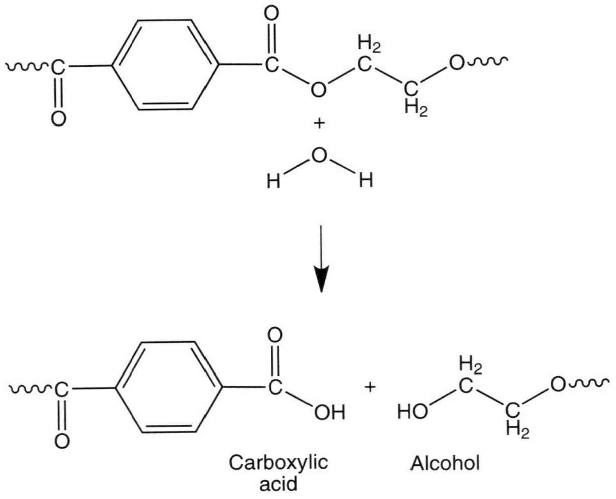

**Figure 24.7**    Hydrolysis of polyester

equipment. In extreme cases, polyethylene terephthalate can overheat to such an extent that the decomposition products volatilize, developing high pressure in the equipment and shooting the low viscosity polymer out of the end of the die. To prevent hydrolysis, water must be removed immediately before processing by drying the resin in high temperature, desiccated air dryers. Effective drying is critical, as even very small amounts water, less than 0.01%, can lead to noticeable viscosity reduction. In addition to drying, some polyester processors incorporate hydrolysis stabilizers, such as esters and epoxides. Although these additives help prevent hydrolysis, nothing replaces the need for efficient drying.

## 24.3    Structure and Morphology

Both polyethylene terephthalate and polybutylene terephthalate exhibit partial crystallinity in the solid state. The molecular weight of the polymer and the time permitted for cooling define the degree of crystallinity of the polymer. Very slow cooling results in high crystallinity and opacity, while fast quenching creates low crystallinity, high clarity material.

### 24.3.1    Polyethylene Terephthalates

Polyethylene terephthalate crystallizes very slowly into only one stable crystalline form, containing monoclinic unit cells. To maximize its physical strength, high crystallinities must be achieved. To achieve this, crystal initiators are often incorporated into the polymer. The initiators allow the formation of many small crystalline regions, creating evenly balanced properties throughout the cooled polymer.

Because polyethylene terephthalate crystallizes slowly, it can readily be produced in its amorphous state. This is especially true when it is used in packaging materials, such as thin films and carbonated drink bottles. The final products exhibit high clarity and directionally balanced properties because they lack crystalline regions.

### 24.3.2    Polybutylene Terephthalates

Polybutylene terephthalate crystallizes much more quickly than polyethylene terephthalate. The four carbon chain derived from 1,4 butanediol creates a more flexible structure than the smaller, more rigid ethylene glycol unit in polyethylene terephthalate. This allows the chains to rearrange easily, allowing very fast crystallization. For this reason, polybutylene terephthalate is rarely found as an amorphous polymer. Polybutylene terephthalate forms two different types of crystalline regions, the alpha and beta forms. The alpha form is the thermodynamically stable one and is observed when slow cooling takes place. It can also be generated from the beta crystalline form by annealing the polymer at elevated temperatures. We observe the beta form when the polymer is quenched rapidly or after axial stretching.

## 24.4    Structure/Property Relationships

Polyesters exhibit excellent physical properties. They have high tensile strength, high modulus, they maintain excellent tensile properties at elevated temperatures, and have a high heat distortion temperature. They are thermally stable, have low gas permeability and low electrical conductivity. For these reasons, polyesters are considered engineering polymers.

### 24.4.1    Tensile Properties

Polyesters have excellent tensile strength, approximately double that of low density polyethylene and polypropylene. The elongation at break is relatively low for both polyethylene terephthalate and polybutylene terephthalate, ranging from 15 to 50% for crystalline materials and reaching as high as 300% for amorphous polyethylene terephthalate. By comparing polyethylene terephthalate and polybutylene terephthalate of identical molecular weight distributions and degrees of crystallinity, we can see the effect of structure on the tensile properties. The relatively inflexible ethylene chain unit in polyethylene terephthalate creates a polymer that is more brittle, has a higher tensile strength, and lower elongation to break than polybutylene terephthalate. Polyethylene terephthalate also exhibits a higher heat distortion temperature and a higher short time exposure use temperature of approximately 200 °C relative to 150 to 165 °C for polybutylene terephthalate.

## 24.4.2    Impact Properties

The impact strength of polyesters is approximately the same as that of pure polystyrene. Crystalline materials shatter via brittle failure under impact, while amorphous materials undergo ductile failure. Though these materials are strong, their strength is often inadequate for impact sensitive applications, such as automotive body panels. For this reason, the resins are often compounded with glass fibers and impact modifiers to increase impact strength. The notched impact resistance of unmodified polybutylene terephthalate is approximately doubled by the addition of 20 to 30% by weight of glass fibers. The cumulative effect of glass fiber and impact modifier can effectively triple the material's notched impact strength.

## 24.4.3    Gas Permeability

Polyalkyl terephthalates demonstrate excellent impermeability to gases, including air and carbon dioxide. Polyethylene terephthalate films are approximately 30 times more resistant to gas permeation than a similar thickness of oriented polypropylene film. Carbonated drink bottles and food packaging applications exploit this property of the polymer. In semicrystalline materials, gas impermeability arises from the presence of the extensive crystalline regions in the solid polymer. For amorphous polyethylene terephthalate, the barrier properties arise from orientation processes during manufacturing. The orientation forms dense regions that inhibit passage of gas molecules from one surface to the other. Orientation occurs during the stretch blow molding of carbonated drink bottles or during bi- or uniaxial orientation processes during film production.

## 24.4.4    Optical Properties

Semicrystalline polyalkyl terephthalates are opaque due to diffraction of light as it crosses the interface between crystalline and amorphous regions. Amorphous polyethylene terephthalate has a low refractive index, making it appear glass-like in quenched parts.

## 24.4.5    Electrical Properties

Polyesters exhibit excellent electrical properties. They resist breakdown when exposed to continual electrical loads and have a high electrical resistance. For this reason, they are often used in electrical housings, as insulating films in electrical components, and as wire insulation where high temperatures are likely to be encountered.

### 24.4.6    Solvent Resistance

Polyesters are, in general, organic solvent resistant. They show excellent room temperature resistance to organic solvents, such as hydrocarbons, alcohols, and chlorinated hydrocarbons. At slightly elevated temperatures of approximately 60 °C, alcohols and aromatic solvents can damage the polymer. Strong acids and bases can cause chemical damage to polyesters, as can ketones and phenols.

## 24.5    End Uses

Polyethylene terephthalate is most often extruded into films or fibers, or blow molded into bottles. Polybutylene terephthalate is primarily found in injection molded parts. Such parts are highly crystalline, which makes them opaque. Polybutylene terephthalate is often modified with glass fibers or impact modifiers. Table 24.1 contains applications by processing method and resin.

**Table 24.1**    Applications of polyethylene terephthalate and polybutylene terephthalate by processing method

|  | Polyethylene Terephthalate | Polybutylene Terephthalate |
|---|---|---|
| Injection Molding | • Electrical connector housings<br>• Light bulb sockets<br>• Flashlight housings<br>• Car headlamp bodies | • Circuit breaker boxes<br>• Headlight reflectors<br>• Automotive body components<br>• Computer keys<br>• Iron housings |
| Extrusion | • Clothing fabrics<br>• Packaging films<br>• Extrusion coating films for food packaging<br>• Insulator films<br>• Photographic film base | |
| Blow Molding | • Carbonated beverage bottles<br>• Cooking oil bottles<br>• Detergent bottles<br>• Drinking water bottles | |

### 24.5.1    Injection Molded Parts

Injection molding grades of polyethylene terephthalate and polybutylene terephthalate have low melt viscosities. Because of this, they can be used to manufacture intricate parts within complicated die cavities. Additionally, they solidify rapidly, allowing short cycle times. High mold temperatures are required for parts having thin walls to prevent premature solidification of the polymer as it fills the mold. Both resins exhibit good mold release, facilitating their removal from the mold. Since both resins are semicrystalline unless quenched, part shrinkage can be considerable as crystallization occurs both in the mold and after removal from the mold. To address in-mold shrinkage, processors increase the hold pressure and the hold time, while also reducing the mold temperature. To reduce post-mold shrinkage, the processor can promote crystallization in the mold by increasing the mold temperature.

Typically, polyester resins are used for high-end applications that require excellent electrical and thermal resistance. When dimensional stability under load is more critical, glass fibers are incorporated to increase the heat distortion temperature and the stiffness of the part. Examples of glass fiber reinforced parts include electrical housings, electrical adapters, computer components, telephone housings, and light bulb sockets. When impact modified, polybutylene terephthalate can be injection molded to make car bumpers.

### 24.5.2    Extruded Films

Polyethylene terephthalate is used to produce a wide variety of different films. These include photographic and X-ray film backing, and electrically insulating films, as well as food and medical packaging films. Polyethylene terephthalate's high heat resistance, chemical inertness, and impermeability make it an excellent choice for packaging films. Additionally, it is food contact safe. Examples of polyethylene terephthalate films in food packaging include boil-in-bag applications and the film covering on microwave meals that remain on the package during preparation. Sometimes polyethylene terephthalate comprises one layer of a multilayer film inhibiting gas permeation. Examples of these applications include stand-up pouches (such as drink pouches) and individual condiment packets. Additionally, polyethylene terephthalate can be extrusion coated onto cardboard to reduce moisture and odor permeability. An example of this is the lining of cardboard potato chip cans.

### 24.5.3    Extruded Fibers

Fabrics manufactured from polyester fibers have several advantages over natural fibers. They have excellent resistance to solvents, are hydrophobic, have good dye receptivity, and are of no interest to moths and mold. They also can take a permanent set and resist wrinkles. Often, fabrics incorporate both natural fibers and polyester to gain a balance of the above mentioned properties with the hand (the feel of the fabric) and drape (the way it falls into folds) of the natural fabric. Most fibers are manufactured via a melt-spun process. In this process, the fiber is extruded and then drawn, after heating above the $T_g$, to introduce orientation, thus further strengthening the fiber, as shown in Fig. 11.7. Beyond fabrics, polyester yarns are woven to reinforce rubber automotive hoses. Additionally, they can be used to make rope and strapping for securing loads on pallets. Crimped polyester fibers are used to manufacture knitting yarns.

### 24.5.4    Blow Molded Bottles

Consumers encounter polyester terephthalate most frequently in the supermarket as carbonated drink bottles, clear plastic juice bottles, shampoo bottles, and spice containers. Its use arises from its gas impermeability, high clarity when quenched, and its chemical and physiological inertness. Additionally, it is one of the easiest resins to recycle, making it an environmentally conscious choice. The blow molding of these containers can be achieved through extrusion blow molding, injection blow molding and injection-stretch blow molding, which are shown in Figs. 14.1, 14.3 and 14.4 respectively. The orientation introduced during the stretching and inflation steps creates a strong bottle with high clarity and low permeability.

## 24.6    Conclusions

Polyesters exhibit excellent high temperature strength and electrical properties making them a good choice for many demanding applications. They also are physiologically inert allowing them to be used in food contact applications. The two common polyesters, polyethylene terephthalate and polybutylene terephthalate, are both used in injection molded products. Polyethylene terephthalate is often used in both extrusion and blow molded processes also.

## Review Questions for Chapter 24

1.  How do polyethylene terephthalate and polybutylene terephthalate differ from one another chemically? How do these differences affect their properties?

2.  When manufacturing very high molecular weight polyesters, a high molecular weight polyester is cooled down to the solid state and then maintained at an elevated temperature just below the melting point. How does this process create a higher average molecular weight than the starting polymer?

3.  Why is it so important to remove water from polyesters during the polymerization processing as well as during melt processing?

4.  Describe the commercial manufacture of polyesters.

5.  Why does the incorporation of ethylene glycol in polyethylene terephthalate reduce the crystallinity of the final polymer?

6.  Polyethylene terephthalate is the dominant material for the manufacture of carbonated beverage bottles? Why are the bottles clear despite the tendency for this polymer to form crystalline domains?

7.  What properties of polybutylene terephthalate make it amenable to injection molding processing?

8.  Polyethylene terephthalate is often used to produce films for food packaging. What properties of this material are critical when designing food packaging?

9.  Propose a mechanism which explains how strong acids damage the chemical structure of polyesters.

10. What properties make polyesters excellent for electrical housing applications?

# 25 Polyurethanes

## 25.1 Introduction

The polyurethane family of polymers encompasses a broad range of molecular compositions with a correspondingly wide range of properties. In general, any polymer that contains a urethane linkage, as shown in Fig. 25.1, is classified as a polyurethane. Urethane links are created by the reaction of organic isocyanates with organic hydroxyls, as illustrated in Fig. 25.2. Polymers are formed when diisocyanates react with diols to form extended chains. In addition, polymers that have linkages other than urethanes that are derived from isocyanates are also called polyurethanes. This classification is based on chemical similarities and the comparable physical characteristics of the resulting polymers. The most common polymers in the polyurethane family consist of urethane linkages connecting polyether or polyester chains, examples of which are illustrated in Fig. 25.3 a) and b), respectively. Polyurethanes

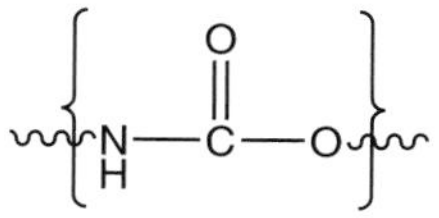

**Figure 25.1**   Urethane linkage

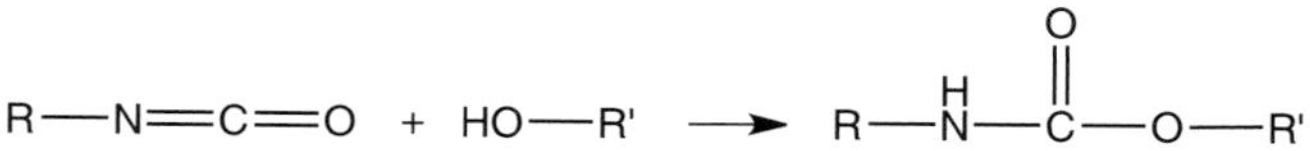

**Figure 25.2**   Reaction of isocyanate and organic hydroxyl to form a urethane linkage

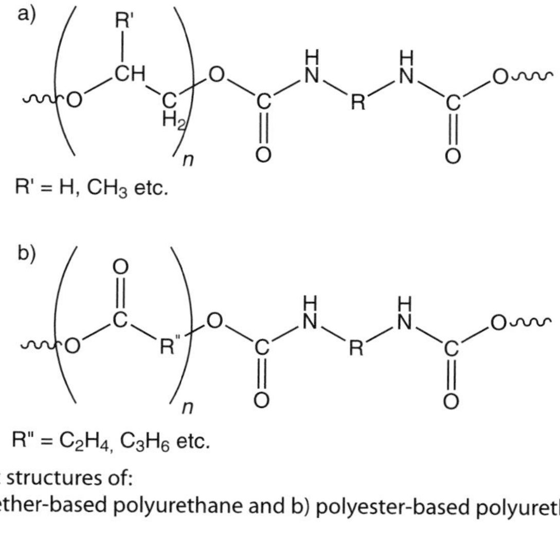

R' = H, CH$_3$ etc.

R" = C$_2$H$_4$, C$_3$H$_6$ etc.

**Figure 25.3**   Generic structures of:
a) polyether-based polyurethane and b) polyester-based polyurethane

can be thermoplastics or thermosets, depending on whether their constituent chains are linear or crosslinked.

We can tune the physical characteristics of polyurethanes over a wide range by careful selection of their components. Polyurethanes are used extensively as foams, which range from soft and resilient to rigid. Solid polyurethanes range from soft moldings and elastic fibers to rigid injection molded items. Other uses of polyurethanes include coatings, sealants, and adhesives.

## 25.2    Chemical Structure

The majority of polyurethanes are based on either polyester or polyether chains connected by urethane or urea linkages. Within these classes there is a host of variants. There are also polyurethanes consisting of polyetherureas, polyureas, polyisocyanates, and polycarbodiimides. The one thing they all have in common is that they are produced by the reaction of organic diisocyanates with molecules containing groups that have one or more exchangeable hydrogen atoms, such as alcohols, glycols, amines, carboxylic acids, and water. In most polyurethanes, the isocyanate residues comprise a relatively small proportion of the polymer chains. In this chapter, we will concentrate on polyurethanes based on polyethers and polyesters, as these are the most widely used.

Unlike other polymers, we have no universally accepted chemical nomenclature for polyurethanes. Instead, we typically refer to polyurethanes in terms of their general characteristics. Thus we refer to "flexible polyurethane foams", "polyurethane elastomers", "integral skin polyurethane foams", and others.

### 25.2.1    Polyether-Based Polyurethanes

The general molecular structure of polyether-based polyurethanes is illustrated in Fig. 25.3 a). Typical polyether sequences include polyethylene glycol and polypropylene glycol. The length of the polyether sequences between urethane links can vary from one or two ether groups up to several hundred. As the length of the polyether sequences between urethane links increases, the polymer exhibits more of the properties normally associated with polyethers.

### 25.2.2    Polyester-Based Polyurethanes

Figure 25.3 b) shows a generic polyester-based polyurethane. The most common polyester repeat units are derived from the polycondensation of adipic acid and a diol, such as ethylene glycol, which results in the structure shown in Fig. 25.4. The average molecular weight of the polyester sequences between urethane links commonly ranges between 400 and 6,000 g/mol.

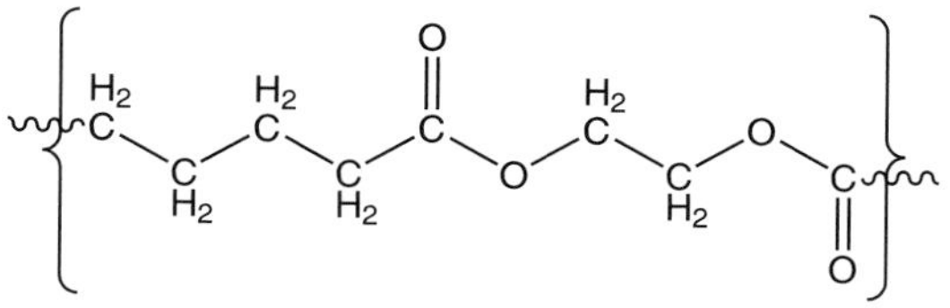

**Figure 25.4** Example of a polyester sequence used in polyurethanes, derived from the condensation of adipic acid and ethylene glycol

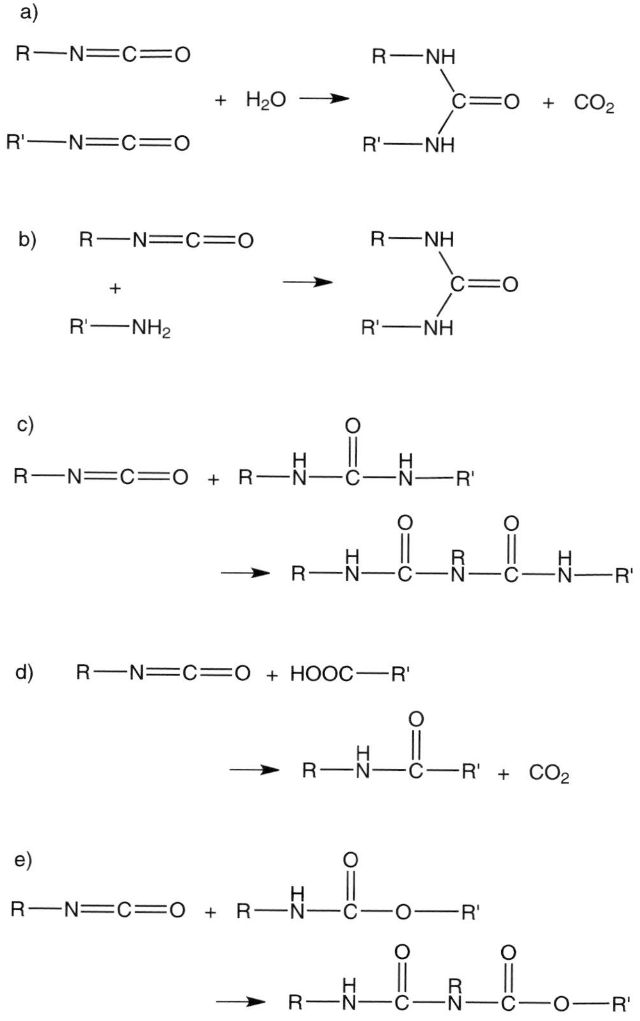

**Figure 25.5** Examples of non-urethane linkages derived from isocyanates: a) urea, b) urea, c) biuret, d) amide, and e) allophanate

### 25.2.3    Non-Urethane Linkages Derived from Isocyanates

In addition to the urethane linkages that are found in the majority of polyurethanes, numerous other linkages derived from isocyanates are also included within the family of polyurethanes. A partial listing of these linkages and the reactions that lead to their formation is presented in Fig. 25.5. The most important of these reactions is that of two isocyanate groups with a water molecule, which yields a urea linkage and generates carbon dioxide. As we shall see in the section on manufacturing, we take advantage of the carbon dioxide evolved by this reaction as a blowing agent to create polyurethane foams. To observe this reaction first hand, it is only necessary to lift the cover off a tank of isocyanate and spit into it. The isocyanate will bubble vigorously where the saliva lands. While providing plant operators with a source of entertainment, this practice results in blocked spray heads and other processing problems. The authors in no way endorse this practice.

### 25.2.4    Diisocyanate Precursors

The reaction rates of diisocyanates are strongly influenced by their molecular structure. The reactivity of isocyanate groups is enhanced by adjacent electron-withdrawing substituents. Aromatic rings are very effective electron withdrawing groups, and it is for this reason that the majority of commercial diisocyanates are aromatic. Many of the diisocyanates used commercially consist of mixtures of isomers. Some of the more important commercial diisocyanates are illustrated in Fig. 25.6. Diisocyanates must be handled carefully to avoid exposing workers to their hazardous vapors.

## 25.3    Manufacture

We can make polyurethanes via one- or two-step operations. In the single-stage process, diols and isocyanates react directly to form polymers. If we wish to make thermoplastic linear polymers, we use only diisocyanates. When thermosets are required, we use a mixture of diisocyanates and tri- or polyisocyanates; residues of the latter becoming crosslinks between chains. In the first step of the two-stage process, we make oligomers known as prepolymers, which are terminated either by isocyanate or hydroxyl groups. Polymers are formed in the second step, when the isocyanate terminated prepolymers react with diol chain extenders, or the hydroxyl terminated prepolymers react with di- or polyisocyanates.

Polyurethanes differ from most other polymers in that polymerization frequently takes place at the same time that we are molding or forming them into a usable shape. The three most common processes of this type are reactive foaming, reactive injection molding (RIM), and reactive spray coating.

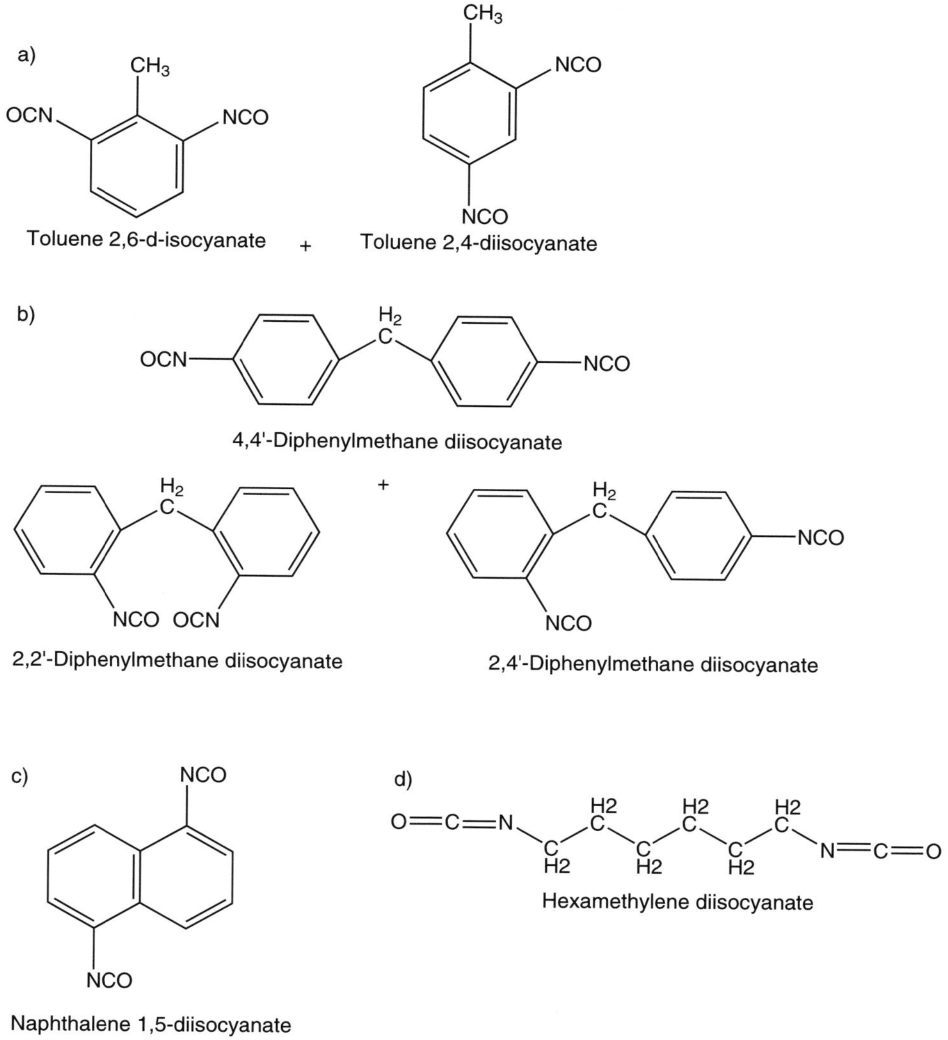

a)

Toluene 2,6-d-isocyanate + Toluene 2,4-diisocyanate

b)

4,4'-Diphenylmethane diisocyanate

+

2,2'-Diphenylmethane diisocyanate          2,4'-Diphenylmethane diisocyanate

c)                                    d)

Hexamethylene diisocyanate

Naphthalene 1,5-diisocyanate

**Figure 25.6**  Examples of diisocyanates used in the manufacture of polyurethanes:
a) TDI, b) MDI, c) NDI, and d) hexamethylene diisocyanate

## 25.3.1  Single-Stage Polymerization

In the single-stage production of polyurethanes, di-, tri-, or polyisocyanates are mixed directly with molecules containing two or more reactive hydrogen atoms, such as diols, diamines, and dicarboxylic acids. We typically increase the reaction rate by adding amine or organotin based catalysts. The reaction normally occurs rapidly, evolving much heat. In order to ensure a uniform product it is important that we rapidly and efficiently mix the reactants in the correct proportions. We normally do this in a mixing head into which liquid feedstreams are metered and vigorously agitated.

The feedstreams can consist of either neat reactants or their solutions. When the feedstreams consist of solutions, the reaction mixture is pumped into a polymerization vessel where the reaction that started in the mixing head proceeds to its conclusion. The polymer is subsequently precipitated from solution, separated, dried, and pelletized. Solvent-free mixtures of reactants are pumped directly to a mold where polymerization proceeds. In this case, other additives, such as, fillers or fire retardants, are co-mixed with the reactants in the mixing head. These additives are permanently incorporated into the finished molding.

## 25.3.2  Two-Stage Polymerization

Two-stage polymerization consists of the steps of prepolymer formation and chain extension. A prepolymer is an oligomer with reactive chain ends. The two basic types of prepolymer are hydroxyl- and isocyanate-terminated, examples of which are shown in Fig. 25.7. We make hydroxyl-terminated prepolymers by reacting an excess of a diol with a diisocyanate. The average length of the prepolymer is governed by the ratio of diol to diisocyanate. A 2 : 1 ratio results in the structure shown in Fig. 25.7 a) where $n = 1$. As the ratio is reduced towards 1 : 1, the length of the prepolymer increases. We can control the length of isocyanate-terminated prepolymers in a similar fashion. Prepolymers are typically viscous liquids with a low vapor pressure, which makes them much easier to handle than their monomeric precursors (especially the diisocyanate).

We create polyurethanes from prepolymers by chain extension. In the case of hydroxyl-terminated prepolymers the chain extender is an isocyanate. If we use a diisocyanate, the resulting polymer is linear. If we substitute some or all of the diisocyanate with a tri- or

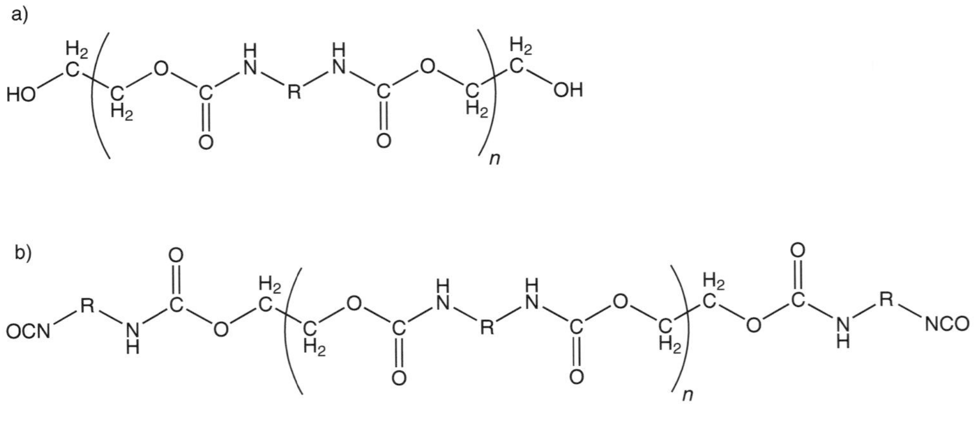

a)

b)

**Figure 25.7**  Examples of prepolymers:
a) hydroxyl-terminated and b) isocyanate-terminated

polyisocyanate, the resulting polymer is crosslinked. Isocyanate-terminated prepolymers are chain extended using diols or triols. The heat generated by the chain extension step of the two-stage process is much less than the heat evolved during single-stage polymerization. The relatively low heat generated by chain extension is preferred for the in-mold polymerization of solid polyurethanes. We also employ this process during reactive spray coating, which is used to apply a protective coating to various materials that are exposed to hostile environments.

### 25.3.3   Reactive Foaming

Foams are one of the most important outlets for polyurethanes. All polyurethane foams are produced by the process of reactive foaming, during which polymerization and expansion proceed simultaneously. There are two basic variants of this process. The first produces large slabs of foam that are subsequently cut or otherwise shaped to meet end use requirements. During the second variant foams expand within a mold of some type that determines their final shape.

#### 25.3.3.1  Production of Foam Slabstock

The process for making foamed slabstock is illustrated schematically in Fig. 25.8. Reactants are sprayed from a mixing head downwards into a moving trough lined with release paper. Bubbles within the foam are created by the evolution of carbon dioxide, which is the major byproduct of the reaction of water and isocyanate to form urea linkages. The mixing head traverses back and forth across the width of the trough, which is moving lengthways underneath it at a rate of approximately 3 to 10 m/min. The relative rates of motion of the mixing head and trough are controlled in such a way as to produce a uniform layer of foam within the channel. The trough is tilted slightly downwards to help create a uniform layer of reactants. The height of the resulting foam "bun" is controlled by the pumping rate, line speed, and angle of the channel. The foam is supported within the paper-lined channel until it has hardened sufficiently to support itself. Typically, foams require 60 to 120 s to rise and another 40 to 100 s to allow them to harden sufficiently for them to be handled. Rectangular slabs are cut from the bun by a traversing saw that moves at the same speed as the foam.

The slabs of foam are transferred to a curing area for up to 12 hours, during which time their final properties develop. Due to the exothermic nature of the foaming reaction and the insulating qualities of the foam, the interior of the slabs can become quite hot. Under extreme circumstances, the temperature rise within polyether-based buns can result in self-ignition. Self-ignition is the source of much unnecessary excitement, for which reason it is generally frowned upon. To minimize the hazards associated with self-ignition, the curing area is separated from the foaming area by fireproof walls.

Most slabstock foams are open-celled, that is, the walls around each cell are incomplete. Towards the end of the foaming process, the polymer migrates from the membranes between cells to the cell struts, which results in a porous structure. In some cases, cells near the surface of the foam collapse to form a continuous skin, which may be trimmed off later.

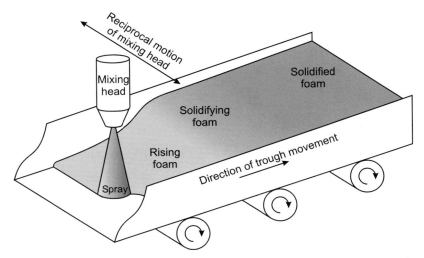

**Figure 25.8**    Schematic diagram outlining the general process for making foamed polyurethane slabstock

Once fully cured, the foam slabs are further sliced to the desired size. Thin slices can be shaved lengthwise from long flexible foam slabs and rolled up to provide foam roll stock for use in upholstery and other industries.

Polyether-based foams account for more than 90% of all flexible polyurethane foams. The properties of foams are controlled by the molecular structure of the precursors and the reaction conditions. In general, density decreases as the amount of water increases, which increases the evolution of carbon dioxide. However, the level of water that can be used is limited by the highly exothermic nature of its reaction with isocyanate, which carries with it the risk of self-ignition of the foamed product. If very low density foams are desired, additional blowing agents, such as butane, are incorporated within the mixing head.

### 25.3.3.2  In-Mold Foaming

The other major polyurethane foaming process involves injecting the mixed reactants into a mold or form that defines the final shape of the product. We use this type of reactive foaming to make a variety of products including shoe soles, furniture cushions, refrigerator insulation, and building insulation. Foams made by this process can provide structural rigidity to the lining of appliances, permitting us to reduce their overall weight. *In situ* molding produces less scrap than slabstock production, especially when complex shapes are desired.

Foams made by *in situ* molding can be either hot or cold cured. In the hot curing process, molds containing the reacting mixture are transferred to an oven to complete the curing process. Cold curing proceeds at ambient temperatures. We use components that are more reactive in the cold curing process than we do in the hot curing process.

The hot curing process normally uses polyether diol precursors with molecular weights of 3,000 to 5,000 g/mole. We can control the stiffness of the foam by adjusting the average number of isocyanate groups on the chain extender molecules. The higher the functionality of the isocyanate molecules, the more crosslinked, and hence stiffer, will be the product.

During the foaming process we must balance the rates of crosslinking and expansion. It is important that foaming occur faster than crosslinking in order to allow the foam to rise and stabilize in its expanded form. However, if crosslinking is delayed too long, the foam can collapse. Insufficient crosslinking results in splitting of the foam. Excessive or premature crosslinking prevents the formation of an open cell structure.

Hot curing can occur at temperatures of up to 250 °C. We normally construct molds from a relatively thin metal, which permits the ready transfer of heat. Before injecting the ingredients into a mold, its interior is sprayed with a mold release agent, which facilitates removal of the foam product. Mold release agents typically consist of a wax dispersion in water. Molds are vented to prevent pressure build up. The total cycle time is about 30 to 40 minutes, of which curing accounts for approximately half. The remainder of the time is devoted to removing the product from the mold, cleaning, and preparing the mold for the next cycle.

Crosslinking occurs rapidly during cold cure foaming, which results in a closed cell structure. In order to create an open cell structure we crush the closed-cell foam after removing it from the mold, which bursts the cells (naturally this only works for resilient products). Crushing can be performed between rollers or in a press. Unlike hot cure foaming, molds have only a few small vents. Cold cure molds must be relatively sturdy in order to withstand internal pressures of 1 to 2 atm. We can control the density of the foam by adjusting how much of the reactive mixture we inject into the mold. The larger the amount of mixture injected, the higher the product density. An advantage of the cold cure process is that its cycle times are typically 6 to 10 minutes, which is much faster than the hot curing process.

Cold curing permits the formation of dual density foam products. We either sequentially inject different formulations or we mold the first foam then mold the second around or on top of it. When making dual density foams, it is important that the first component develop a continuous skin prior to the injection of the second reactive mixture.

An important variant of the *in situ* process is foaming within a flexible cover. As the foam expands, it forces the cover outwards to conform to the interior shape of a mold. We make extensive use of this process to make seat cushions for automobiles and office furniture.

By the appropriate choice of reactants, blowing agents, reaction conditions, mold temperature, etc., we can make a product known as integral skin foam. This consists of a foamed core within an integral solid skin. Such products can be rigid, semi-rigid, or flexible. Applications include shoe soles, automobile arm rests, and snowboards. These products are lightweight, abrasion resistant, and shock absorbing. We generally use polyester-based polyols for their superior mechanical properties. Polyether-based foams are used in special applications where microbial attack is common, such as golf shoes and hiking boots, which are exposed to moisture and biologically active environments.

Polyurethane foams are widely used for carpet backing. In the direct coating method, the pre-heated carpet passes directly underneath spray heads, which deliver the required amount of the reactive mixture. The polyurethane backing is subsequently passed between embossing rolls to give it a uniform thickness. Alternatively, the foaming reactant mixture can be sprayed onto a drum against which the pre-heated carpet is pressed, while the polyurethane is still fluid. Carpets that experience a lot of traffic or high wear, such as in corridors, on stairs, and under chairs equipped with rollers are generally given a denser coating of foam to handle the repeated abuse. High elasticity and low permanent set are valued to prevent permanent deformation under the feet of furniture.

### 25.3.4    Reaction Injection Molding

We employ reaction injection molding to make crosslinked elastomeric polyurethane products. Liquid feedstocks are mixed and injected into molds where the polymerization and cross-linking reactions take place. We typically use a hydroxyl-terminated polyester prepolymer, which reacts within the mold with a diisocyanate such as naphthalene 1,5-diisocyanate. The resulting elastomers are abrasion resistant and impervious to oils and other solvents. This process is widely used to make solid tires, such as those found on fork lift trucks and grocery carts. By injecting a small amount of water we can generate elastomer foams with densities in the range of 0.25 to 0.65 $g/cm^3$. These foams have excellent shock absorbing properties, which suits them for railroad buffers, bridge suspension components, and building foundations.

### 25.3.5    Fiber Spinning

Elastic polyurethane fibers are widely used in apparel to improve shape retention and fit. We can spin polyurethane fibers by several techniques. Dry spun fibers are formed by spinning a solution of polyurethane in a polar solvent, such as dimethylformamide or dimethylacetamide that is subsequently flashed off at reduced pressure. In wet spinning, fibers of polyurethane solution are spun into a non-solvent, typically water, in which they coagulate. Wet spun fibers are typically coarser than dry spun fibers and have a rougher surface. Reactive spinning involves simultaneous polymerization and spinning into a coagulation bath. Of these options, direct reactive spinning is the most economical process for making elastic polyurethane fibers, such as Spandex.

## 25.4    Morphology

The morphology of polyurethanes varies widely depending on the molecular characteristics of their components. We take advantage of this variability by selecting components that give us the properties that we desire for specific applications. In general, the lower the fraction of isocyanate residues in a polyurethane, the closer its morphology and properties will match those of the polyether or polyester sequences between the urethane linkages. The properties of polyurethane foams are further controlled by their density, degree of crosslinking, and whether the cells are open or closed. The area where we see the most direct relationships between molecular structure, morphology, and physical properties is in thermoplastic polyurethane rubbers.

## 25.4.1    Thermoplastic Polyurethane Rubber Morphology

The elasticity of thermoplastic polyurethane rubbers (which are also known as thermoplastic urethanes or TPUs) is a function of their morphology which comprises hard and soft phases. The hard phases consist of hydrogen bonded clusters of chain segments, which are linked by flexible chain segments that make up the soft phase. The hard blocks, which are the minor phase, exist as separate domains within a continuous matrix of the majority soft phase, as shown schematically in Fig. 25.9.

Each hard block consists of a bundle of chain segments containing urethane units that are linked to each other by hydrogen bonds, as shown in Fig. 25.10. Hard blocks comprise bundles of chains with dimensions on the order of a few nanometers. These blocks are paracrystalline, that is, their chains are not perfectly ordered into true crystalline matrices, instead they are approximately straight and parallel to each other. The precise size and nature of the hard phases is controlled by the chemical structure and length of their chain segments. Hard blocks are connected to their neighbors by meandering amorphous chains, which are flexible when they are above their glass transition temperature. The molecular weights of the connecting chains can be in the region of 1,000 to 4,000 g/mole, i.e., approximately 8 to 30 nm. The hard blocks act as crosslinks that prevent the flexible polymer chains from flowing when they are subjected to deformational forces. Once the deforming force is removed, the flexible chains relax to their original state, causing the material to retract.

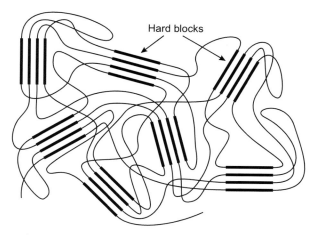

**Figure 25.9**   Schematic representation of hard blocks connected by flexible chains

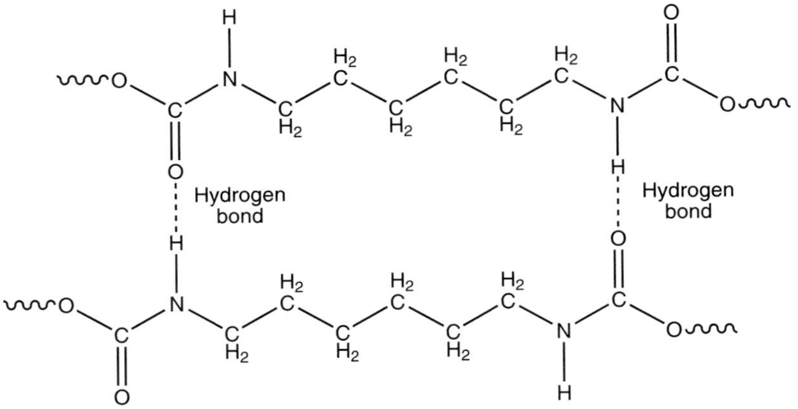

**Figure 25.10**    Example of hydrogen bonding between urethane linkages in a hard block of a thermoplastic polyurethane elastomer

When a thermoplastic polyurethane elastomer is heated above the melting point of its hard blocks, the chains can flow and the polymer can be molded to a new shape. When the polymer cools, new hard blocks form, recreating the physical crosslinks. We take advantage of these properties to mold elastomeric items that do not need to be cured like conventional rubbers. Scrap moldings, sprues, etc. can be recycled directly back to the extruder, which increases the efficiency of this process. In contrast, chemically crosslinked elastomers, which are thermosetting polymers, cannot be reprocessed after they have been cured.

Thermoplastic polyurethane elastomers are normally based on polyester prepolymers. The properties of these polymers can be systematically varied by tailoring the nature and ratio of the hard and soft segments. The stiffness of a polyurethane elastomer increases as the proportion of hard blocks increases. As the stiffness increases, the extensibility of the material decreases.

## 25.5    Structure/Property Relationships and End Uses

Given the wide variety of chemical structures within the polyurethane family, it should come as no surprise that polymer scientists can tailor compositions and structures to meet a correspondingly broad range of end use requirements. In order to better understand the nature of these relationships, we have divided this section into subsections based on some of the more characteristic forms in which we encounter these materials.

## 25.5.1    Foams

Polyurethane foams fall into three basic categories, flexible, semi-rigid, and rigid. As a general rule, we make the softer and more flexible products from long polyether diols. As the concentration of isocyanate linkages increases, the foam becomes stiffer, especially when tri- or polyisocyanates are used as crosslinks. We can make semi-rigid and rigid foams either with or without integral skins.

### 25.5.1.1  Flexible Foams

The two principal methods by which we control a flexible foam's properties are via the chemical characteristics of the prepolymer on which it is based and the amount of water used in the foaming reaction. Longer prepolymers yield softer products. Polyether-based foams are more flexible than those made with polyester prepolymers, due the inherently more flexible polyether chain. Polyether-based foams account for more than 90% of all flexible polyurethane foams. In general, foam density decreases as the amount of water increases. The level of water that we can use is limited by the highly exothermic nature of its reaction with the isocyanate, which carries with it the risk of self-ignition. If lower density foams are desired, we use supplementary physical blowing agents, such as butane. Typical densities for resilient foams fall in the range of 0.2 to 0.6 g/cm$^3$. The vast majority of flexible foams are open celled.

The largest single use of flexible foams is for upholstery; this includes residential, commercial, automotive, bus, and airplane seating. There is a fair chance that you are supported by polyurethane foam as you read this sentence. Other major uses include mattresses and protective packaging. Smaller scale uses include waist bands in disposable diapers, acoustic damping, and backing for automobile headliners. Open cell foam is preferred for acoustic insulation, in which sound is absorbed by the friction of air moving through the interconnected cells.

Some of the key factors that lead us to select polyurethane foam for upholstery are its durability, resilience, and controllable hardness (or softness, depending on your point of view). Vibration dampening and shock absorbance are important attributes in automobile and public transportation seating. Open cell foams are preferred for these applications because they allow for air and moisture transport, which improve the comfort of passengers who may occupy a seat continuously for several hours. This inherent breathability is also a valuable attribute in mattresses. Shock absorbance plays a key role in selecting flexible polyurethane foams for the packaging of fragile items.

When rectangular blocks are cut from foam buns, there is inevitably a relatively large amount of scrap, which can amount to as much 25%. These irregular off cuts are ground down to make uniform size chips, which can be glued together to form a continuous product, such as carpet underlay. When used in carpet underlay, polyurethane provides mechanical and acoustic damping, which improve our comfort. Other uses for foam chips include pillow filling and stuffing in plush toys.

### 25.5.1.2  Semi-Rigid Foams

Semi-rigid foams take much longer to return to their original dimensions after deformation than flexible foams. They offer exceptional shock absorbance, which suits them for many protective applications, especially in automobiles. In general, we use polyester diols in semi-rigid foams because of their superior mechanical properties.

Automotive applications include head rests, dashboard covers and bumpers. Typically, such applications involve foaming of the polyurethane behind a pre-formed outer cover. Integral skin applications include spoilers, wind dams, and body trim. Steering wheels are often constructed of a semi-rigid foam surrounding a metal frame. Arm rests and similar items routinely incorporate metal inserts to permit fastening using screws. Other foams are reinforced with inorganic fillers or fibers to improve their stiffness. Such foams can have densities in the range of 1.0 to 1.3 $g/cm^3$.

Another major use for semi-rigid polyurethane foams is as shoe soles. In this application, their light weight, abrasion resistance, and shock absorbing properties are important. In safety shoes, chemical and oil resistance are important. Polyether-based foams are used in applications where microbial attack is common, such as shoes intended primarily for use on soil or grass.

We also use semi-rigid polyurethane foams to package irregularly shaped objects. Liquid reactants are injected into a plastic bag, which swells to form a custom fit that immobilizes items within an outer box.

### 25.5.1.3  Rigid Foams

Rigid polyurethane foams can be made from either polyester or polyether prepolymers, which are crosslinked with polyfunctional isocyanates. The resulting foams are largely closed cell, with only about 5 to 10% of cells being open. Rigid polyurethane foams are widely used as insulation in commercial, residential, and industrial settings.

When used purely as an insulator, foam densities can be as low as 0.02 to 0.08 $g/cm^3$. In structural applications the foam's density can rise to 0.4 to 0.7 $g/cm^3$. The use of fluorocarbons as blowing agents has largely given way to more environmentally friendly agents, such as low molecular weight hydrocarbons. We can impart flame retardancy by incorporating chlorine- or phosphorus-based compounds.

Rigid polyurethane foams are often polymerized *in situ*. That is, the foaming and polymerization reaction takes place within the cavity where the product will perform its end use. Products of such *in situ* foaming include ice chests, refrigerators, refrigerated fishing boat holds, insulated shipping containers, and cavity insulation within houses and automobiles. In many of these applications the foam insulation acts as a structural component, stiffening a relatively flexible outer skin, as it does in refrigerator doors. Foam within automobile cavities provides multiple benefits: it insulates, quiets the interior, and eliminates penetration by water, thus reducing rusting.

Rigid integral skin polyurethane foams are used to make seat pans, office equipment housings, tanning beds, water skis, and surfboards, all of which must be light and stiff.

### 25.5.2    Elastic Fibers

Elastic polyurethane fibers are widely used in apparel to improve fit and reduce sagging. The elastic strands are woven or knitted into fabrics of other yarns while in their extended state. When tension is released, they return to their original length. The resulting fabrics exhibit improved fit relative to normal fabrics. Woven fabrics and knitted textiles incorporating polyurethane fibers can be stretched up to 25% and 200%, respectively, without causing permanent damage. We encounter polyurethane fibers, in the form of Spandex or Lycra™, in swim wear, hosiery, sports clothing, and many other types of apparel.

Another widespread use of polyurethane elastic fibers is in disposable diapers and adult incontinence garments. Elastic strands are incorporated into waistbands and side panels made of non-woven fabrics where they improve fit and reduce leakage.

### 25.5.3    Thermoplastic Elastomers

Polyurethane-based thermoplastic elastomers are extensively used in applications requiring physical resilience and chemical resistance. In addition to their elasticity, they also exhibit vibration damping, abrasion, tear, and cut resistance.

The elasticity in polyurethane elastomers comes from their semicrystalline morphology in which amorphous chains above their glass transition temperature are connected via hydrogen bonded crystallites, as shown in Fig. 25.9 and discussed in Section 25.4.1. Being largely amorphous, these polymers are typically flexible with modest tensile strength. As their crosslink density increases, their hardness increases and their extensibility diminishes. We can vary the properties of thermoplastic urethanes by tailoring the nature and ratio of the hard and soft segments. The principal diisocyanate that we use is diphenylmethane diisocyanate, with 1,4-butane diol as the chain extender. Thermoplastic polyurethane elastomers are normally based on polyesters, which have higher tensile strengths than polyether-based elastomers, but lower resistance to hydrolysis and microbial attack.

The most important physical characteristics of polyurethane elastomers are their elasticity, low permanent set, high tear strength, and good abrasion resistance. Polyester-based elastomers have higher tear strength than polyether-based analogues. Polyether-based elastomers have better recovery. Important chemical characteristics include stability when exposed to the elements and resistance to oil and grease.

Wheels and tires are one of the major uses of cast polyurethane elastomers. We commonly see these on fork lift trucks and shopping carts, where their excellent abrasion resistance, resistance to oil, and good elasticity are valued. In industrial settings we find polyurethane covers on rollers used for paper, steel, and textile conveyor systems. In such applications, their excellent cut and abrasion resistance help prolong their useful life. If used in hot and humid conditions, polyether-based polyurethanes are preferred.

Other common uses include gaskets and seals for hydraulic systems in harsh environments, such as mining operations, where abrasion is a constant hazard. We take advantage of the flexibility of thermoplastic urethane elastomers and their resistance to chemical and bacterial attack in gaskets connecting sewer pipes, seals in water treatment plants, and expansion joints in bridges and airport runways. Automotive applications include gaskets and boots in suspension and drive systems. Oil resistance and good tear strength are valuable characteristics in these applications.

We take advantage of the shock absorbing characteristics of thermoplastic urethane rubbers in automotive, motorcycle, and bicycle suspensions and power transmission systems. Industrial shock absorbing applications include buffers for elevators and cranes.

### 25.5.4   Sprayed Coatings

An important industrial use of polyurethanes is as sprayed coatings. These coatings can be applied *in situ* in areas subject to high abrasion or chemical attack. Applications include coatings on screens for separating aggregates by size, pipes used to handle slurries in mining operations, pump impellers, and chemical storage tanks.

### 25.5.5   Miscellaneous Uses

Polyurethanes are used in many other applications, which we do not have room to discuss. These applications include paints, varnishes, adhesive binders for chipboard and plywood, hoses, timing belts, athletic tracks and playing surfaces, and the encapsulation of electrical components.

## 25.6   Conclusions

In this chapter we have discussed some of the extensive variety of polyurethanes. In doing so, we have barely scratched the surface of the available compositions. The wide range of chemical structures available within the polyurethane family enables us to tailor their properties to a seemingly endless variety of applications. The versatility of polyurethane chemistry allows us to create products with specific characteristics for use in many commodity and specialty applications. As flexible foams they improve the comfort of the furniture on which we sit and the carpet on which we walk. Semi-rigid foams help protect us in our automobiles. Rigid foams are used for insulation that reduces our energy consumption. As fibers, polyurethanes improve the fit of apparel on all ages of people, from disposable baby diapers to sports clothing, and adult incontinence products. Polyurethane rubber characteristics can be tuned to provide a range of elastic moduli suitable for a wide variety of automotive, commercial, industrial and residential uses. In short, polyurethanes contribute greatly to our comfort, safety, and convenience.

# Review Questions for Chapter 25

1.  What makes a polymer a polyurethane?

2.  Single stage polymerization generates much more heat than the polymerization of a prepolymer in the chain extension stage of the two stage polymerization process. Why is this so, and what benefit is there in having a lower heat of polymerization?

3.  What differentiates a thermoset polyurethane from a thermoplastic one?

4.  Describe the differences in properties between polyester and polyether based polyurethanes.

5.  Why does the density of a polyurethane foam generated from a slab process generally decrease as the water content of the system increases?

6.  How does the number of multifunctional isocyanate units in a foamed polyurethane influence the rigidity of a foam?

7.  Why is it necessary to have a balance between the rates of crosslinking and expansion when manufacturing a foamed polyurethane? What variables affect these rates?

8.  What properties of polyurethanes make them an excellent choice for foamed seat cushions?

9.  How does the elasticity of thermoplastic urethanes arise from the structure of the polyurethane?

10. What is the difference between paracrystalline regions and crystalline regions?

11. How does the length of the prepolymer units affect the physical properties of a polyurethane material?

# Additional Reading

## *General Organic Chemistry*

Wade, L. G. *Organic Chemistry, 5th Edition.* (2002) Prentice-Hall, Inc. Upper Saddle River, NJ.

## *General Polymer Science Texts*
### Materials

Ehrenstein, G. W. *Polymeric Materials: Structure-Properties-Applications.* (2001) Hanser Gardner Publications, Cincinnati.

Charrier, J. M. *Polymeric Materials and Processing: Plastics, Elastomers and Composites.* (1991) Hanser Publishers, New York

Billmeyer, F. W. Jr. *Textbook of Polymer Science, 2nd Edition.* (1971) John Wiley and Sons, Inc. New York.

Painter, P. C.; Coleman, M. M. *Fundamentals of Polymer Science.* (1997) Technomic Publications. Lancaster, PA.

### Alloys, Blends and Additive Mixtures

Utracki, L. A. *Polymer Alloys and Blends: Thermodynamics and Rheology.* (1990)Hanser Publishers, New York.

Zweifel, H. *Plastics Additives Handbook 5th Edition.* (2001) Hanser Gardner Publications, Cincinnati.

Wool, R. P. *Polymer Interfaces: Structure and Strength.* (1995) Hanser Gardner Publications, Cincinnati.

### Properties and Testing

Ferry, J. D. *Viscoelastic Properties of Polymers, 3rd Edition.* (1980) Wiley, New York.

Brostow, W.; Corneliussen, R. D. *Failure of Plastics.* (1986) Hanser Gardner Publications, Cincinnati.

Ehrenstein, G. W. *Thermal Analysis of Plastics: Principles and Applications.* (2004) Hanser Gardner Publications, Cincinnati.

Carreau, P. J.; DeKee, D. C. R.; Chhabra, R. P. *Rheology of Polymeric Systems: Principles and Applications.* Hanser Gardner Publications, Cincinnati.

Hylton, D. C. *Understanding Plastics Testing.* (2004) Hanser Gardner Publications, Cincinnati.

## *Polymer Chemistry*

Allcock, H. R.; Lampe, F. W. *Contemporary Polymer Chemistry, 3rd Edition.* (2002) Prentice-Hall, Inc. Upper Saddle River, NJ.

Flory, P. J. *Principles of Polymer Chemistry.* (1953) Cornell University Press. Ithaca, New York.

Odian, G. *Principles of Polymerization, 3rd Edition.* (1991) John Wiley and Sons, Inc. New York.

Sandler, R. S.; Karo, W. *Polymer Syntheses Volume 1, 2nd edition.* (1992) Academic Press, Inc. San Diego.

Schnabel, W. *Polymer Degradation: Principles and Practical Applications.* (1992) Hanser Publishers, New York.

## Chain Growth Polymerization:

Kennedy, J. P.; Marechal, E. *Carbocationic Polymerization*. (1982) John Wiley and Sons, Inc. New York.

Moad, G.; Solomon, D. H. *The Chemistry of Free Radical Polymerization*. (1995) Elsevier Science Inc. Tarrytown, New York.

Morton, M. *Anionic Polymerization, Principles and Practice*. (1983) Academic Press. New York.

## Step Growth Polymerization:

Gupta, S. K.; Kumar, A. (1987) *Reaction Engineering of Step Growth Polymerization*. Plenum Press, New York.

Solomon, D. H. (ed) *Step Growth Polymerizations*. (1972) Marcel-Dekker, Inc. New York.

## *Polymer Processes*

Osswald, T. A. *Polymer Processing Fundamentals*. (1998) Hanser Gardner Publications, Cincinnati.

## Extrusion

Giles, H. F., Jr.; Wagner, J. R., Jr.; Mount, E. M. *Extrusion: The Definitive Processing Guide and Handbook – PDL Handbook Series*. (2005) William Andrew Publishing, Norwich, NY.

Vlachopoulos, J.; Wagner, J. R. Jr. (Eds). *The SPE Guide on Extrusion Technology and Troubleshooting*. (2001) The Society of Plastics Engineers, Brookfield, CT.

## Compounding

Todd, D. B. (ed) *Plastics Compounding: Equipment and Processing*. (1998) Hanser Gardner Publications, Cincinnati.

## Injection Molding

Poetsch, G.; Michaeli, W. *Injection Molding: An Introduction*. (1995) Hanser Gardner Publications, Cincinnati.

Gramann, P.; Osswald, T. A.; Turng, T. *Injection Molding Handbook*. (2001) Hanser Gardner Publications, Cincinnati.

## Blow Molding

Rosato, D. V.; Rosato, A. V.; Domatha, D. P. *Blow Molding Handbook*. (2003) Hanser Gardner Publications, Cincinnati.

Lee, N. C. *Blow Molding Design Guide*. (1998) Hanser Gardner Publications, Cincinnati.

## Rotational Molding

Beall, G. L. *Rotational Molding: Design, Materials, Tooling and Processing*. (1998) Hanser Gardner Publications, Cincinnati.

## Thermoforming

Throne, J. L. *Technology of Thermoforming*. (1996) Hanser Gardner Publications, Cincinnati.

## Recycling

Brandrup, J.; Bittner, W.; Michaeli, W.; Menges, G. *Recycling and Recovering Plastics*. (1996) Hanser Gardner Publications, Cincinnati.

## *Polymer Specific Texts*

### Polyethylene

Peacock, A. J. *Handbook of Polyethylene: Structures, Properties and Applications*. (2000) Marcel Dekker, New York.

### Polypropylene

Moore, E. P., Jr. (ed) *Polypropylene Handbook 1st Edition*. (1996) Hanser Gardner Publications, Cincinnati.

Pasquini, N. (ed) *Polypropylene Handbook 2nd Edition*. (2005) Hanser Gardner Publications, Cincinnati.

Karian, H. G. (ed) *Handbook of Polypropylene and Polypropylene Composites 2nd Edition*. (2003) Marcel Dekker, New York.

### Polycarbonates and Polyesters

Bottenbruch, L. (ed) *Engineering Thermoplastics: Polycarbonates, Polyacetals, Polyesters, Cellulose Esters*. (1996) Hanser Gardner Publications, Cincinnati.

LeGrand, D. G.; Bendler, J. T. (eds) *Handbook of Polycarbonate Science and Technology*. (2000) Marcel Dekker, New York.

Fakirov, S. (ed) *Transreactions in Condensation Polymers*. (1999) Wiley-VCH, New York.

Schultz, J. M.; Fakirov, S. (eds) *Solid State Behavior of Linear Polyesters and Polyimides*. (1990) Prentice Hall, Edgewood Cliffs, N. J.

### Polystyrene

Choi, D. D.; White, J. L. *Polyolefins: Processing, Structure, Development and Properties*. (2004) Hanser Gardner Publications, Cincinnati.

Skodchdopole, R. E. in *Encyclopedia of Polymer Science and Engineers, V16 2nd Edition*. "Styrene Polymers". (1989) John Wiley and Sons.

### Polyvinyl Chloride

Daniels, C. A.; Summers, J. W.; Wilkes, C. E. (eds) *PVC Handbook*. (2005) Hanser Gardner Publications, Cincinnati.

Matthews, G. *PVC: Production, Properties and Uses*. (1996) Institute of Materials V. 587, Great Britain.

Titow, W. V. *PVC Plastics: Properties, Processing and Applications*. (1990) Elsevier Applied Science. New York.

### Nylons

Kohan, M. I. *Nylon Plastics Handbook*. (1995) Hanser Gardner Publications, Cincinnati.

### Polyurethanes

Oertel, G. (ed) *Polyurethane Handbook: Chemistry, Raw Materials, Processing, Applications and Properties*. (1993) Hanser Gardner Publications, Cincinnati.

Uhlig, K. *Discovering Polyurethanes*. (1999) Hanser Gardner Publications, Cincinnati.

# Subject Index

*Andrew Peacock* is a Development Associate with Tredegar Film Products, Richmond, Virginia. Previously he worked as a Senior Research Chemist with Exxon Chemical Company, Baytown, Texas. Publications include the "Handbook of Polyethylene – Structures, Properties and Applications", nine patents in the field of polymer science, and numerous journal articles. Dr. Peacock received a B. Sc. in Chemistry from the University of London, England, an M. Sc. in Polymer Science and Technology from Lancaster University, England and a Ph. D. in Chemistry from the University of Southampton, England.

*Allison Calhoun* is an Assistant Professor of Chemistry at Whitman College, Walla Walla, Washington. Prior to accepting an academic position, she worked at Imerys Pigments and Additives as a Research and Development Scientist and Application Development Leader focusing on the interfacial properties of mineral loaded polymeric systems. In addition to three US patents and numerous journal articles, she co-edited the "Plastics Technician's Toolbox: Extrusion" and portions of the "Plastics Technician's Toolbox: Injection Molding" published by the Society of Plastics Engineering. Dr. Calhoun received a B. S. ED in Biological Sciences and her Ph. D. in Physical Chemistry with a specialization in surface chemistry from the University of Georgia.

# HANSER

# Polymers and Their Performance.

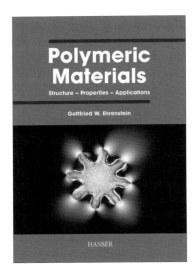

Ehrenstein
**Polymeric Materials**
295 pages. 205 fig.
ISBN 3-446-21461-5

This book focuses on the relationships between the chemical structure and the related physical characteristics of plastics, which determine appropriate material selection, design, and processing of plastic parts. The book also contains an in-depth presentation of the structure-property relationships of a wide range of plastics, including thermoplastics, thermosets, elastomers, and blends.

One of the special features is the extensive discussion and explanation of the interdependence between polymer structure, properties, and processing. The book contains numerous application-oriented examples and is presented at an intermediate level for both practicing plastics engineers and advanced engineering students.

More Information on Plastics Books and Magazines:
**www.kunststoffe.de** or **www.hansergardner.com**

## HANSER

# "... fewer headaches right from the start." INJECTION MOLDING

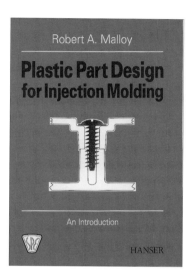

Malloy
**Plastic Part Design
for Injection Molding**
472 pages. 427 fig. 11 tab.
ISBN 3-446-15956-8

This overview assists the designer in the development of parts that are functional, reliable, manufacturable, and aesthetically pleasing. Since injection molding is the most widely used manufacturing process for plastic parts, a full understanding of the integrated design process presented is essential to achieving economy and functional design.

Contents: Introduction to Materials. Manufacturing Considerations for Injection Molded Parts. The Design Process and Material Selection. Structural Design Considerations. Prototyping and Experimental Stress Analysis. Assembly of Injection Molded Plastic Parts. Conversion Constants.

More Information on Plastics Books and Magazines:
**www.kunststoffe.de** or **www.hansergardner.com**

## HANSER

# Plastics Processing 101.

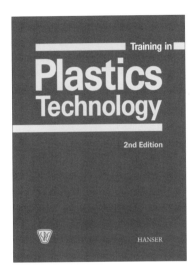

Michaeli/Greif/Wolters/Vossebürger
**Training in Plastics Technology**
180 pages. 143 fig.
ISBN 3-446-21344-9

This book is a clearly written, comprehensive introduction to the major aspects of plastics processing. It is divided into independent educational units, each covering a distinct subject area. Review questions at the end of each lesson help determining whether the individual educational goals were reached.

Partial Contents: Raw Materials and Polymer Synthesis. Methods of Polymer Synthesis. Bonding Forces. Classification. Deformation Behavior of Plastics. Time-Dependent Behavior of Plastics. Physical Properties. Plastics Fabrication and Processing. Compounding. Extrusion. Injection Molding. Fiber-Reinforced Plastics. Foams. Thermoforming. Welding. Machining. Adhesive Bonding

More Information on Plastics Books and Magazines:
**www.kunststoffe.de** or **www.hansergardner.com**

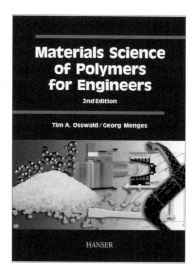

# HANSER

# Plastics in Engineering Practice.

Osswald/Menges
**Materials Science of Polymers
for Engineers**
640 pages. 600 fig.
ISBN 3-446-22464-5

This unified approach to polymer materials science is divided into three major sections: Basic Principles, Influence of Processing on Properties, and Engineering Design Properties.

The first edition of this textbook was praised for its vast number of graphs and data that can be used as reference. The new, second edition further strengthens this attribute with a new appendix containing material property graphs for the commonly used polymers. However, the most important change implemented in this edition is the introduction of real-world examples and a variety of problems at the end of each chapter.

More Information on Plastics Books and Magazines:
**www.kunststoffe.de** or **www.hansergardner.com**